T0173919

An Introduction to
Ceramics and
Refractories

An Introduction to
Ceramics and
Refractories

A. O. Surendranathan

CRC Press
Taylor & Francis Group
Boca Raton London New York

CRC Press is an imprint of the
Taylor & Francis Group, an **informa** business

CRC Press
Taylor & Francis Group
6000 Broken Sound Parkway NW, Suite 300
Boca Raton, FL 33487-2742

First issued in paperback 2020

© 2015 by Taylor & Francis Group, LLC
CRC Press is an imprint of Taylor & Francis Group, an Informa business

No claim to original U.S. Government works

ISBN-13: 978-1-4822-2044-5 (hbk)
ISBN-13: 978-0-367-73872-3 (pbk)

This book contains information obtained from authentic and highly regarded sources. Reasonable efforts have been made to publish reliable data and information, but the author and publisher cannot assume responsibility for the validity of all materials or the consequences of their use. The authors and publishers have attempted to trace the copyright holders of all material reproduced in this publication and apologize to copyright holders if permission to publish in this form has not been obtained. If any copyright material has not been acknowledged please write and let us know so we may rectify in any future reprint.

Except as permitted under U.S. Copyright Law, no part of this book may be reprinted, reproduced, transmitted, or utilized in any form by any electronic, mechanical, or other means, now known or hereafter invented, including photocopying, microfilming, and recording, or in any information storage or retrieval system, without written permission from the publishers.

For permission to photocopy or use material electronically from this work, please access www.copyright.com (http://www.copyright.com/) or contact the Copyright Clearance Center, Inc. (CCC), 222 Rosewood Drive, Danvers, MA 01923, 978-750-8400. CCC is a not-for-profit organization that provides licenses and registration for a variety of users. For organizations that have been granted a photocopy license by the CCC, a separate system of payment has been arranged.

Trademark Notice: Product or corporate names may be trademarks or registered trademarks, and are used only for identification and explanation without intent to infringe.

Visit the Taylor & Francis Web site at
http://www.taylorandfrancis.com

and the CRC Press Web site at
http://www.crcpress.com

To my beloved parents, Late Sreedharan Nair and Sarojini Devi, and to

my brothers and sisters, Sreenivasan, Unnikrishnan (now deceased),

Girija, Sreenarayanan, and Ajitha

Contents

Section II   Ceramics

Section III   Refractories

Preface

This book is the result of my intention to produce a book containing all aspects of ceramics and refractories. To date, I have had to refer to more than two books to teach the course "Ceramics and Refractories" to my undergraduate students. In addition, Dr. Gagandeep Singh of the Taylor and Francis Group happened to contact me when I was thinking of writing such a book.

This volume on ceramics and refractories is divided into three sections. The first section, "Ceramics and Refractories," is divided into seven chapters. Apart from an introductory section, Chapter 1 mainly details the applications of ceramics and refractories. Chapter 2 is on selection of materials and it describes the two stages in selection with a case study. Chapter 3 is on new developments in the ceramic and refractory fields. Chapter 4 describes the phase equilibriums in ceramic and refractory systems and outlines the three important systems, namely, unary, binary, and ternary. Corrosion of ceramics and refractories is the subject of Chapter 5. Chapter 6 discusses failures in ceramics and refractories. Design aspects are covered in Chapter 7.

Section II, which focuses entirely on ceramics, is divided into nine chapters (Chapters 8–16). Each chapter contains problems to be solved. Chapter 8 deals with bonding and Chapter 9 is on structures of ceramics. Chapter 10 deals with defects in ceramics. Ceramics' microstructures are covered in Chapter 11. Chapter 12 covers the production of ceramic powders starting from the raw materials. It also includes powder characterization. Four forming methods are described in Chapter 13. Chapter 14 discusses three types of thermal treatments. Mechanical properties are the subject matter of Chapter 15. Chapter 16 addresses thermal and thermo-mechanical properties.

Section III, which deals entirely with refractories, is divided into 10 chapters (Chapters 17–26). Chapter 17 classifies refractories. Principles of thermodynamics as applied to refractories are covered in Chapter 18. Chapter 19 is on properties and testing. Production of refractories is discussed in Chapter 20. Chapters 21 and 22 deal with the most important two single-phase refractories, silica and alumina. The combination of these two refractories is described in Chapter 23. Chapter 24 describes the chrome-magnesite system in terms of one and two components. The pure element refractory carbon is dealt with in Chapter 25, and the concluding chapter is on insulating refractories.

Solutions to the problems are given in Appendix 1. Appendix 2 contains the conversion from SI units to FPS units for the benefit of U.S. and U.K. readers.

A. O. Surendranathan

Acknowledgments

I would like to express my heartfelt thanks to Dr. Gagandeep Singh of the publishing company Taylor & Francis Group for inviting me to write this book. He was in constant touch with me via telephone and e-mails after my acceptance of his invitation. His perseverance enabled me to complete this volume.

Ms. Joselyn B. Kyle, the project coordinator, patiently reviewed my manuscript to find out whether it fulfills all the criteria required for its publication. I am very grateful to her.

My seven months' sabbatical at Deakin University allowed me to devote the time required to complete this manuscript. I am grateful to Peter Hodgson, the director of the Institute for Frontier Materials (IFM), for extending an invitation to me to stay at Deakin University. The excellent library facilities available there helped a great deal in the completion of my book. I would like to acknowledge the help rendered by the librarian in residence, who came to the IFM every week, for locating the necessary literature.

Next, I should thank my wife Latha (though she does not want me to) for drawing most of the figures in the book. She used her free time very effectively when I was at the university and was careful to check each drawing's accuracy with me on a daily basis.

My sons, Nikhil and Akhil, inquired about the progress of the book whenever we were in contact. I acknowledge their support.

At my parent institution, the National Institute of Technology Karnataka (NITK), Mr. B. K. Mahesh, foreman of the mechanical engineering department, and Mr. Ronak N. Daga, my own student, helped me in completing several more print quality figures. I would like to express my gratitude to them here.

Finally, I should thank my NITK colleagues for sharing the course instruction I was supposed to offer to our students during the time of this project's undertaking.

A. O. Surendranathan

Author

Dr. A. O. Surendranathan was born on October 15, 1954, in the village of Thenhippalam in the Malappuram District of Kerala, India. He is a professor in the Department of Metallurgical and Materials Engineering, National Institute of Technology Karnataka, India.

His academic qualifications are a BSc, AMIE, ME, and a PhD. He has 7 years of research and 29 years of teaching experience. During his academic career, he was awarded a summer research fellowship by the Indian Academy of Sciences, Bangalore, India. He has also undergone industrial training at the Eicher Tractors Engineering Centre, Ballabgarh, Haryana.

Dr. Surendranathan has conducted research in powder metallurgy, superplasticity, batteries, and corrosion. In addition to these areas, he has guided advanced diploma, undergraduate, postgraduate, and doctoral students in the fields of electroplating, welding, heat treatment, polymers, composites, and nanotechnology encompassing more than 80 projects.

Dr. Surendranathan has presented and published more than 100 papers and has received numerous medals, awards, prizes, and honors. In addition to teaching a variety of subjects and guiding his advanced degree students in the preparation of seminar topics, he has conducted laboratory courses and given more than 140 seminars.

Numerous seminars, workshops, conferences, programs, conventions, and functions have been organized by Dr. Surendranathan, who has completed four sponsored projects in addition to delivering guest lectures on various topics. He has chaired technical sessions at conferences and has been a member or office holder in eight professional organizations. Dr. Surendranathan's foreign assignments have included travel to Ethiopia, Australia, and the United States.

Section I

Ceramics and Refractories

Section 1

Ceramics and Refractories

1

General Aspects

1.1 Introduction

Materials constitute the universe. Anything having a mass is a material. All materials fall into 4 types. They are metals, polymers, ceramics and composites. This book deals with the *ceramics*. The word ceramics has its origin from the Greek word *keramos*, which roughly means *burned stuff*. Ceramic products are subjected to heat and then cooled while manufacturing, hence the name.

Refractories are a class of ceramics with high fusion points. The word *refractories* means materials that are difficult to fuse. The melting point of refractories is normally above 2000°C.

The first man-made ceramic product was earthenware, the making of which dates back to about 24000 BC. The raw material for this product was clay, which was mixed with water, then molded into a shape, dried, and fired. During 7000–6000 BC, lime mortar was used for construction. The gaps between stones were filled with this material. It was also used to make thick-walled containers. In the period from 7000 to 5000 BC, floors were laid with a plaster-like cementitious material. The interiors of walls were given a coating of this material, and they were later decorated.

By about 4500 BC, the kilns used for smelting copper were lined with ceramics. This period, therefore, also marks the beginning of the Chalcolithic period, or the Copper Age. Until about 4500 BC, bricks were only dried before use, whereas during 4500–3500 BC, they were also fired before use. Such fired bricks were found to be stronger and more durable. During this period, ancient arts using ceramics were developed, and white porcelain ceramics with bright blue glassy coatings were made. The first writings were in the form of pictures and symbols. The initial symbols were wedge-shaped, and inscribed on ceramic tablets. Because they were wedge-shaped, the symbols were called *cuneiform symbols*. The term *cuneiform* owes its origin to the Latin words *cuneus*, meaning *wedge*, and *forma*, meaning *shape*. Evolution of pictographs and cuneiform symbols using ceramic tablets took place during 3500–2800 BC. Earthenware pots were made available to all after the invention of the potter's wheel, by about 3300 BC.

During 2500–1600 BC, glassmaking developed into a craft. From 1750 to 1150 BC, a number of glazes were developed, including lead and colored glazes. During 1025–750 BC, stoneware was developed in China. These were lighter in color, stronger, and less porous than earthenware. It was during 500–50 BC that concrete was developed by the Greeks and the Romans. Glassblowing was invented during 100–50 BC. Thus, glass was commonly available. The period AD 600–800 saw the evolution of high-quality porcelain in China.

The use of white, tin-based glazes and lustrous overlay glazes were refined by Persians during AD 800–900. During the 1740s, *transfer printing* was invented. This technique increased the production rate of tiles. After about another 100 years, *dust pressing* was invented. This further increased the production rate of tiles. This method of pressing was improvised to uniaxial pressing. Today, all the high-volume production of ceramic products is carried out using uniaxial pressing. The late 1800s saw the synthesis of single-crystalline ruby and sapphire. In the early 1900s, kilns were developed that could generate up to 1600°C of heat. During the mid-1900s, alumina-based ceramics were developed and many new engineering ceramics were synthesized. In the late 1900s, ceramics and ceramic matrix composites with high-strength and high-toughness that could withstand ultra-high temperature were developed.

The importance of ceramics and refractories can be gauged from the wide variety of their applications, which are dealt with in the following section.

1.2 Applications of Ceramics and Refractories

Even though refractories are also ceramics, they have special applications. Hence, they are dealt with separately in this section, subsequent to a discussion on ceramics and their applications.

1.2.1 Applications of Ceramics

Ceramics both natural and synthetic ceramics find applications in various fields-traditional and engineering applications.

1.2.1.1 Natural Ceramics

Man has been surrounded by ceramics from the very beginning. The natural tendency of humans is to put to use any and all available materials to satisfy their needs. Thus, the earliest humans used the natural ceramics around them—the boulders provided them shelter, and stones were used for defense and hunting. As time passed, man learned to chip rocks to make tools, leading to the evolution of hand-held axes. About 70,000 years ago, the Neanderthals emerged, and they developed small spearheads, hand-held axes, long knives with one edge sharp and the other blunt, and notched blades for shredding food.

The Neanderthals disappeared about 35,000–30,000 years ago. Modern man appeared later on, and made advanced tools out of stone, bone, and ivory. A natural ceramic called red ochre was used to cover dead bodies. A later development was the use of pigments obtained by mixing colored soil with water. These were used to decorate the body and paint images. Prehistoric artisans used clay to mold animal images. They found that naturally occurring clay, when mixed with water, could be shaped.

1.2.1.2 Synthetic Stone

The first use of fired clay was in making female sculptures. These sculptures represent the first man-made ceramic products. The firing temperatures were low to medium, and these products are now referred to as *earthenware*. Earthenware is used even today. After its initial use in sculpture making, with further development in ceramics in the later years, earthenware pots were used for storage and cooking. Dwellings were also made using fired clay.

The other type of ceramic material that came to be used is gypsum. Today, gypsum is used to make wallboards, molds, sculptures, and casts. Plaster of Paris, as the set gypsum is called, was used in ancient times in construction. It was also used to mold the front of the skull to depict the features of a dead person in the family.

The use of lime mortar was another application of ceramics in old days. This was done by mixing lime with the ash of salty grasses, with water added later. This mixture was used, for example, to fill the spaces between stones in a wall construction. This process increased the strength of the wall on natural drying.

The next stage in development was the making of clay bricks. Larger buildings were built by cementing with mortar made with waterproof bitumen. Earlier, pots and bricks were porous, because the temperature of firing was low. Later, kilns were built, facilitating an increase in temperature and enabling the production of higher quality vessels. For painting, slips of colored clay were used. The surface was polished with a stone by rubbing. Decoration was also done by incising before firing.

As Neolithic people started trading, to keep track of exchanges, they started incising damp clay tokens. As written language evolved, writing was also incised in the same way. Written language transformed from pictographs into a wedge-shaped script (cuneiform). Figure 1.1 shows a cuneiform tablet.

1.2.1.3 Traditional Ceramics

During the Chalcolithic period, copper was produced in clay-lined furnaces. Ceramic containers were made to hold and refine molten metal, and ceramic molds were used to cast the metal. Colored ceramic glaze was made

FIGURE 1.1
Cuneiform tablet.

from quartz, malachite, and natron. Glaze has the same structure as glass. When the temperature of kilns could be increased, complete melting of glass could be achieved.

1.2.1.4 Application of Ceramics in Buildings

Ceramics have been used in buildings since early civilizations. During the Greek civilization, ceramic bricks, plaster, and decorative tiles were used. Romans used mortar. The quality of their concrete was marvelous. The traditional cement industry started in the 1750s. After AD 632, Islamic mosques built in the Islamic territories in the East contained colorful tiles.

1.2.1.5 Whiteware

The Chinese were able to produce whiteware by AD 618 as they, by then, had kilns that attained a temperature of about 1300°C. Porcelain was manufactured in Europe only around 1770.

1.2.1.6 Modern Ceramics

When electrical devices were produced, porcelain came to be used as an insulator. One such device was the spark plug. As furnace technology improved, spark plugs with more alumina content could be produced. These had better properties. When alumina is mixed to the extent of 95%, the firing temperature has to be increased to about 1620°C. The better properties of alumina were used later to make abrasives, laboratory ware, wear-resistant parts, and refractories.

1.2.1.6.1 Engineering Ceramics

Engineering ceramics are those modern-day ceramics that have been engineered and produced to suite specific present-day applications.

1.2.1.6.1.1 High-Temperature Ceramics   High temperatures are involved in processes such as metal smelting, glass production, petroleum refining, and energy conversion.

- *Metal Processing*: Metal processing requires furnaces that can reach high temperatures. These furnaces are coated inside with refractory linings. The crucibles, in which the metals are melted, and the molds used for shaping the molten metal are made of ceramics. In the investment casting process, a ceramic is used to form a coating on the wax pattern. Later, this coating holds the molten metal. The cores used are also made of ceramics. All the other parts of the complete mold set, such as the fibrous insulation wrapped around the mold and the filter at the mold entrance, are also made of ceramics. The cutting tools are exposed to high temperatures while working. There are cermets, such as WC-Co, which are used as tools. The inserts of these tools are made of ceramics. The ceramics used in the iron and steel industries should withstand temperatures in the range of 1590°C–1700°C and possess sufficient hot strength.
- *Glass Tanks*: Glass tank linings, made of ceramics, should withstand the molten reactive glass apart from the high temperature involved. They are made up of zirconium oxide (zirconia) or a mixture of zirconia, alumina, and silica.
- *Process Industries*: Furnaces used for the fabrication of spark plug insulators, catalyst supports for automobile emission control systems, dinnerware, magnetic ceramics, and electrical ceramics are lined with ceramics. Ceramic high-temperature furnace linings and thermal insulation are used in petroleum and chemical processing, cement production, heat treatments, and paper production.

    Another application of ceramics in high temperature is as a heat source. In ceramic tubes, fuel is mixed with air and burned at their tips. Ceramics, such as SiC and $MoSi_2$, are used as heating elements. Ceramic heat exchangers are used along with glass-melting furnaces. They are also used in chemical processing. Porous ceramics are used to provide support for catalysts in petrochemical cracking processes, ammonia synthesis, and in polymer precursor processing.

- *Ceramics in Heat Engines*: The thermal barrier coatings on metal parts have increased the efficiency of gas turbine engines. The ceramic used for this purpose is zirconium oxide. The spark plug insulator used in internal combustion engines has a high alumina composition. Silicon nitride is used in turbocharger rotors. The other applications of ceramics in heat engines are silicon carbide glow plugs and swirl chambers for diesel engines, doped zirconia oxygen sensors, cordierite honeycomb substrates for catalytic converters, diesel particle traps, ceramic fiber insulation, and alumina mounting substrates. Other examples are diesel cam follower rollers, diesel fuel injector timing plungers, water pump seals, air conditioner compressor seals, and sensors.

- *Ceramics in Aerospace*: Apart from the use of ceramics in gas turbine engines used in aerospace, the other components made of ceramics are rocket nozzle liners, thruster liners, afterburner components, igniter components, leading edge structures for hypersonic missiles, and thermal protection for space shuttles.

1.2.1.6.1.2 Applications Involving Wear and Corrosion   Examples of these applications are seals, valves, pump parts, bearings, thread guides, papermaking equipment, and liners.

- *Ceramics as Seals*: Ceramic seals are used between two rapidly moving parts. These are called face seals. The compressor seal in an automobile air conditioner is an example of such a seal. A combination of graphite and Al_2O_3 is used for making these seals. Water pump seals for automobiles are also face seals. The materials used are silicon carbide and alumina. The main rotor-bearing seal used in jet engines is made of impregnated graphite or silicon nitride. Gas turbine engine shaft seals are also made of silicon nitride.

- *Ceramics as Valves*: Alumina or silicon carbide is the material for valves. Gate valves used in an irrigation ditch are made of metal or concrete. Slide-gate valves in ladles holding molten steel are made of ceramics. Alumina is the material used to make oxygen valves of respirators used in hospitals. Ball-and seat valves, made of WC-Co, transformation-toughened zirconia, or silicon nitride, are used in downhole pumps required for deep oil wells.

- *Ceramics in Pumps*: In addition to ceramic seals and valves in pumps, ceramics are also used as bearings, plungers, rotors, and liners. A pump completely lined with ceramic materials is shown in Figure 1.2.

- *Ceramics as Bearings*: Single crystal alumina have been used in watches where the load is low. The ceramic that has surpassed metal bearings in properties for use in high-load, high-speed applications is silicon nitride. Figure 1.3 shows silicon nitride–bearing balls.

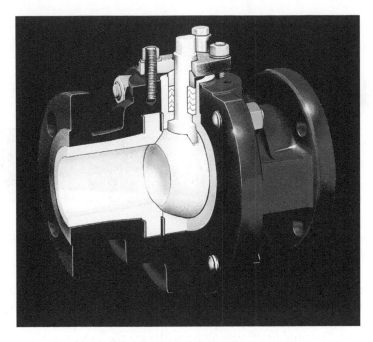

FIGURE 1.2
Pump completely lined with ceramic materials.

FIGURE 1.3
Silicon nitride ball bearings.

- *Ceramics as Thread Guides*: Ceramics used as thread guides are alumina, porcelain, and glass-coated stoneware. Alumina thread guides are shown in Figure 1.4.
- *Papermaking*: For sawing of logs, the saws are tipped with WC-Co cermet. The chipper used to chop a log into small chips is made of a composite consisting of TiC ceramic particles in a metal matrix.

FIGURE 1.4
Aluminum oxide thread guides. (From D. W. Richerson, *Modern Ceramic Engineering— Properties, Processing, and Use in Design*, CRC Press, Boca Raton, 2006, pp. 7–86.)

For chip screening and conveying, alumina tiles and block-lining screens, pneumatic conveyor pipes, and saw dust–removing equipment are used. For chip digestion, a ceramic-lined pressure cooker is used. Ceramic-lined vats along with ceramic valves, seals, cyclone separator liners, and pump parts are used for pulp cleaning and bleaching. Pulp is stored and refined in concrete tanks lined inside and outside with ceramic tiles. In paper machines, water reduction is achieved with the help of rows of alumina suction boxes and rows of foils made of segments of alumina, transformation-toughened alumina or zirconia, or silicon nitride. Initial roll pressing is carried out with the help of granite rock. Ceramic heaters are also used for drying. Slitting of the prepared paper is carried out with the help of ceramic cutters.

1.2.1.6.1.3 Cutting and Grinding   Ceramic is a good choice for cutting and grinding applications.

- *Cutting Tool Inserts*: Composites consisting of very hard tungsten carbide (WC) bonded with about 6% volume fraction of cobalt were developed as cutting tool inserts. These could be operated above a speed of 30 m/min. In the 1950s, hot-pressed alumina was developed, which could be used at speeds of up to 300 m/min. The reliability of these inserts was improved by adding about 35% TiC in the 1960s. In the 1970s, a wide range of ceramic cutting tool inserts were developed. They are hot-pressed Si_3N_4 with particle or whisker reinforcement, alumina with SiC whisker reinforcement, and transformation-toughened alumina. These could reach speeds of 900 m/min. As ceramics are not tough enough, they

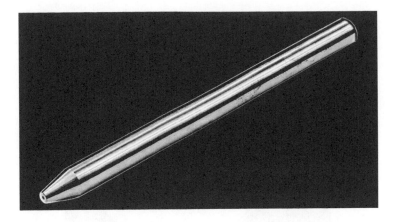

FIGURE 1.5
Waterjet cutting nozzle made from hard, wear-resistant WC.

cannot perform well in machining operations. Hence, the tool life of cermets was increased by giving a thin coating of Al_2O_3, TiC, or TiN.

- *Superhard Abrasives*: Quartz sand, garnet, and corundum have been used as abrasives for many centuries. Another abrasive that has been used is natural diamond. Various superhard abrasives have been synthesized during the twentieth century. They are SiC, B_4C, TiC, cubic BN, polycrystalline, and single-crystalline diamond.

- *Ceramics in Waterjet Cutting*: In waterjet cutting, fine ceramic particles such as boron carbide are mixed with water. The slurry obtained is blasted with about 400 MPa pressure through a small bore ceramic nozzle (Figure 1.5).

1.2.1.6.1.4 Ceramics for Electrical Applications Ceramics are used in varied electrical applications in the form of insulators, semiconductors, and conductors.

- *Electrical Insulators*: Alumina insulators (Figure 1.6) are used for high-voltage applications such as in x-ray tubes and spark plugs.
  Integrated circuits use very thin films of SiO_2 and Si_3N_4.
- *Ceramics as Dielectrics*: Barium titanate is used as a miniature capacitor because of its dielectric properties, which include ferroelectricity and piezoelectricity. Figure 1.7 shows ceramic piezoelectric parts and assemblies.
- *Ceramic Semiconductors*: SiC and $MoSi_2$ semiconductors are used as heating elements. CdS, InP, TiO_2, $SrTiO_3$, and Cu_2O are used in solar cells. Cu_2O is used in rectifiers. $BaTiO_3$ and $Fe_3O_4/MgAl_2O_4$ solid

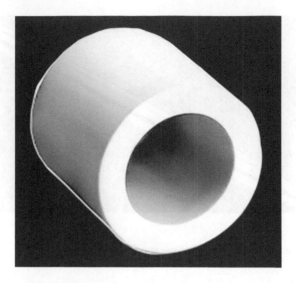

FIGURE 1.6
About 12 cm long alumina insulator.

FIGURE 1.7
Ceramic piezoelectric parts and assemblies.

solution are used as thermistors. Varistors use ZnO and SiC. SnO_2, ZnO, and Fe_2O_3 are useful as sensors. GaAs and GaP are found in LEDs and lasers.

- *Conductors*: TiO, ReO_3, and CrO_2 possess conductivity comparable to metals. VO_2, VO, and Fe_3O_4 have good conductivity at elevated temperatures. NiO, CoO, and Fe_2O_3 have less conductivity than the previous group of ceramics at elevated temperatures. All these ceramics possess conductivity due to the movement of electrons.

  The next group of ceramic conductors are ionic ones. Examples are doped ZrO_2, beta alumina, NaCl, AgBr, KCl, BaF_2, and various lithium compounds. Ionic conductors are used in lithium batteries, oxygen sensors, and solid oxide fuel cells.

- *Superconductors*: A series of ceramic compositions were identified between 1986 and 1988 that exhibited superconductivity between 90 and 120 K. Figure 1.8 shows ceramic superconductor wires wound into a cable.

1.2.1.6.1.5 Ceramics as Magnets The first magnet used was naturally occurring magnetite. Ceramic magnets are termed as ferrites. There are *hard* and *soft* ferrites. Speakers, linear particle accelerators, motors for tooth brushes, and electric knives are some applications of hard ferrites. Soft ferrites are used in television, radio, touch-tone telephones, electrical ignition systems, fluorescent lighting, high-frequency welding, transformer cores, and high-speed tape and disk recording heads. Composite magnets are produced by combining ceramics with other materials. Recording tape is an example of a composite magnet in which a ferrite is combined with a polymer.

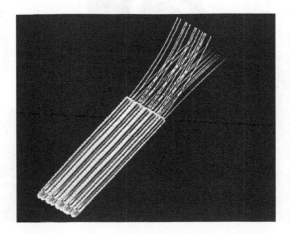

FIGURE 1.8
Ceramic superconductor wires wound into a cable.

1.2.1.6.1.6 Ceramics in Optics   Ceramics in optics are based on their optical properties, such as transparency, phosphorescence, and fluorescence.

- *Transparency*: The primary source of glass before the 1900s was a soda-lime-silica composition. High-lead glass, high-silica and boro-silicate glasses, optical fibers, and doped laser glasses are special types of glass. Glass is used for making windows, containers, optical fibers, and lenses.

- *Phosphorescence and Fluorescence*: Many ceramics phosphoresce and fluoresce. These ceramics find applications in fluorescent lights, oscilloscope screens, television screens, electroluminescent lamps, photocopy lamps, and lasers. Lasers include tungstates, fluorides, molybdates, and garnet compositions doped with chromium, neodymium, europium, erbium, praseodymium, and so on. Ceramic phosphor is $Ca_5(PO_4)_3(Cl,F)$ or $Sr_5(PO_4)_3(Cl,F)$ doped with Sb and Mn. Figure 1.9 shows some configurations of fluorescent lights.

1.2.1.6.1.7 Composites   Fiberglass is a composite in which glass fibers are used for reinforcement. Other ceramic fibers that are used as reinforcements are of carbon, silicon carbide, alumina, and aluminosilicate. Lightweight structures for aircraft and stiff and light sporting equipment use carbon fibers.

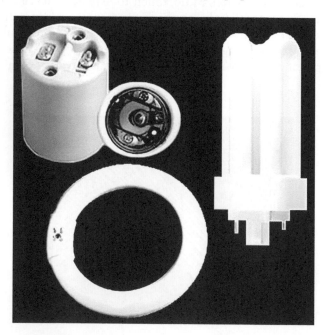

FIGURE 1.9
Some configurations of fluorescent lights.

In ceramic–ceramic composites, both the reinforcement and the matrix are ceramic materials. Examples are carbon fibers in acarbon matrix, SiC fibers in an SiC matrix, and alumina fibers in an alumina matrix. Apart from the fiber form, ceramic reinforcements are also used in the form of particles and whiskers.

1.2.1.6.1.8 Medical Applications There are three categories of medical applications: replacement and repair, diagnosis, and treatment and therapy.

- *Ceramics for Replacement and Repair*: Ceramics are used as implants, for repairs and restorations in the human body. Hydroxyapatite-based compositions stimulate bone and tissue growth. Other ceramics that could be used for replacement and repair are alumina and zirconia. Teeth replacement has been the earliest use of ceramics in the human body. Prior to 1774, teeth of animals and those extracted from human donors were used, along with other materials such as ivory, bone, and wood. In 1774, porcelain was used. In recent years, ceramics have become the standard materials for tooth replacement, dentures, and restoration techniques. A composite of ceramic particles in a polymer is used to fill cavities. Jawbones are restored by hydroxyapatite.

  Metal balls in hip implants are replaced by ceramic balls made of alumina, transformation-toughened alumina, or zirconia (Figure 1.10).

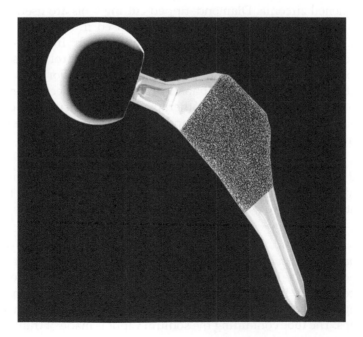

FIGURE 1.10
Hip replacement consisting of a ceramic ball on a metal stem.

Ceramic *bioglass* implants could solve the hearing problems associated with damaged ossicles. Transparent ceramic lenses can replace damaged eye lenses. *Orbital implants* are artificial eyes. Pyrolytic carbon is the material used to make mechanical heart valve. Bone replacement has also been done with the same carbon material.

- *Medical Diagnoses*: Many ceramic items are used for medical diagnoses. Laboratory ware such as petri dishes and test tubes have been in use for many years. These are useful for sterile handling of medical samples for analyses. Lenses are used in microscopes, and for examining eyes and ears. Ceramics are also used to make electrical and optical components for x-ray machines, oscilloscopes, spectroscopes, and computers, all of which are used in medical diagnosis. Ceramics have been of use in CT equipment, endoscope, and ultrasound scanning. In CT equipment, ceramic *scintillator* slides are used, which collect the x-rays passing through the body. In an *endoscope*, ceramic is the optical fiber. Piezoelectric ceramic transducers are used in ultrasound scanning.

- *Medical Treatment and Therapy*: Endoscopes, which contain ceramic optical fiber, are used in gall bladder surgery, in arthroscopic knee surgery, and also in the detection of colon cancer. For treating cancer, endoscope built with a tool containing a tiny ceramic rod is used. Ultrasharp scalpels used in surgery are made of transformation-toughened zirconia. Diamond–tipped cutting tools are used to saw bone and drill teeth. Ceramic lasers are used in operating rooms and are also used to vaporize sun-damaged skin. In delicate surgeries, piezoelectric transducers are used to control the scalpel. Piezoelectric transducers are also used in dialysis. For liver cancer treatment, tiny glass microspheres are used. Piezoelectric ceramics also find application in physiotherapy. Electrical insulators, capacitors, and glass seals are the components of pacemakers. The valves for respirators are made of ceramics.

1.2.1.6.1.9 Ceramics for Energy Efficiency and Pollution Control   Ceramics have addressed energy and pollution challenges. They have contributed to energy savings at home and are used for power generation and in the transportation sector.

- *Home Energy Savings*: Fiberglass thermal insulation has been in use in homes since 1938 and has helped save much energy. Use of fluorescent lights has contributed to further energy savings. In street lights, incandescent bulbs have been replaced with sodium vapor lamps. The tube containing the sodium vapor is made of transparent polycrystalline alumina.

By 1980, ceramic igniters replaced the pilot lights of natural gas appliances. These pilots were wasting 35%–40% of the natural gas. These igniters were made using SiC.

- *Ceramics in Power Generation*: Coal, atoms, natural gas, and water have been the major sources of power. Ceramics are widely used in these plants in the form of concrete, wear- and corrosion-resistant parts, and electrical components. Dams for hydroelectric power generation are constructed using concrete. Nuclear plants use ceramic fuel pellets, control rods, and seals and valves. Ceramic thermal barrier coatings on superalloy turbine engine components have increased the efficiency of gas turbine engines. Turbine blades for harnessing wind energy are made from ceramics, fiber-reinforced polymers. Ceramic insulators are needed for safe transport of the electricity generated.

- *Ceramics in the Transportation Sector*: *Catalytic converters* reduced the pollution from automobiles by 90% by the mid-1970s. The ceramics used in these devices were in the form of a honeycomb.

  In other uses of ceramics in pollution control, oxide fibers are woven into filter bags to remove ash particles resulting from coal burning. For water purification, porous ceramic filters are used. *Firebooms* are used to contain oil spills. They contain high-temperature fibers and a floatable porous ceramic core, in addition to other materials.

1.2.1.6.1.10 Military Applications Electromagnetic windows, armor, composite structures, stealth technology, lasers, ring laser gyroscopes, rocket nozzles, afterburner seals, sensors, fuel cells gas turbine engines, communication systems, smart weapons, night vision devices, and flash-blindness prevention goggles are used in military. All of these contain ceramics.

WC core was inserted into the bullet. High hardness ceramic was developed against this bullet. This was made with an outer layer of boron carbide, aluminum oxide, or silicon carbide backed with multiple layers of fiberglass. Figure 1.11 shows boron carbide armor with bullets.

1.2.1.6.1.11 Ceramics in Recreation There are many ceramic articles on which an artist can work. Some of them are pottery, decorating, enameling, and glass blowing.

Ceramics also play a role in the field of music. Many speakers, electronic musical instruments, and microphones use piezoelectric ceramics. Magnetic ceramics are used in tape recording heads and audio tapes. Radios, CD players, computers, and televisions use ceramics. Lenses are used in projectors.

Shafts for golf clubs, kayaks and paddles, boat hulls, hockey sticks, skis, and tennis rackets use ceramic fibers as reinforcement in polymer matrixes.

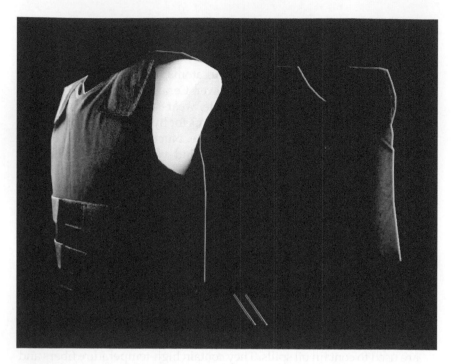

FIGURE 1.11
Boron carbide armor with bullets.

Golf putters have been constructed with ceramic heads. Some drivers contain ceramic inserts and some golf cleats are also made of ceramics. The ceramics used for all these are transformation-toughened zirconia or alumina. Some bowling balls contain a ceramic core. Some race cars are equipped with ceramic valves and composite ceramic brakes.

1.2.1.7 Widely Used Ceramics

The following is a brief summary of applications of some of the more widely used ceramic materials.

1.2.1.7.1 Alumina (Al_2O_3)

Al_2O_3 is used to contain molten metal or in applications where a material must operate at high temperatures but where high strength is also required. Alumina is also used as a low dielectric constant substrate for electronic packaging that houses silicon chips. One classical application is for insulators in spark plugs. Some unique applications are also being found in dental and medical use. Chromium-doped alumina is used for making lasers. Fine particles of alumina are used as catalyst supports.

1.2.1.7.2 Diamond (C)

Diamond is the hardest naturally occurring material. Industrial diamonds are used as abrasives for grinding and polishing diamonds. Diamond and diamond-like coatings prepared using chemical vapor deposition processes are used to make abrasion-resistant coatings for many different applications (e.g., cutting tools). They are, of course, also used in jewelry.

1.2.1.7.3 Silica (SiO₂)

SiO_2 is probably the most widely used ceramic material. Silica is an essential ingredient in glasses and many glass ceramics. Silica-based materials are used in thermal insulation, refractories, abrasives, fiber-reinforced composites, laboratory glassware, and so on. In the form of long continuous fibers, silica is used to make optical fibers for communication. Powders made using fine particles of silica are used in tires, paints, and many other applications.

1.2.1.7.4 Silicon Carbide (SiC)

SiC provides outstanding oxidation resistance even at temperatures higher than the melting point of steel. SiC often is used as a coating for metals, carbon–carbon composites, and other ceramics to provide protection at these extreme temperatures. SiC is also used as an abrasive in grinding wheels and as particulate and fibrous reinforcement in both metal matrix and ceramic matrix composites. It is also used to make heating elements for furnaces. SiC is a semiconductor and is a very good candidate for high-temperature electronics.

1.2.1.7.5 Silicon Nitride (Si₃N₄)

Si_3N_4 has properties similar to those of SiC, although its oxidation resistance and high-temperature strength are somewhat lower. Both silicon nitride and silicon carbide are likely candidates for components for automotive and gas turbine engines, permitting higher operating temperatures and better fuel efficiencies with less weight than traditional metals and alloys.

1.2.1.8 Why Ceramics?

Like any material, ceramics are applied because of the required properties they possess in that application. The following discussion is based on the applications by virtue of the particular properties that ceramics possess.

1. Compressive strength makes ceramics good structural materials (e.g., bricks in houses, stone blocks in the pyramids).
2. High-voltage insulators and spark plugs are made from ceramics due to the electrical conductivity properties of ceramics.
3. Good thermal insulation results from ceramic tiles being used in ovens.

4. Some ceramics are transparent to radar and other electromagnetic waves and are fused in radomes and transmitters.

5. Chemical inertness makes ceramics ideal for biomedical applications such as orthopedic prostheses and dental implants.

6. Glass ceramics, due to their high-temperature capabilities, are of use in optical equipment and fiber insulation.

7. Ceramic materials that contain alumina and silicon carbide are extremely rigid and are used to sand various surfaces, cut metals, polish, and grind.

1.2.2 Applications of Refractories

The word *refractory* means *difficult to be fused*: thus the applications of refractories involve usage of these hard-to-be-fused materials in circumstances where higher temperatures are involved. Typically, the equipment in which these are used are called ovens, kilns, and furnaces. There are other high-temperature equipment that use refractories, but these are not named by any of the terms mentioned earlier. Examples are steam boilers, incinerators, stills, retorts, reactors, cracking towers and so on. Most ovens are operated at relatively lower temperatures with certain exceptions such as the coke oven. Coke ovens operate at temperatures up to 1100°C. The next higher temperature operating equipment is the kiln, and the highest temperatures are involved in furnaces. Again, some of the high-temperature operating equipment are not termed as *furnaces*. Examples are converters used for converting iron to steel or copper matte to blister copper.

Refractory articles used as shelves, supports, trays, setters, spacers, baffles, and enclosing muffles or saggers may be classified as kiln furniture. These are used for the heating or sintering of ceramic ware. Chutes, conveyors, slides, scoops, cars, and so on are again refractory parts, which are used to convey hot solid products. Liquid products from furnaces are transported through refractory troughs and runners. Ladles containing refractories are used to transport molten charges such as liquid steel. A tundish feeds a continuous casting device. The inside of all furnaces are lined with refractory materials.

The largest of all furnaces is the blast furnace. This is used to produce pig-iron. Apart from its main shaft structure, it has other accessories. One of them is a blast furnace stove. It is a tall chamber filled with an open checker work of refractory heat-exchanger bricks and with associated ducting. All the parts of blast furnace—hearth, bosh, waist, stack, and port—are lined inside with refractory bricks. The other associated parts of this furnace are downcomer checkers, blast ducts, and bustle pipes. As these parts also operate at high temperatures, they also contain refractory linings. Taphole refractories and the refractories used as plugs must resist fast-flowing liquid metal and slag. Both these liquids are intensely corrosive and erosive. Troughs and runners and their associated devices are refractory-lined ditches through which metal flows from the blast furnace taphole to the point of loading of ladles. Torpedo ladles and teeming ladles transport hot metal to a steelmaking area and to a pouring pit, respectively, for the casting of pig-iron.

Cupola lining materials are ordinarily limited to oxide refractories. In lead blast furnaces, refractory linings are employed in the upper shaft or charging area for abrasion resistance, and in the hearth or crucible area below the tuyeres for resistance to collected molten lead and slag. Refractory heat exchangers are built as adjuncts to the glass-melting tank. Continuous circulating pebble-bed heaters are a smaller capacity heat exchangers that employ uniformly sized refractory spheroids. In chemical reactors, the typical refractory application is as the lining in steel shells. The numerous ports and tubulations, internal heaters or stirrers, instruments, and accessories for maintenance may call for refractory gasketing and seals.

In an aluminum electrolytic smelter, the collector base, cell bottom, and its sidewalls are made up of a conducting refractory. In the shaft kiln, the shaft is refractory lined. Refractories are used to line hearths, roofs, and the outer walls of a multiple hearth furnace. Refractory vessel linings are used to protect the shell and for thermal insulation in a calciner. The continuous rotary kiln is a long, slow rotating, refractory-lined steel tube. The glass-melting tank is a rectangular refractory box of about 90–120 cm high. Oxide refractories are used for the construction of the tank. The checkers used along with a glass-melting tank are rectangular latticeworks of refractory brick.

In tunnel kilns used for ceramic sintering, each car top is a square or rectangular refractory platform. The kiln walls are of free-standing refractory brick construction. Some kiln designs are modular, with refractory seals between segments. Burner blocks are also made of refractories. Refractory gasketing is also needed. The heat-treatment furnaces used for metals and glass contain refractories in the roof walls, floor, and interior equipment. Refractory linings for drying ovens and tunnels have the sole purpose of thermal insulation of the peripheral walls and roof. Fiber refractories, even glass wool, serve the purpose.

Steam boiler walls contain refractory insulation. In basic oxygen furnace (BOF) used for steelmaking, a water-cooled, refractory-armored tube or lance is inserted down through the mouth of the furnace, after charging it. In the bottom blown BOF, two types of bottom plugs are used: (1) dense refractory, through which an array of vertical tuyeres penetrate, and (2) porous, permeable refractory, which serves as an oxygen diffuser. Refractory zones in oxygen furnaces are lane, hood, taphole area, barrel, trunnion area, charge pad, tuyere areas, and the bottom. Electric arc furnaces (EAFs) used for alloy steelmaking consist of refractory-lined vessels and their tops. In EAFs, the refractory materials make up their roofs, electrode coatings, side walls, slag lines, bottoms, tapholes and plugs, and spouts.

The coreless induction furnace (CIF), used widely as a melting, alloying, and foundry furnace, consists of a thermally insulating, nonsuscepting, and impermeable refractory crucible. The various refractory parts in bottom pouring steelmaking ladles are its orifice, slide-gate valve, nozzle, taphole, stopper, and plug. The transfer ladle used in continuous casting of steel uses a *ladle shroud*, which is nothing but a vertical refractory tube. The mold powder, used in the same process, is a refractory material.

The copper converter is a closed-ended refractory-lined tube. Arc furnaces are also used for oxide melting. In this water-cooled refractory, insulators are used around the electrodes. In coke ovens, the oven overhead, consisting of the volatile by-products of coking, is taken by refractory-lined manifolding above the entire line. The oven roofs contain charging doors. Sidewalls are common for the adjacent ovens. The entire array is of an intricate architecture, calling for a host of brick shapes and monolithic and sealing refractories. All external walls are made of thermally insulating refractories.

Carbon baking furnaces can withstand temperatures as high as 1800°C. The principal difference among the refractories employed relates to the maximum temperature of use. In batch kilns used for ceramic sintering, the front wall is closed by moveable refractory-lined doors. Beta-spodumene, zirconia, and silicon nitride perform duties in heat engines. The nose cone of a guided missile is refractory. The launching pad of a space vehicle installation and the protective tiles on the vehicle are also refractory. Ablative hypersonic-vehicle coatings are assuredly refractory. The selective-reflectance coatings on space satellite skins are made of a refractory material.

Ceramic abrasives and machine tools are refractory as they glow orange-hot on the job. In advanced, high-strength filament-reinforced composites, there is at least one refractory component, such as carbon or silicon carbide. The legs of a high-temperature thermocouple are also made of these two materials. Nuclear reactors use graphite, uranium oxide, and carbide. Military armor is made of refractory materials.

The oxygen content of automotive exhaust gases is measured by zirconia. Alumina is the carrier for combustion catalysts. Alumina also insulates the electrodes of a hot spark plug. Chromia prevents superalloys from burning up in an aircraft turbojet.

Figures 1.12–1.18 show photographs of some of the equipment in which refractories are used.

Applications of refractories are growing as new developments continue to take place.

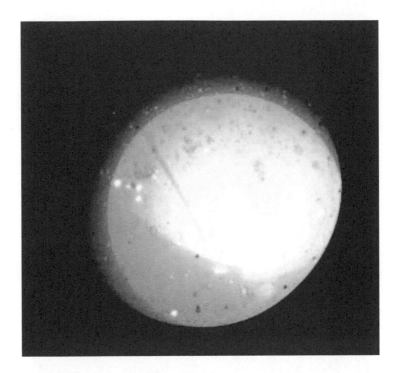

FIGURE 1.12
Converter (steel industry).

FIGURE 1.13
Ladle (steel industry).

FIGURE 1.14
Electric arc furnace (steel industry).

FIGURE 1.15
Tundish (steel industry—flow control).

FIGURE 1.16
Rotary kiln (cement industry).

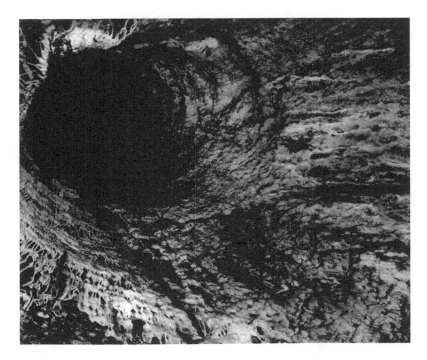

FIGURE 1.17
Lime shaft kiln (lime industry).

FIGURE 1.18
Glass tank (glass industry).

Bibliography

T. H. Benny, *Applications of Ceramics*, http://www.articlesbase.com/education-articles/applications-of-ceramics-6266693.html?utm_source=google&utm_medium=cpc&utm_campaign=ab_paid_12&gclid=CPz9v7rjsLgCFYVKpgodhjIApw (accessed October 2, 2014).

S. C. Carniglia and G. L. Barna (Eds.), *Handbook of Industrial Refractories Technology—Principles, Types, Properties and Applications*, Noyes Publications, Westwood, NJ, 1992, Chapter III: Foundations of Refractory Application, pp. 67–133.

Radex-Heraklith Industriebeteiligungs AG, *Refractory Applications*, http://www.rhiag.com/internet_en/media_relations_en/mr_photo_archives_en/mr_photo_archives_applications_en/ (accessed October 2, 2014).

D. W. Richerson, *Modern Ceramic Engineering—Properties, Processing, and Use in Design*, CRC Press, 2006, Boca Raton, pp. 7–86.

2

Selection of Materials

2.1 Introduction

There are about 70,000 engineering materials available. Selecting the most suited material for a given application out of this large variety is a daunting task. Only a systematic selection method will help in this, and a number of such methods have been developed [1–11].

The selection of materials involves two stages.

2.2 First Stage

In this stage, the performance requirements for the part that has to be met by the material are first specified. Also, the processing requirements of the part are also specified. Once these are specified, some of the available materials that do not meet these requirements get eliminated from the group.

2.2.1 Performance Requirements

Performance requirements consist of functional requirements, processability requirements, cost, reliability, and resistance to service conditions.

2.2.1.1 Functional Requirements

These requirements are those needed for the component to work in the application. To illustrate, if the component will be subjected to a bending load, then a functional requirement for the material of the component will be to have sufficient bending strength.

2.2.1.2 Processability Requirements

Processability requirements are those by which the components may be processed to the required shape. If the process is casting, then the material should be castable. Cast iron possesses better castability than steel. The processing of a material affects its properties. For example, a casting will contain dendritic grains. Hence, the properties become anisotropic. That means, along the length of the dendrite material, strength will be different from that perpendicular to it. Thus, processing affects a material's functional requirements.

2.2.1.3 Cost

Material cost adds to the total cost, and there are limitations to the total cost of a component. Thus, material cost has to be considered for its selection. If all other criteria are satisfied for two materials, then the material of lower cost will be selected.

2.2.1.4 Reliability

A material is said to be reliable if a component made from it fulfills its intended application without failure during its life period. Reliability depends not only on the material properties, but also on the processing. Processing can incorporate certain defects in the component, which will affect its functioning. Failure analysis is carried out on the failed components, and the inference from this analysis can be used in the selection of substituting components to avoid future failures. Failure can take place not only from defects, but also from wrong procedures adopted in processing, faulty design of the component, service conditions, and misuse of the component.

2.2.1.5 Service Conditions

Service conditions can adversely affect the performance of a part. The conditions could be the composition of the environment in which the component is used, its temperature, and pressure. If the composition of the environment is such that the environment corrodes the material, then the performance of the component is adversely affected. Higher temperatures generally decrease the strength of a material. Increasing the pressure of the environment may not decrease the life of a component as long the material is ceramic. This is because ceramics possess good compressive strength.

2.2.2 Quantitative Methods

We have to now consider the available materials meeting the performance requirements. The materials may be metals, ceramics, or plastics. Under the performance requirements, there can be *essential* ones and *desirable* ones. These are to be weighed against the different materials available.

For example, in the case of a load-bearing component, the essential requirement is its load-bearing capacity. This is given by its compressive strength. Its processability and resistance to environment are the desirable characteristics.

Once a group of materials satisfying the performance characteristics are formed, the number in the group is narrowed down to a few by applying quantitative methods.

2.2.2.1 Material Properties

Material properties can be rigid ones or soft. Rigid properties should be possessed by all the materials in the group. Those materials that do not possess the rigidity requirements will automatically get eliminated from the group of available materials. For example, for a structural material, a minimum compressive strength is specified. If any candidate material does not have this minimum strength, it goes out of the group. Also, we should consider the processability. If the component can be fabricated by various methods, then those materials that can be processed by the available method will be initially selected. The next consideration will be on the service conditions. If the service conditions involve high temperatures, then refractory materials will be present in the initially selected group. Thus, the selected materials group can be truncated.

Soft properties can be compromised on. For example, in the case of a structural material, its density can be considered to be a soft property.

2.2.2.2 Cost per Unit Property

Cost per unit property can be initially used to group the materials. This method of initial selection is applicable if one property is critical in the application. If the strength of the material is the critical property, then we can calculate the cost per unit strength in the following manner:

Stress on the material in the application is given by:

$$S = F/A \tag{2.1}$$

Here, F is the force applied in the application and A is the cross-sectional area of the part. The material is supposed to possess at least this much strength, though the actual strength requirement is higher after incorporating the factor of safety. Let us say that the length of the part is L. Now, the cost of the material, let us say, is C per unit mass. Then the cost of the part becomes:

$$C_p = AL\rho C \tag{2.2}$$

Here, L is the length of the part and ρ is the density of the material. Substituting for A from Equation 2.1, we get:

$$C_p = FL\rho C/S \tag{2.3}$$

For a given application, both F and L are fixed. Hence, the different materials can be compared with respect to their ρC/S values. Fix an upper limit for this value, and make a group of materials having this upper limit value or lower. Materials from this group are taken for further analysis in the selection process.

2.2.2.3 Ashby's Material Selection

Ashby's material selection method involves the comparison between two properties. This is also used for the initial selection. To illustrate, we can consider strength and density as the two properties. Ashby plot is available between strength and density for a variety of materials. A straight line having a constant slope is drawn in this plot. For uniaxial loading, the ratio between the yield strength and density value is fixed for the given application, and this ratio is taken for drawing the straight line. Materials falling above this line are grouped for further analysis.

2.2.2.4 Computer-Based Selection

Because of the difficulty in manual handling of large amounts of data available for the initial selection, several computer-based methods [12–15] have been devised. Out of these many methods, let us consider Dargie's method. A part classification code is given in this method. This code is called the MAPS 1 code.

MAPS 1 code consists of eight digits. Based on the first five digits, the manufacturing processes are eliminated. Based on the last three digits, the materials are eliminated. Table 2.1 gives the illustration for the different digits.

Further, two databases are used in this selection method. They are suitability and compatibility matrices. In the suitability matrix, each digit has a matrix. The column in the matrix contains the digit in the code, and the rows are made up of the processes and materials. The elements in the matrix contain 0 or 2, where 2 represents suitability and 0 unsuitability.

As the name suggests, compatibility matrix says whether a process and a material is compatible or not. In this, the process is represented by rows and

TABLE 2.1

MAPS 1 Code

Position of the Digit	Illustration
First	Batch size
Second	Major dimension
Third	Shape
Fourth	Tolerance
Fifth	Surface roughness
Sixth	Service temperature
Seventh	Corrosion rate
Eighth	Environment

the materials by columns. The elements in these matrices are 0, 1, or 2, where 0 says incompatibility, 1 is difficult compatibility, and 2 is compatibility. On running the program, a list of combinations of the process and the material for producing the component is obtained.

2.3 Second Stage

In the first stage, we obtain a group of materials for a particular application. In the second stage, the size of this group is further cut down. Again, there are a number of methods used for this. We shall consider the method based on weighted properties.

2.3.1 Weighted Properties

In this method, each property is given a weight (α) based on its importance in the service of the component. Then, a weighted property value is obtained by multiplying the value of the property of the material by α. At the end, the property values for all the properties required for the application of the material are summed up. This sum gives the performance index for that material. Whichever material has got the highest value will be the most preferred one for that application.

2.3.1.1 Digital Logic Method

Digital logic method is adopted when there are numerous properties to be considered, and their relative importance is not very clear. In this method, only two properties are considered at a time. Every possible combination of the properties are compared. For each comparison, a yes-or-no evaluation is given. A table (see Table 2.2) is constructed to determine the relative importance of each property.

TABLE 2.2

Determination of Relative Importance of Each Property

| Property | Number of Positive Decisions, N | | | | | | | | | | Positive Decisions | Relative Emphasis Coefficient, α |
	1	2	3	4	5	6	7	8	9	10		
1	0	1			1				1		3	0.3
2		0		1		0		1			2	0.2
3			1		0		0			0	1	0.3
4	1	0		1					0		2	0.2
5			1		0			1		0	2	0.3
Total number of positive decisions											**10**	$\Sigma\alpha = 1.0$

Note: n = number of properties; $N = n\,(n-1)/2$; α = positive decisions/10.

In order to prevent the influence of the property having higher numerical value, a scaling factor is introduced. The property after incorporating the scaling factor is called the *scaled property*. It is given by:

B = (Numerical value of the property/Maximum value in the list) × 100

(2.4)

After obtaining all the scaled properties for a material, its performance index (γ) is obtained as:

$$\gamma = \sum_{i=1}^{n} B_i \alpha_i$$ (2.5)

In the case of a space-filling material, its cost per unit weight (C) and density (ρ) can be incorporated to its performance index. Thus, a figure of merit (M) can be defined as:

$$M = \gamma/C\rho$$ (2.6)

In the case of large number of properties and materials, a computer will simplify the selection process. The material having the highest M value is selected for the given application.

2.4 A Case Study

Now we shall run through a material selection process example [16]. In this process, we shall select the material for a sailing boat mast. The mast will be in the form of a hollow cylinder having a length of 1 m. This structural component will be subjected to 153 kN compressive axial force. The other specifications for the mast are 10 cm outer diameter and 84 mm inner diameter. The weight should not exceed 3 kg. The mast will also be subjected to impact force and water spray. There should be small holes for fixing the neighboring components.

2.4.1 Material Performance Requirements

Material performance requirements can be listed as follows:

    a. Resistance to catastrophic fracture under impact load
    b. Resistance to yielding

   c. Buckling resistance

   d. Resistance to corrosion

   e. Reliability

2.4.2 Initial Selection

The cross-section of the mast can be found from the given inner and outer diameters:

   Cross-sectional area of a tube $= \pi(R^2 - r^2)$, where R is the outer radius and r is the inner radius.

   In the present case, R = 5 cm and r = 4.2 cm.

   Therefore, cross-sectional area of the mast $= 3.14 \times (25 - 17.64) = 3.14 \times 7.36 = 23.11$ cm^2

   ~ 2300 mm^2

   The mast will be subjected to 153 kN compressive axial force.

   Therefore, the compressive stress on the mast will be (153/2300) kPa $\sim$ 67 MPa

   Applying a factor of safety of 1.5 to establish the reliability, the minimum yield strength required for the mast will be $67 \times 1.5 \sim 100$ MPa.

   Looking into Ashby's Materials Selection Charts [10] it is clear that engineering polymers, wood, and some of the lower-strength engineering alloys will be excluded as the candidate materials for the mast. Corrosion resistance can be considered as a soft requirement, and therefore, it will not be considered in the initial selection.

2.4.3 Alternate Solutions

The next step in the selection process is to compare the various materials that satisfy the initial selection criteria. Here, we have to consider the other requirements such as the elastic modulus, density, corrosion resistance, and cost, apart from the yield strength. Elastic modulus value indicates the mast's buckling resistance, which is one of the performance requirements. Table 2.3 gives a list of the materials satisfying the initial selection criteria.

   Since the mast is the component of a sailing boat, which is designed for movement, the weight of the mast is important. That is why specific gravity is considered. It will be more relevant if we consider the specific yield strength and specific modulus. Hence, the weighting factors are fixed for these two properties plus corrosion resistance and cost. These are shown in Table 2.4. Table 2.5 gives the performance indices of the candidate materials. From this, those materials possessing performance index greater than 45 are selected. These happen to be 7 out of the 15 in the list.

TABLE 2.3

List of Materials Satisfying Initial Selection Criteria with Performance
Requirements

Material	Yield Strength, MPa	Elastic Modulus, GPa	Specific Gravity	Corrosion Resistance[a]	Cost[b]
UNS G10200	280	210	7.8	1	5
UNS G10400	400	210	7.8	1	5
UNS K11510	330	212	7.8	1	5
UNS G41300	1520	212	7.8	4	3
UNS S31600	205	200	7.98	4	3
UNS S41600	440	216	7.7	4	3
UNS S43100	550	216	7.7	4	3
UNS A96061	275	69.7	2.7	3	4
UNS A92024	393	72.4	2.77	3	4
UNS A92014	415	72.1	2.8	3	4
UNS A97075	505	72.4	2.8	3	4
Ti-6Al-4V	939	124	4.5	5	1
Epoxy-70% glass fabric	1270	28	2.1	4	2
Epoxy-63% carbon fabric	670	107	1.61	4	1
Epoxy-63% aramid fabric	880	38	1.38	4	1

[a] 5—excellent, 4—very good, 3—good, 2—fair, 1—poor.
[b] 5—very inexpensive, 4—inexpensive, 3—moderately expensive, 2—expensive, 1—very
    expensive. From M. M. Farag and E. El-Magd, *Mater. Design*, 13, 323–327, 1992.

TABLE 2.4

Weighting Factors

Property	Specific Yield Strength, MPa	Specific Elastic Modulus, GPa	Corrosion Resistance	Cost
α	0.3	0.3	0.15	0.25

2.4.4 Optimum Solution

The optimum solution in the selection of a material is based on its resistance
to fracture. Hence, we now have to consider the different ways by which the
mast may fail. The mast can fail by four ways

1. Yielding
2. Local buckling
3. Global buckling
4. Internal fiber buckling in the case of composites

The material will not yield, if

$$F/A < \sigma_y \qquad (2.7)$$

TABLE 2.5

Performance Indices of the Candidate Materials

Material	Scaled Specific Yield Strength, MPa	Scaled Specific Elastic Modulus, GPa	Scaled Corrosion Resistance	Scaled Cost	γ
UNS G10200	1.7	12.3	3	25	42
UNS G10400	2.4	12.3	3	25	42.7
UNS K11510	2	12.3	3	25	42.3
UNS G41300	9.2	12.3	6	15	42.5
UNS S31600	1.2	11.3	12	15	39.5
UNS S41600	2.7	12.7	12	15	42.4
UNS S43100	3.4	12.7	12	15	43.1
UNS A96061	4.8	11.6	9	20	45.4
UNS A92024	6.7	11.8	9	20	47.5
UNS A92014	7	11.6	9	20	47.6
UNS A97075	8.5	11.7	9	20	49.2
Ti-6Al-4V	9.8	12.5	15	5	42.3
Epoxy-70% glass fabric	28.4	12.6	12	10	63
Epoxy-63% carbon fabric	19.6	30	12	5	66.6
Epoxy-63% aramid fabric	30	12.4	12	5	59.4

Here, F is the axial applied force on the material, and σ_y is the yield strength of the material. The component will have local buckling resistance as long as:

$$F/A \le 1.21ES/D \qquad (2.8)$$

Here, E is the elastic modulus of the material, S is the wall thickness of the mast, and D is its outer diameter. Global buckling will not take place if:

$$F/A \le \pi^2 EI/AL^2 \qquad (2.9)$$

Here, I is the second moment of area and L is the length of the component. Assuming that an eccentricity (ξ) exists between the load and component axes at the beginning of loading, the condition to avoid both yielding and global buckling can be shown to be:

$$\sigma_y \ge (F/A)[1 + (\xi DA/2I)\sec\{(F/EI)^{1/2}(L/2)\}] \qquad (2.10)$$

Eccentricity is the distance between the centerlines of cylindrical component and axial compressive load. Internal buckling resistance is obtained if:

$$F/A < [E_m/4(1 + \upsilon_m)(1 - V_f^{1/2})] \qquad (2.11)$$

Here, E_m is the elastic modulus of the matrix, v_m is the Poisson's ratio of the matrix, and V_f is the volume fraction of the fibers parallel to the direction of the applied force.

The next stage in material selection is to find the optimum dimensions for the mast if it is made of each of the seven materials shortlisted. This can be done by plotting the outer diameter versus thickness of the mast, which satisfy Equations 2.7 through 2.10 for the four metallic materials, and Equations 2.7 through 2.11 for the composites. This plot is given for the UNS A97075 alloy in Figure 2.1. The optimum value of the outer diameter and thickness will be that at which all the three conditions are satisfied. From Figure 2.1, it is the point represented by (e), which is a point below the line marking an outer diameter of 100 mm. This value is the maximum diameter allowed for the mast. This way, the optimum dimensions for the mast are found for all the other six materials. The results are tabulated in Table 2.6. After getting the optimum values for D and S, the internal diameter can be calculated. This is also shown for the different materials. From the table, it is seen that only the UNS A96061 alloy satisfies the criterion of minimum 84 mm as the internal diameter. Thus, this alloy is selected for making the mast. In case there are more than one material that satisfy all the criteria, then the selection will be based on the total cost of the material. The one having the least cost will be selected.

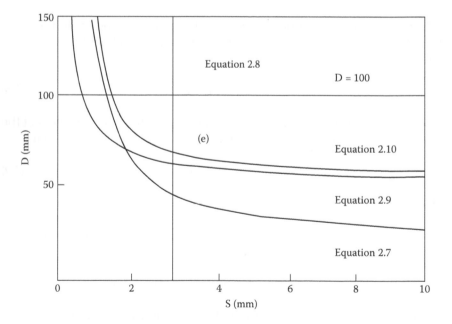

FIGURE 2.1
Design range as predicted by Equations 2.7 through 2.10.

TABLE 2.6

Designs Using Candidate Materials

Material	D^{16} (mm)	S^{16} (mm)	d (mm)
UNS A96061	100	3.4	93.2
UNS A92024	88.3	2.89	82.5
UNS A92014	85.6	2.89	79.8
UNS A97075	78.1	2.89	72.3
Epoxy-70% glass fabric	78	4.64	68.7
Epoxy-63% carbon fabric	73.4	2.37	68.7
Epoxy-62% aramid fabric	75.1	3.99	67.1

References

1. M. M. Farag, *Materials Selection for Engineering Design*, Prentice Hall Europe, London, 1997.
2. G. Dieter, Overview of the material selection process, in *ASM Metals Handbook*, Vol. 20, *Materials Selection and Design*, ASM, Materials Park, OH, 1997, pp. 243–254.
3. J. Clark, R. Roth, and F. Field, III, Techno-economic issues, in *ASM Metals Handbook*, Vol. 20, *Materials Selection and Design*, ASM, Materials Park, OH, 1997, pp. 255–265.
4. M. F. Ashby, Materials selection charts, in *ASM Metals Handbook*, Vol. 20, *Materials Selection and Design*, ASM, Materials Park, OH, 1997, pp. 266–280.
5. M. F. Ashby, Performance indices, in *ASM Metals Handbook*, Vol. 20, *Materials Selection and Design*, ASM, Materials Park, OH, 1997, pp. 281–290.
6. D. Bourell, Decision matrices in materials selection, in *ASM Metals Handbook*, Vol. 20, *Materials Selection and Design*, ASM, Materials Park, OH, 1997, pp. 291–296.
7. T. Fowler, Value analysis in materials selection and design, *ASM Metals Handbook*, Vol. 20, *Materials Selection and Design*, ASM, Materials Park, OH, 1997, pp. 315–321.
8. F. A. Crane and J. A. Charles, *Selection and Use of Engineering Materials*, Butterworths, London, 1994.
9. M. F. Ashby, *Materials Selection in Mechanical Design*, Pergamon, London, 1992.
10. M. F. Ashby, Materials selection in conceptual design, *Mater. Sci. Tech.*, 5: 517–525, 1989.
11. R. Sandstorm, An approach to systematic materials selection, *Mater. Design*, 6: 328–338, 1985.
12. V. Weiss, Computer-aided materials selection, in *ASM Metals Handbook*, Vol. 20, *Materials Selection and Design*, ASM, Materials Park, OH, 1997, pp. 309–314.
13. P. A. Gutteridge and J. Turner, Computer-aided materials selection, and design, *Mater. Design*, 3: 504–510, 1982.
14. L. Olsson, U. Bengston, and H. Fischmeister, Computer-aided materials selection, in *Computers in Materials Technology*, T. Ericcson (Ed.), Pergamon, Oxford, 1981, pp. 17–25.

15. P. P. Dargie, K. Parmeshwar, and W. R. D. Wilson, MAPS 1: Computer-aided design system for preliminary materials and manufacturing process selection, *Trans. ASME, J. Mech. Design*, 104: 126–136, 1982.
16. M. M. Farag and E. El-Magd, An integrated approach to product design, materials selection, and cost estimation, *Mater. Design*, 13: 323–327, 1992.

Bibliography

J. F. Shackelford and W. Alexander, *CRC Practical Handbook of Materials Selection*, Taylor & Francis, New York, 1995.

3

New Developments in Ceramic and Refractory Fields

3.1 Introduction

There are a number of new developments in the ceramic and refractory fields, which are covered separately in this chapter.

3.2 New Developments in the Ceramic Field

One of the major developments in the twentieth century took place at the CSIRO, Melbourne, thanks to the work by Ron Garvie et al. [1]. They developed the partially stabilized zirconia (PSZ), which was later toughened by phase transformation. Phase transformations, alloying, quenching, and tempering techniques were applied to a range of ceramic systems. Significant improvements to the fracture toughness, ductility, and impact resistance of ceramics were realized, and thus the gap in physical properties between ceramics and metals began to close. More recent developments in nonoxide and tougher ceramics (e.g., nitride ceramics) have narrowed the gap even further.

New or *advanced* ceramics, when used as engineering materials, possess several properties that can be viewed as superior to metal-based systems. These properties place this new group of ceramics in a most attractive position, not only in the area of performance but also in cost effectiveness. These properties include high resistance to abrasion, excellent hot strength, chemical inertness, high machining speeds (as tools), and dimensional stability.

Engineering ceramics can be classified into three—oxides, nonoxides, and composites. Examples for oxides are alumina and zirconia. Carbides, borides, nitrides, and silicides come under nonoxides. Particulate-reinforced oxides and nonoxides are examples for composites. Oxide ceramics are characterized by oxidation resistance, chemical inertness, electrical insulation, and low thermal conductivity. Low oxidation resistance, extreme hardness, chemical inertness, high thermal conductivity, and electrical conductivity are the properties of nonoxide ceramics. Ceramic-based composites possess toughness, low and high oxidation resistance (depends upon the type), and variable thermal and electrical conductivity. The manufacturing processes are complex, and finally result in high cost.

3.2.1 Production

The production of technical or engineering ceramics, as compared to earlier traditional ceramic production, is a much more demanding and complex procedure. High-purity materials and precise methods of production must be employed to ensure that the desired properties of these advanced materials are achieved in the final product.

3.2.1.1 Oxide Ceramics

High-purity starting materials (powders) are prepared using mineral processing techniques to produce a concentrate, followed by further processing (typically wet chemistry) to remove unwanted impurities and to add other compounds to create the desired starting composition. This is a most important stage in the preparation of high-performance oxide ceramics. As these are generally high-purity systems, minor impurities can have a dynamic effect—for example, small amounts of MgO can have a marked effect upon the sintering behavior of alumina. Various heat treatment procedures are utilized to create carefully controlled crystal structures. These powders are generally ground to an extremely fine or *ultimate* crystal size to assist ceramic reactivity. Plasticizers and binders are blended with these powders to suit the preferred method of forming (pressing, extrusion, slip casting, etc.) to produce the *raw* material. Both high- and low-pressure forming techniques are used. The raw material is formed into the required *green* shape or precursor (machined or turned to shape if required) and fired to high temperatures in air or a slightly reducing atmosphere to produce a dense product.

3.2.1.2 Nonoxide Ceramics

The production of nonoxide ceramics is usually a three-stage process involving the following steps: first, the preparation of precursors or starting powders; second, the mixing of these precursors to create the desired compounds (Ti + 2B, Si + C, etc.); and third, the forming and sintering of the final component. The formation of starting materials and firing for this group require carefully controlled furnace or kiln conditions to ensure the absence of oxygen during heating, as these materials will readily oxidize during firing. This group of materials generally requires quite high temperatures to effect sintering. Similar to oxide ceramics, carefully controlled purities and crystalline characteristics are needed to achieve the desired final ceramic properties.

3.2.1.3 Ceramic-Based Composites

This group can be composed of a combination of oxide ceramics–nonoxide ceramics; oxide–oxide ceramics; nonoxide–nonoxide ceramics; ceramics–polymers, and so on—and an almost infinite number of combinations are possible. The reinforcements can be granular, platy, whiskers, and so on. When it is a combination of ceramics and polymers, the objective is to improve the toughness of the ceramics, which otherwise is brittle. When we combine two ceramics, the intention is to improve the hardness, so that the combination becomes more suited to a particular application. This is a somewhat new area of development, and compositions can also include metals in particulate or matrix forms.

3.2.1.4 Firing

Firing conditions for new tooling ceramics are somewhat diverse, both in temperature range and equipment. In general, these materials are fired to temperatures well above the melting points of most of the metals. The range is 1500°C–2400°C and even higher. Attaining these high temperatures requires very specialized furnaces and furnace linings. Some materials require special gas environments, such as nitrogen, or controlled furnace conditions such as a vacuum. Others require extremely high pressures to achieve densification. Hot isostatic pressing (HIP) is a high-pressure process. The furnaces are quite diverse, both in design and concept. The typical methods of heating the furnaces are by using gases, resistance heating, or induction heating. The gases used for heating can be fuel gas plus oxygen or heated air. Resistance heating is achieved by using metallic, carbon, and ceramic resistors. Radio frequency or microwave is used for induction heating.

3.2.1.5 Firing Environments

Gas heating is generally carried out in normal to low pressures. Resistance heating is carried out in pressures ranging from vacuum to 200 MPa. Induction heating can also be done over the same range as resistance. In both resistance and induction heating, the systems do not have to contend with high volumes of ignition products, and thus can be contained. The typical furnace types used in gas heating are box, tunnel, and bell. The HIP furnace can be heated by gas and resistance. When carbon is the heating element, the furnace is an *autoclave* sealed-type one. This is to prevent any oxidation of the carbon at high temperatures. Again, for water-cooled-type furnace for RF-heated one, sealed special designs are used. Open design is used for microwave-heated furnace for heating small items.

3.2.1.6 Finishing

One of the final stages in the production of advanced materials is the finishing to precise tolerances. These materials can be extremely hard, with hardness approaching that of diamonds, and thus finishing can be quite an expensive and slow process. Finishing techniques can include: laser, water jet and diamond cutting, diamond grinding, and drilling. However, if the ceramic is electrically conductive, techniques such as electrical discharge machining (EDM) can be used. As the pursuit of hardness is one of the prime developmental objectives, and as each newly developed material increases in hardness, the problems associated with finishing will also increase. The development of CNC grinding equipment has lessened the cost of final grinding by minimizing the labor content; however, large runs are generally required to offset the set-up costs of this equipment. Small runs are usually not economically viable. One alternative to this problem is to *net form*, or form to predictable or acceptable tolerances to minimize machining. This has been achieved at Taylor Ceramic Engineering by the introduction of a technique called *near to net shape forming*. Complex components can be formed by this unique Australian development, with deviations as low as ±0.3%, resulting in considerable savings in final machining costs.

In many applications today, the beneficial properties of some materials are combined to enhance and, at times, support other materials, thus creating a hybrid composite. In the case of hybrid composites, it is the availability and performance properties of each new material that sets the capability of the new material. To determine the long-term durability of a new composite before actually committing to service, field evaluation tests will have to be carried out.

3.2.2 Future Developments

Advanced ceramic materials are now well established in many areas of everyday use. The improvements in performance, service life, savings in operational costs, and savings in maintenance are clear evidence of the benefits of advanced ceramic materials. Life expectancies, now in years instead of months, with cost economics in the order of only double the existing component costs, give advanced ceramic materials a major advantage. The production of these advanced materials is a complex and demanding process with high equipment costs and the requirement of highly specialized and trained people. The ceramic materials of tomorrow will exploit the properties of polycrystalline phase combinations and composite ceramic structures—that is, the coprecipitation or inclusion of differing crystalline structures having beneficial properties working together in the final compound.

Tomorrow (even today), the quest will be to pack the highest amount of bond energy into the final ceramic compound and to impart a high degree of ductility or elasticity into those bonds. This energy level has to be exceeded to cause failure or dislocation generation. The changing pace of technology and materials also means that newer compounds precisely engineered to function will be developed. Who can tell just how this will be achieved and when the knowledge becomes public. Ceramics, an old class of material, still present opportunities for new material developments.

Another interesting development has been robocasting [2]. It relies on robotics for computer-controlled deposition of ceramic slurries. Mixtures of ceramic powder, water, and trace amounts of chemical modifiers are injected through a syringe. The material, which flows like milkshake even though the water content is only about 15%, is deposited in thin sequential layers onto a heated base. Figure 3.1 shows the syringe and the product formed. It may result in cheaper and faster fabrication of complex parts. Because the new method allows a dense ceramic part to be free-formed, dried, and baked in less than 24 hours, it is perfect for rapid prototyping. Engineers can quickly change the design of a part and physically see if it works. Besides prototyping, Joseph Cesarano III, an engineer at the U.S. Department of Energy's Sandia National Laboratories, who originated the

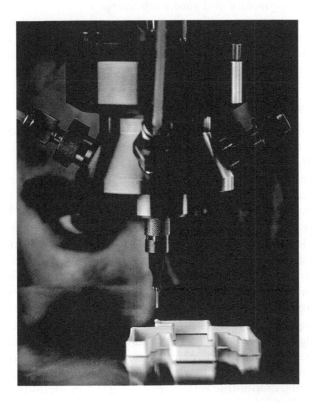

FIGURE 3.1
The syringe and the product formed.

concept, sees the new method also being used in the actual manufacturing of ceramics. Other ceramic processing techniques, such as slip casting, gel casting, and injection molding, can yield complicated ceramic parts, but require the design and manufacture of molds prior to fabrication. After the part is formed by layering and completely dried, it is sintered in a furnace at very high temperatures, so that the particles can densify.

3.3 New Developments in the Refractory Field

The use of monolithic refractories has been a new development in the refractory field.

3.3.1 Applications of Monolithic Refractories

Monolithic linings are a relatively recent development and consist of unshaped refractory products [3]. There are two basic types of monolithic linings: *castable refractory* and *plastic refractory*.

The following sections outline the applications of monolithic refractories, and the types of refractories that are used in specific parts of the operation [4]. However, the type of refractory used and even the kind of materials used in a particular application will differ from plant to plant.

3.3.1.1 Ferrous Metallurgy

Gunned insulating castables are used to line the outer middle section of modern blast furnaces; also, the uptake sections of the top of the blast furnace are lined with gunned insulating material.

Dense castable is used to line the blowpipe section, which joins the tuyeres to the bustle main pipe. In addition to this, dense castables are used for reinforcing the middle section of the blast furnace linings. The blast furnace trough where molten iron exits the furnace is lined with a high-temperature dense castable formulation. The covering area of the blast furnace trough is lined with high-strength castable.

The blast furnace stove, which heats the air that will be blown into the furnace from the tuyeres and burns the coke, is insulated with an acid-resistant gunned material. The stack section is also lined with a gunned castable. A high-alumina monolithic refractory is used around the spout of the torpedo ladle. Torpedo ladles are used for transferring molten iron from the iron-making section of the plant to the steel section. Often when short-term repairs are done to this item, monolithic refractories are used.

In steel plants, monolithic refractories are used in sections of the linings of teeming ladles, converters, molten steel ladles, vacuum degassers, and tundishes. In tundishes, the majority of refractory lining is made from a dense castable that is coated with magnesia. The outer lining of the converter is lined with a gunned castable. The injection lance, which is used to blow gas into the steel in the converter, is fabricated from a dense castable. The flooring in molten steel ladles is cast from high-strength magnesia spinel castable materials.

3.3.1.2 Petrochemical Applications

Refractories in applications such as petroleum refineries are used primarily with insulating and abrasion-resistant properties in mind. Sections such as cyclone walls and reactors are lined with dense high-alumina castables. Various areas of chemical plants, such as reactors and high-pressure vessels, are commonly lined with high-grade alumina monolithics.

3.3.1.3 Cement Plant Applications

Monolithic refractories are used in sections of cement kilns, typically at the outlet, where wear rates are very high. At these zones, dense silicon-carbide-based castables are often used. The cyclones in the preheater section of cement plants are lined with dense castables due to the complexity of their shape and the wear resistance required. The shell of the exhaust gas duct is made from both dense and insulating castable materials. In the cooler section of the plant, monolithic refractories are used to line the ceiling.

3.3.1.4 Incinerators

Incinerators used for the disposal of municipal waste rely heavily on various monolithic linings. In the primary combustion chamber, the entrance section is lined with an abrasion-resistant castable. The main section is usually lined with a high-alumina plastic gunned material, which is backed with an insulating castable. The ducting that connects the primary and secondary combustion chambers is lined with a dense castable material, and so is the secondary chamber. The ducting that connects the dust collectors to the stack is made from an acid-resistant castable.

3.3.1.5 Nonferrous Metallurgy

For the majority of nonferrous applications, monolithics tend to be used only as repair materials and backing insulation.

In the aluminum industry, melting furnaces are made entirely out of monolithic refractories. The heating chamber is lined with high-alumina castable, while sections that are in direct contact with the metal are made from phosphate-bonded plastic refractory.

Now we shall consider the applications of some individual refractory materials. These are summarized in Table 3.1.

Refractory products are used in all industrial high-temperature processes in the basic industries and cannot be substituted [5].

TABLE 3.1

Some Individual Refractory Materials and Their Uses

Refractory Material	Uses
Fireclays	Furnaces, regenerators, ovens, and kilns
High-alumina refractories	Blast furnace stoves, cement and lime rotary kilns, electric arc furnace roofs, ladle, glassmaking furnaces, etc.
Chromite—Magnesite	Inner lining of basic oxygen steelmaking vessels, side walls of soaking pits, etc.

References

1. D. A. Taylor et al., Advanced ceramics—The evolution, classification, properties, production, firing, finishing and design of advanced ceramics, *Mater. Aust.*, 33(1): 20–22, 2001.
2. Sandia National Laboratories, *Robocasting*, http://www.sandia.gov/media/robocast.htm (accessed October 2, 2014).
3. National Programme on Technology Enhanced Learning, *Application of Refractory Materials*, http://nptel.iitm.ac.in/courses/113104058/mme_pdf/Lecture15.pdf (accessed October 2, 2014).
4. CERAM Research Ltd, *Refractories - Some Examples of Applications of Monolithic Refractories*, http://www.azom.com/article.aspx?ArticleID=1410 (accessed October 2, 2014).
5. Carborundum Universal Limited, *Industrial Ceramics*, http://www.cumi-murugappa.com/useful-articles/industrial-ceramics-products.html- (accessed October 2, 2014).

4

Phase Equilibria in Ceramic and Refractory Systems

4.1 Introduction

Phase equilibria in ceramic and refractory systems can be studied with the help of phase diagrams. Phase diagrams are based on Gibb's phase rule, which can be mathematically represented by the following equation:

$$P + F = C + 2 \tag{4.1}$$

Here, P represents the number of phases present at equilibrium at the point in a phase where the system is considered. F is the degree of freedom (the number of independently variable factors affecting the range of states in which a system may exist), representing the number of variables to be changed in order to change the state of the system. C is the number of components (the number of chemically independent constituents of the system, i.e. the minimum number of independent species necessary to define the composition of all the **phases** of the system) of the system. The number 2 stands for the environmental variables, pressure, and temperature.

Depending on the number of components, the phase diagrams can be classified into unary, binary, and ternary, when the number of components is one, two, and three, respectively.

4.2 Unary-Phase Diagrams

Unary-phase diagrams will contain various polymorphic forms of the solid (if the component exhibits polymorphism), liquid, and vapor as the phases. The largest number of phases that will be at equilibrium will be when the degrees of freedom is zero, as evident from Equation 4.1.

Rearranging this equation to find the number of phases, we get:

$$P = C - F + 2 \tag{4.2}$$

Substituting the values for C and F, we get:

$$P = 1 - 0 + 2 = 3 \tag{4.3}$$

There are a number of applications of one-component phase diagrams in ceramics. One such application is the development of commercial production of synthetic diamonds from graphite. For this, high temperatures and high pressures are necessary. This is evident from the phase diagram of carbon shown in Figure 4.1 [1].

Phase diagrams do not give any idea about the kinetics of the phase transitions. In the present case of transition of graphite into diamond, a liquid metal catalyst or a mineralizer such as Ni is required for the reaction to proceed at a useful rate.

Another system, which has been studied at high pressures and temperatures, is that of silica. At pressures above 3–4 GPa, a new phase of silica called *coesite* is formed. This phase is also found in nature. This forms in earth's crust as a result of meteorite impacts. When the pressure on silica is increased to above 10 GPa, another phase called *stishovite* is formed. The low-pressure phases of silica are shown in Figure 4.2.

At low pressures, there are five condensed phases for silica. They are α-quartz, β-quartz, β$_2$-tridymite, β-cristobalite, and liquid silica. The transition of α-quartz to β-quartz takes place at 573°C. This transition is rapid

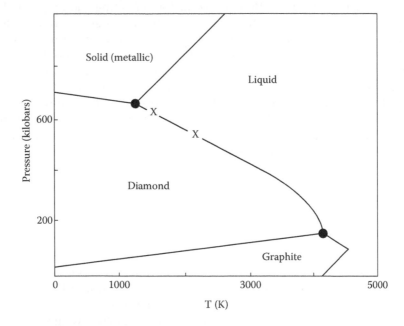

FIGURE 4.1
High-pressure, high-temperature phase equilibrium for carbon.

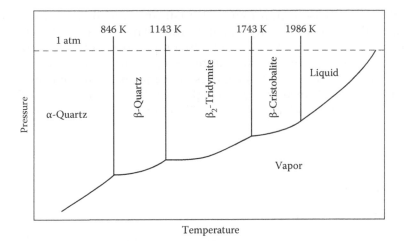

FIGURE 4.2
Equilibrium diagram for SiO$_2$.

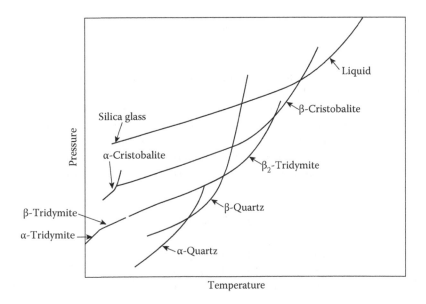

FIGURE 4.3
Phase diagram including metastable phases occurring in the system SiO$_2$.

and reversible. The other transitions, as shown in Figure 4.3, are sluggish. Apart from stable forms of silica, it also forms metastable silica allotropes. These are shown in Figure 4.3.

Transition between cristobalite and quartz is a sluggish process. Hence, β-cristobalite transforms on normal cooling rates to α-cristobalite.

For the same reason, β_2-tridymite transforms into β- and α-tridymite. Also, on the same line of argument, liquid forms silica glass. The transition of silica glass to stable form at room temperature being too sluggish, it remains indefinitely in this state at this temperature. Thus, glass becomes applicable for all practical purposes. It requires long-time heating at about 1100°C for transforming silica glass to the crystalline stable form of silica. This form is called cristobalite. This kind of crystallization is called devitrification.

4.3 Binary Systems

Compared to unary-phase diagrams, binary ones contain an additional composition variable. Hence, the C increases by 2 in the phase rule. If the system has only one phase, then the degree of freedom becomes three:

$$1 + F = 2 + 2 \quad \text{or} \quad F = 3 \tag{4.4}$$

This implies that, in order to represent pressure, temperature, and composition region of stability of a single phase, a three-dimensional diagram must be used. While dealing with condensed systems involving only solids and liquids, the effect of pressure is small. Also, most often, we are concerned with systems at or near atmospheric pressure. Hence, diagrams at constant pressure can be drawn. The result is that binary diagrams can be represented as two-dimensional figures. The axes in these figures will be temperature and composition. The phase rule in such a representation is given by:

$$P + F = C + 1 \tag{4.5}$$

A simple binary diagram is shown in Figure 4.4.

Point A represents a single-phase region. In this region, both temperature and composition can be varied, and the system still does change its state of equilibrium.

In areas in which two phases are present at equilibrium, the composition of each phase is indicated by lines on the diagram. Point B in Figure 4.4 is a point in such an area. Intersection of a constant-temperature *tie line*, shown, with the phase boundaries gives the compositions of the phases in equilibrium at temperature T. In Figure 4.4, the compositions are the pure components involved.

With two phases present, either temperature or composition needs to be specified to define the system in this area. That means the variance is one.

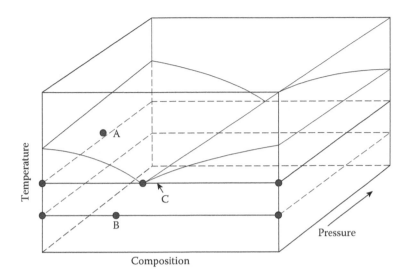

FIGURE 4.4
Simple binary diagram showing the independence of the system on pressure.

This value of variance is obtained from Equation 4.6. The maximum number of phases, when $F = 0$, is obtained from Equation 4.7.

$$P + F = C + 1, 2 + F = 2 + 1, F = 1 \qquad (4.6)$$

$$P + F = C + 1, P + 0 = 2 + 1, P = 3 \qquad (4.7)$$

When three phases are present, the composition of each phase and temperature of the system are fixed. Figure 4.4 shows the tie line that passes through such an invariant point C.

4.3.1 Binary-Phase Diagrams

A hypothetical binary-phase diagram is shown in Figure 4.5. The pure components are A and B.

The right-hand side of the diagram is not completed as it is a repetition of the left-hand side in terms of the nature of the plot. The complete diagram will have three single-phase regions: α, β, and liquid-phase regions. α and β are solid solutions of B in A and A in B, respectively. There are also three two-phase regions. They are $\alpha + L$, $\beta + L$, and $\alpha + \beta$.

When a second component is added to a pure material, the freezing point is often lowered. This is evident in the lowered liquidus curves for both end members. This diagram shows a eutectic point, k. It is an invariant point, as

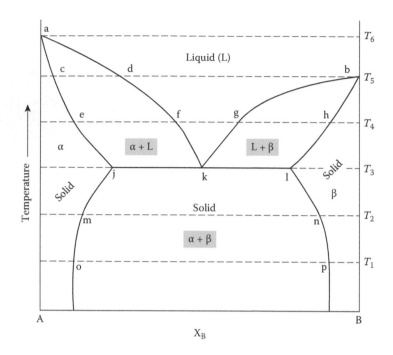

FIGURE 4.5
A hypothetical binary-phase diagram.

the three phases are at equilibrium at this point, and the phase rule gives that, when three phases are formed from two components in a condensed system, the value of F becomes zero. The temperature at which this happens is called *eutectic temperature*. This temperature can be defined as that at which liquidus curves intersect, and it is the lowest temperature at which liquid occurs. An eutectic composition can also be defined, which is the composition of the liquid at this temperature, with the liquid coexisting with two solid phases. The temperature cannot change at the eutectic point, unless one phase disappears.

BeO–Al_2O_3 (Figure 4.6) can be taken as an example for a binary system. This system can be divided into three simpler two-component systems (BeO–$BeAl_2O_4$, $BeAl_2O_4$–$BeAl_6O_{10}$, and $BeAl_6O_{10}$–Al_2O_3), in each of which the freezing point of pure material is lowered by addition of a second component.

The BeO–$BeAl_2O_4$ subsystem contains a compound, $Be_3Al_2O_6$, which melts incongruently. In only one single-phase region, the composition comprises 100% of the system. Point A is present in such a region. In two-phase regions, the phases are indicated. Considering point B in one of such regions, the composition of each phase is represented by the intersection of a constant temperature line (tie line) and phase-boundary lines. The amount of each phase

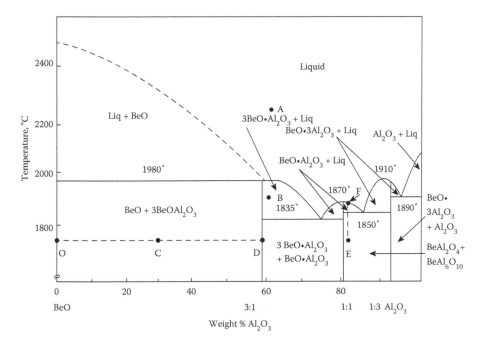

FIGURE 4.6
The binary system BeO–Al$_2$O$_3$.

can be determined from the fact that the sum of the composition times the amount of each phase present equals the composition of the entire system. This fact can be represented graphically by the *lever principle*, which states that the distance from one phase boundary to overall system composition, divided by the distance between the boundaries. The result is the fraction of the second phase.

For example, consider point C, which is another point in a two-phase region. Applying the lever principle, we get:

$$(OC/OD)100 = \text{Percent } 3BeO \cdot Al_2O_3 \tag{4.8}$$

That is,

$$[(29 - 0)/(58 - 0)]100 = 50 = \text{Percent } 3BeO \cdot Al_2O_3 \tag{4.9}$$

$$DC/OC = (58 - 29)/(29 - 0) = 1 = BeO/3BeO \cdot Al_2O_3 \tag{4.10}$$

Thus, the entire system is composed of 29% Al$_2$O$_3$ and consists of two equal amounts of phases, BeO and 3BeO·Al$_2$O$_3$ (which contains 58% Al$_2$O$_3$).

There must be 50% of each phase for a mass balance to give correct overall composition.

Now consider the changes that occur in the phases present on heating a composition represented by the point E, which consists of a mixture of $BeAl_2O_4$ and $BeAl_6O_{10}$. These phases remain until 1850°C; at this eutectic temperature, the reaction, $BeAl_2O_4$ + $BeAl_6O_{10}$ = Liquid (85% Al_2O_3), continues at constant temperature to form the eutectic liquid until all $BeAl_6O_{10}$ is consumed. On further heating, more of $BeAl_2O_4$ dissolves in the liquid and the liquid composition changes along G-F until, at 1865°C, all $BeAl_2O_4$ disappears. On cooling this liquid, the reverse reaction occurs.

One of main features of eutectic systems is the lowering of the temperature at which the liquid is formed. In the $BeO-Al_2O_3$ system, the pure end members melt at 2500°C and 2040°C, respectively. In this two-component system, the liquid is formed at temperatures as low as 1835°C. This can be an advantage or a disadvantage. For getting the maximum temperature use, there should not be any liquid. In Figure 4.6, it can be seen that the addition of a small amount of BeO to Al_2O_3 results in the formation of a substantial amount of liquid at 1890°C. This composition then becomes useless as a refractory above this temperature.

The existence of a liquid becomes advantageous in the case of firing during the processing of ceramics. It is desirable to form a liquid as an aid to firing at lower temperatures, since the liquid increases the ease of densification.

Lowering the melting point is made use of in the Na_2O-SiO_2 system. Liquidus is lowered from 1710°C in pure SiO_2 to 790°C for eutectic composition (75% SiO_2–25% Na_2O). Thus, eutectic composition can be fired at a temperature as low as less than half the melting point of SiO_2. Sintering is achieved at a very low temperature, thereby economizing the processing.

Now, let us consider *incongruent melting*. Mention was made of this melting for the compound $Be_3Al_2O_6$ present in the $BeO-Al_2O_3$ system. A solid does not melt to form its own composition; rather, it dissociates to form a new solid and a liquid.

At this *incongruent melting point* or *peritectic temperature*, three phases (two solids and a liquid) are at equilibrium. The temperature remains fixed until the reaction is completed.

Another phenomenon that can take place in a binary system is *phase separation*. When a crystalline solid solution is cooled, it separates into two separate phases. This is evident in the CoO–NiO system (Figure 4.7). The same system has a complete series of *solid solutions*. Limited solid solutions occur in some other systems. For example, in the MgO–CaO system, two partial solid solutions are seen (Figure 4.8). They are MgO ss and CaO ss.

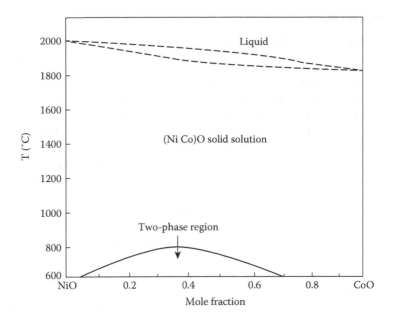

FIGURE 4.7
The binary system NiO–CoO.

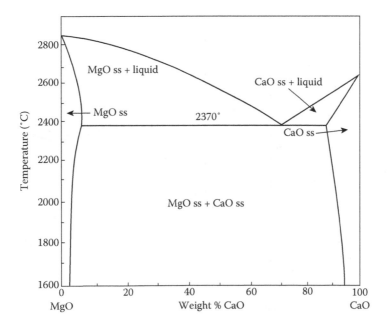

FIGURE 4.8
MgO–CaO system.

4.4 Ternary-Phase Diagrams

Ternary-phase diagrams, consisting of three components, possess four independent variables: pressure, temperature, and the concentrations of any two components out of three present in a ternary system. If pressure is fixed, then the presence of four phases gives rise to an invariant system. Compositions are represented on an equilateral triangle and the temperature on the vertical ordinate (Figure 4.9).

For two-dimensional representation, the temperature can be projected on an equilateral triangle, with liquidus temperatures represented by isotherms. The diagram is divided into areas representing the equilibrium between a liquid and a solid phase. The boundary curves represent the equilibrium between two solids and a liquid, and the intersections of the three boundary curves represent the points of four phases in equilibrium.

In another method of two-dimensional representation, a constant-temperature plane is cut through the diagram. This plane indicates the phases at equilibrium at that temperature. The interpretation of ternary diagrams is not different from that of binary diagrams. The composition of each

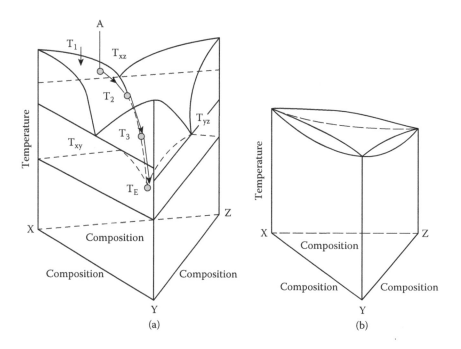

FIGURE 4.9
Representation of ternary diagram: (a) system in which ternary eutectic is present, (b) system forming a complete series of solid solutions.

phase is given by the phase boundary surfaces or intersections. The relative amounts of each phase is determined by the principle that the sum of the individual phase compositions must equal the total composition of the entire system.

In Figures 4.9 and 4.10, the composition A falls in the primary field of X. If we cool the liquid A, X begins to crystallize when the temperature reaches T_1. The composition of the liquid changes along AB. Along this line, the lever principle applies. At any point, the percentage of X is given by 100(BA/XB). When the temperature reaches T_2 and the crystallization path reaches the boundary representing the equilibrium between the liquid and two solid phases X and Z, Z begins to crystallize also, and the liquid changes in composition along CD. At L, the phases in equilibrium are the liquid of composition L and the solids X and Z, whereas the overall composition of the entire system is A. As shown in Figure 4.10b, the mixture of L, X, and Z that gives a total corresponding to A is (xA/xX)100 = percent X, (zA/zZ)100 = percent Z, and (lA/lL)100 = percent L.

Very often, a constant temperature diagram is useful. Such a diagram is illustrated for subsolidus temperatures by the lines between the forms that exist at equilibrium in Figure 4.11 [2].

These lines form composition triangles in which three phases are present at equilibrium.

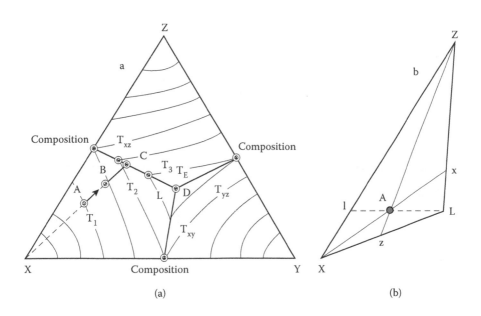

FIGURE 4.10
(a) Cooling path illustrated in Figure 4.9a, (b) application of the center-of-gravity principle to find the proportion of phases.

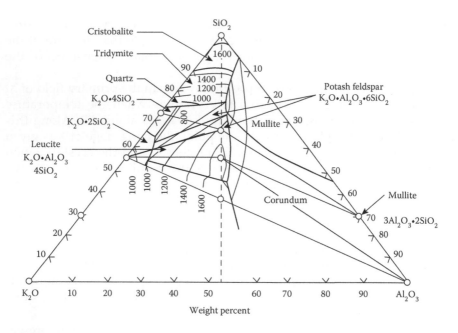

FIGURE 4.11
The ternary system $K_2O–Al_2O_3–SiO_2$.

References

1. G. Suits, Pressure-temperature phase diagram for carbon, *Am. Sci.*, 52: 395, 1964.
2. J. F. Schairer and N. L. Bowen, The ternary system $K_2O-Al_2O_3-SiO_2$, *Am. J. Sci.*, 245: 199, 1947.

Bibliography

W. D. Kingery, *Introduction to Ceramics*, 2nd Edn, John Wiley, New York, 1991, pp. 269–317.

5

Corrosion of Ceramics and Refractories

5.1 Introduction

Ceramics are favorite materials for applications under severe conditions [1].

The reason for their extreme resistance can be explained as follows. Ceramics are compounds between metallic and non-metallic elements. Corrosion products are also ceramics. Hence, ceramic may be thought of as having already been corroded [2]. Contrary to the corrosion of metals, corrosion of ceramics, if at all they corrode, involves chemical dissolution. Metals corrode by electrochemical processes. Another form of corrosion is oxidation. Oxidation takes place in an oxidizing environment. In this respect, the oxidation resistance of materials is to be considered. When we consider ceramics for application in an oxidizing environment, the nonoxide ceramics may not be that resistant. Members of the ceramic spectrum of borides, carbides, nitrides, silicides, and so on tend to get oxidized when exposed to an oxidizing atmosphere. Herein, oxide ceramics are the most stable. In general, ceramics are more stable than metals in terms of their oxidation.

5.2 Corrosion of Ceramics

The excellent corrosion resistance of ceramics can be gauged from the following. Household cutlery, pottery, century-old vases, and so on are made of ceramics. Corrosion resistance arises due to their chemical stability and the high covalent bonding. Often hydrofluoric acid (HF), one of the strongest chemicals, is required to etch the microstructure of engineering ceramics.

Ceramics are generally electrically insulating. They have few charge carriers. Hence, electrochemical corrosion is negligible. They corrode mainly by acid–base type of reactions. Ceramics undergoes dissolution in acids and bases. The same composition can exist as crystalline or amorphous ceramics. There is a difference in corrosion behavior between the two forms. Also, there are glass-ceramics, in which both the phases coexist.

5.2.1 Mechanisms of Dissolution

In a polyphasic ceramic, the total rate of dissolution (R_{tot}) is the sum total of the rates of individual phases. The rates of individual phases are also a function of their areas exposed to the corrosive medium. Thus, we have to consider their weighted surface areas. The weighted surface area of the i-th phase is given by its surface area (S_i) divided by the total surface area of the ceramic body (S) under consideration. Therefore, R_{tot} is given by:

$$R_{tot} = (1/S)\Sigma S_i R_i \qquad (5.1)$$

During the dissolution of the bulk ceramic, there could be three types of secondary effects. As the amount of major phases will be more in the solution, this solution can modify the dissolution of the minor phases. Secondly, solvents can induce chemical reactions among the phases. The insoluble reaction products formed out of the major phases can cover the minor phases and thereby reduce their dissolution.

There are five mechanisms for the corrosion of ceramics. The first one is the *congruent dissolution by simple dissociation*. If the ceramic corrodes by this mechanism, then the ratio of the elements dissolved will be the same as that of them in the ceramic. For example, let us consider sodium chloride. If it dissolves by this mechanism, then the ratio of the number of Na^+ ions to the number of Cl^- ions will be 1. This is the same ratio of these ions in the sodium chloride crystal. This kind of dissolution will lead to uniform removal of the surface—that is, the reduction in thickness will be the same at any point on the surface. Hence, the concentration profile on the surface that is exposed will be same before and after the dissolution. This and the solid undergoing dissolution are represented in Figure 5.1. Here, X_o represents the original surface, and X the surface after equilibrium concentration is achieved.

The second mechanism for the corrosion of ceramics is *congruent dissolution by chemical reaction with the solvent*. Here, the dissolution takes place by the attack of the solvent on the surface of the ceramic. The attack is uniform. Hence, the aftereffect of this kind of dissolution is the same as that of the first—that is, there will be uniform retreat of the surface.

Incongruent dissolution with the formation of crystalline reaction products is the third mechanism for the corrosion of ceramics. In this mechanism, on reaction between the ceramic and the solvent, a crystalline reaction product is formed. It happens in the dissolution of strontium titanate in an acid. The reaction is shown in Equation 5.2.

$$SrTiO_3 + 2H^+ \leftrightarrow Sr^{2+} + TiO_2\downarrow \qquad (5.2)$$

As a result of the reaction, strontium goes into solution as Sr^{2+} ions, and the crystalline reaction product TiO_2, being insoluble in the acid, remains on the surface. The mechanism is shown schematically in Figure 5.2.

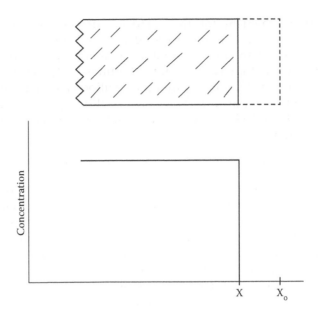

FIGURE 5.1
Concentration profile for congruent dissolution by simple dissociation.

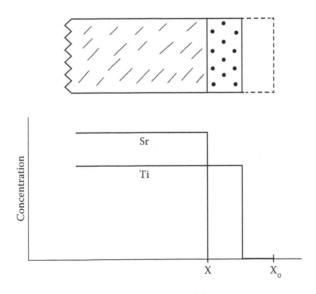

FIGURE 5.2
Incongruent dissolution with the formation of crystalline reaction product.

The fourth mechanism for the corrosion of ceramics is *incongruent dissolution with the formation of noncrystalline layer*. Here, the chemical reaction between the solvent and the solid leaves a noncrystalline layer on the surface. It happens in the case of sodium aluminosilicate. When it is allowed to react with an acid, Na^+ ions go into the solution along with some silicon tetrahedral units. On the surface, a noncrystalline gel layer of combined aluminum octahedral–silicon tetrahedral units are formed. The reaction is shown in Equation 5.3.

$$NaAlSi_3O_8 + H^+ \leftrightarrow Na^+ + Al_2Si_2O_5(OH)_4 + H_4SiO_4 \qquad (5.3)$$

The mechanism is represented in Figure 5.3. As the concentration of silicon and sodium decreases on the surface, aluminum concentration increases.

The fifth mechanism for the corrosion of ceramics is known as *ion exchange*. Here, when a type of ion goes into the solution, another type of ion from the solution gets attached on the bulk phase. Sodium aluminosilicate can be taken as an example here also. From this compound, Na^+ ions go into the water, which contains H_3O^+ ions. These ions go into the bulk solid surface. The reaction is represented by Equation 5.4.

$$NaAlSi_3O_8 + H_3O^+ \leftrightarrow (H_3O)AlSi_3O_8 + Na^+ \qquad (5.4)$$

After the reaction has reached equilibrium, the Si/Al ratio remains the same at the interface between the leached surface and the solvent phase.

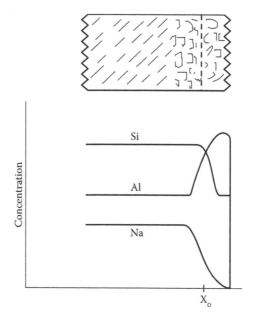

FIGURE 5.3
Incongruent dissolution with the formation of a noncrystalline layer.

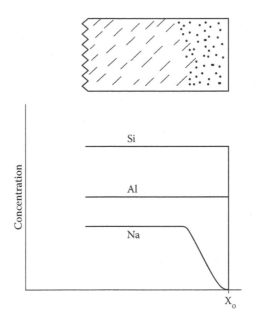

FIGURE 5.4
Ion exchange mechanism.

The mechanism is shown in Figure 5.4. It can be seen that the concentration of Na^+ ions continuously decreases from the interface between the bulk phase and the interface between the leached surface and the solvent.

5.2.2 Corrosion of Glass

5.2.2.1 Introduction

Glasses are used as containers for liquids, and as window glass and optical glass. Another specialized use of glass is for disposing off high-radioactive-level liquid nuclear waste, which is vitrified in borosilicate glass. The fission products after a chain reaction are highly radioactive isotopes. The liquid borosilicate glass is poured into stainless steel canisters. On solidification of the liquid glass, these canisters are placed in geologic repositories. Glass properties are impaired by weathering or by contact with water. When the nuclear waste is contained in glass, this glass should not get corroded for thousands of years, so that mankind is protected from radiation. The actual number of years for which it should remain stable is decided by the half-lives of the radioactive isotopes it contains. The purpose of containment of nuclear wastes in glass is to immobilize them once they are stored in large depths in the earth's crust. The chances of radionuclides coming into the biosphere are from the groundwater flowing near the repositories. When the glass comes into contact with the groundwater, there is a chance of the radionuclide ions getting leached into the groundwater. Thus, it is important that the glass becomes stable toward the groundwater over geologic timescales.

Laboratory tests are compared with the degradation of glasses in nature. Glasses in nature have been in existence for geologic timescales. This comparison is used to determine the stability of man-made glasses. These man-made glasses used for containing nuclear wastes are meant to survive for 10^3–10^5 years. Comparison between these glasses and the naturally occurring glasses such as obsidians, basalts, and tektites is done with the help of thermodynamics. Glass corrosion behavior is described in terms of a mechanism that is based on glass network breakdown. In addition, saturation effects, surface layer formation, diffusion processes, and changes in solution composition are to be taken into account. The theoretical treatment of the glass–water reaction is based on transition state theory in combination with reversible and irreversible thermodynamics.

5.2.2.2 Concepts of Glass Dissolution

The reaction of alkali silicate glasses with aqueous solutions is usually interpreted as the result of a combination of two independent processes: initial diffusion-controlled extraction of alkali ions out of the glass matrix, and the dissolution of glass matrix itself [3–6]. The initial reaction shows a square root of time dependence for the alkali release, and it is described as an ion exchange (interdiffusion Na^+/H_3O^+ or Na^+/H^+) process [6] or as a result of water diffusion into the glass network [7]. As time passes, alkali diffusion goes on taking place, and the depletion depth goes on increasing. With this increasing depth of alkali depletion, the rate of release of alkali to the medium decreases. When this happens, the matrix dissolution becomes the dominant reaction.

5.2.2.3 Glass Durability

The durability of a glass is a function of both its kinetic rate of approach to equilibrium and its final thermodynamic equilibrium state in an aqueous environment [8,9]. Time-dependent corrosion of many glasses has been subjected to kinetic models [5,10–12]. Ion exchange, diffusion, and protective layer formation mechanisms are explained by these models.

Silicates exist both in crystalline and amorphous forms. The stability of these two forms have been predicted with the help of chemical thermodynamics [8,9,13–18]. With this, glass durability could be predicted with the knowledge of glass composition.

5.2.2.4 Corrosion Process for Nuclear Waste Glasses

Corrosion of glasses used for immobilization of nuclear waste takes place in three stages. They are *interdiffusion, matrix dissolution,* and *surface layer formation.* These three stages are shown in Figure 5.5 [19]. Interdiffusion is

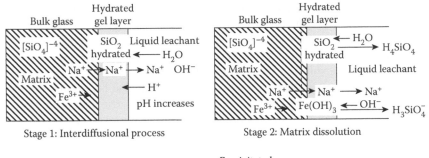

Stage 1: Interdiffusional process   Stage 2: Matrix dissolution

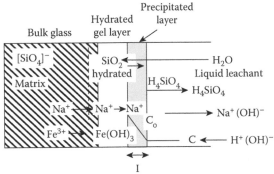

Stage 3: Surface layer formation

FIGURE 5.5
Three-stage process of nuclear waste glass corrosion. (From D. E. Clark and B. K. Zoitos (Eds.), *Corrosion of Glass, Ceramics, and Ceramic Superconductors—Principles, Testing, Characterization, and Applications*, Noyes Publications, 1992, pp. 1–6, 124–125, 153–155, 242–243, 269, 273–274, 285–295, 315–328, 372–381, 432–435, 455–479, 514–529, 548–556, 559, 567–568, 571–573/28; D. G. Chen, Graduate Thesis, University of Florida, 1987, p. 101.)

the predominant phenomenon in the initial stages of nuclear waste glasses. In this stage, water penetrates the glass, and alkali ions enter into the water. Both these processes take place by diffusion. Once the alkali ions enter the water, they form a hydroxide. This reaction increases the pH of the solution. The surface of the glass remains alkali deficient.

In stage two, matrix dissolution takes place. In this stage, the matrix tends to break down. The glass matrix is made up of amorphous silica. This amorphous silica starts dissolving. In this leaching process, the solution formed after leaching is called the leachate. The pH of the leachate controls the dissolution. This is because the solubility of amorphous silica is determined by the pH. For pH less than 5, the dissolution increases as pH decreases. For pH between 5 and 9, the dissolution remains almost constant. As the pH increases beyond 9, dissolution increases as pH increases. Another factor that determines the dissolution rate is whether the system is open or closed. An open system is a once-through system. Therefore, for this system, the pH remains the same. Whatever leached silica is formed is removed from the interface between the glass and the leachate. Therefore, the concentration of the leachate also remains the same. These two factors give rise to a constant dissolution rate. In a closed system, the concentration of the leached products goes on increasing with time. Hence, unless the pH increases with the dissolution, the dissolution rate comes down.

In the third stage, surface layers start forming. These layers are the result of the insoluble corrosion products formed during the leaching process. These products may contain crystalline or amorphous materials. They can be protective or nonprotective to the surface. If they are protective, further dissolution is inhibited by their formation. The extent of protection is decided by three factors—the chemistry of the product, its properties, and the environment.

5.2.2.5 Corrosion of Optical Glasses

Glasses find applications as spectacles, lenses, and filters of ordinary light. As lenses, they are used in magnifying glasses, optical microscopes, telescopes, binoculars, cameras, and projectors. Glasses used for these applications are called optical glasses. This is because optical light is used in these applications, and they possess optical properties. As is the case for all glasses, optical glasses also are not immune from corrosion.

After their manufacture from molten glass compositions, and before their being used, optical glasses are polished. In these polished glasses, three types of defects are possible. They are dimming, staining, and scratches.

When white deposits are formed on a glass surface, then this defect is called dimming. Being white, these deposits are opaque. Opacity is a characteristic of crystalline materials. Hence, dimming is the result of crystalline deposits. Dimming results from the interaction of the glass surface with vapors of water, carbon dioxide, sulfur dioxide, and so on. This interaction can take place when glass is stored after polishing. During polishing, water is used. If this water is allowed to dry on the polished glass, then also dimming results.

When we see different colors on the surface of a polished glass, then we say that staining has taken place. After the lens preparation by polishing, the soluble glass components are removed by water. If this water is allowed to dry on the surface of the lens, it leaves behind a film on the surface. When light falls on this film, there will be refraction of the light coming from the bottom surface of the film as it passes through the film thickness. As it comes out of the film, it will interfere with the light coming from the top surface of the film. This interference gives rise to different colors. The condition for interference is given by Equation 5.5.

$$2\mu t \cos\theta = (n + \tfrac{1}{2})\lambda \tag{5.5}$$

Here, μ is the refractive index, t is the thickness of the film, and n is an integer.

The color of the stain is also dependent on μ and t. Scratches tend to form on lenses during polishing. After polishing, the lenses are cleaned, and then they are examined. If there are scratches on them, such lenses are rejected.

The composition of optical glasses has an effect on their durability. The main effect is from the type of network former that the glass contains. There are three main glass network formers—silicon, boron, and phosphorus.

Silicon is the network former in silicate glasses. Silicate glasses are the most common optical glasses. These glasses are resistant to water at ordinary temperatures. Except hydrofluoric and phosphoric acids, other acids do not attack these glasses. However, alkaline solutions do attack them.

Borate glasses have boron as the network former. These glasses absorb water. They are also attacked by acids and alkalis. Boron trioxide is the main additive in borate glasses. Durability of the borate glass is poor. Because of their excellent optical properties, they are widely used.

Phosphorus is the network former in phosphate glasses. These glasses are soluble in water. Though the durability of these glasses is poor, their attractive optical properties keep them in good demand. They have good ultraviolet transmission.

By changing the composition of optical glasses, their durability can be improved. In order to get resistance to water and acid, a cation should be selected such that the coulombic attraction between this cation and singly bonded oxygen increases. Ge et al. [20] found that increasing the amount of Al_2O_3 in calcium aluminosilicate glass decreased its alkali resistance. Here, the structural change has affected the behavior. When alumina content is more, the coordination number of aluminum ion increases from 4 to 6. For lower alumina content, aluminum ion replaces the silicon ion from the tetrahedron. For higher alumina content, aluminum forms its own polyhedron, which is an octahedron. When there are more than one type of alkali oxide present in the glass, the durability of the glass is increased further. This is because of mixed alkali effect [21].

Microbial corrosion is also seen in optical glasses. Once it happens, the transmissivity of these glasses is affected. Also, the surface can get damaged. The growth of microorganisms is enhanced under tropical weather conditions. In tropical regions, the weather is hot and humid. If organic matter such as dust, grease, fingerprints, and dead microscopic insects are present on the glass surface, plant and animal life can grow on these surfaces under hot and humid weather. In humid weather, water absorption initiates the growth of microorganisms. Fungi and bacteria are the microorganisms found on optical glasses. Fungal growth will not take place if the glass surfaces are perfectly clean [22]. Hermitical sealing of instruments and the use of a fungicide are the preventive measures to inhibit fungal attack [23]. The attack by bacteria is of less significance than that by fungi [24].

5.2.2.6 Heavy-Metal Fluoride Glass Corrosion

The main constituents of heavy-metal fluoride (HMF) glasses are fluorides of certain heavy metals, which have high densities. The heavy metals used for making HMF glasses are zirconium, lanthanum, lead, gadolinium, lutetium, ytterbium, thorium, uranium, yttrium, and iron. Along with the fluorides of these heavy metals, these glasses will also contain other metal fluorides such as fluorides of barium, aluminum, lithium, sodium, zinc, sodium, manganese, and calcium. These were discovered in the year 1975 [25].

The aqueous corrosion of HMF glasses can be considered in two stages. In the first stage, the various reactions are selective leaching, ion exchange, diffusion, and hydrolysis. In selective leaching, the soluble metal fluoride species are leached. In an ion exchange reaction, fluoride ions in the glass are exchanged for hydroxyl ions from the solvent. Each of them carry a negative charge. Hence, an equivalent number of ions is exchanged. In the diffusion step, water molecules diffuse into the surface of the glass. In the final reaction of the first stage, the dissolved species hydrolyze. This reaction tends to decrease the pH of the solution. The first three reactions produce a transform layer on the surface. This layer is porous. As time increases, the thickness of this layer increases. The thickness varies as the square root of time [26]. This shows that the layer formation is diffusion-controlled. When the hydrolyzed products increase and the pH of the solution decreases, the second stage starts to dominate. The various reactions in this stage are dissolution, diffusion, hydrolysis, and precipitation. In the first reaction, additional glass elements dissolve. In the second reaction, further diffusion of water molecules into the transform layer takes place. In the third reaction, hydrolysis takes place, both in the glass and the solution. In the glass, it takes place in the transform layer. As more hydrolysis takes place in the solution, the pH of the solution further decreases. When the solution becomes saturated, the corrosion products start precipitating.

The corrosion behavior of HMF glasses is dependent mainly on their composition. Based on composition, glasses can be classified according to their primary glass formers. For example, they can be classified into fluorozirconate, thorium, and uranium glasses. In fluorozirconate glasses, the primary glass former is thorium, in thorium glasses, it is thorium, and in uranium glasses, it is uranium.

Thorium glasses have greater resistance in acid environments than fluorozirconate ones. However, fluorozirconate (ZBLA) glasses are better in neutral and basic environments. In the case of acidic environments, the lower the pH, the higher will be the leach rate for all glasses. The pH of the solution and its agitation influence the appearance of the leached surface. The lower the pH in the acidic range, the greater will be the crystallization on the transform surface. If the solution is moving, these crystals are washed away. In stagnant solution, the crystals formed remain on the surface. The microstructure of such a glass is shown in Figure 5.6. The experiment was conducted using ZBLA glass. The crystalline precipitates formed were those of hydrated zirconium fluoride and zirconium–barium fluorides.

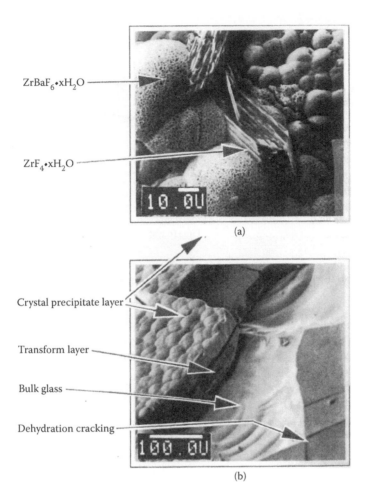

ZrBaF$_6$·xH$_2$O

ZrF$_4$·xH$_2$O

(a)

Crystal precipitate layer

Transform layer

Bulk glass

Dehydration cracking

(b)

FIGURE 5.6
Corrosion layers formed on ZBLA glass (From D. G. Chen, Graduate Thesis, University of Florida, 1987, p. 101). (a) the two kinds of precipitates, (b) the bulk glass, transform layer, and precipitate layer.

The first type is seen as needle-like crystals, and the second type is in spherical shape. The transform layer is amorphous. The bulk glass beneath the transform layer is unaffected.

The rate of growth of the transform layer with exposure time is also strongly influenced by test conditions [19,26]. Figure 5.7 shows the transform layer thickness versus corrosion time for ZBLA glass during stagnant test. It can be observed that the rate of growth of the transform layer decreases as time increases, and finally a constant thickness is reached. When the precipitate completely covers the surface of the glass, the corrosion of glass is stopped and the transform layer thickness remains constant. In the case of

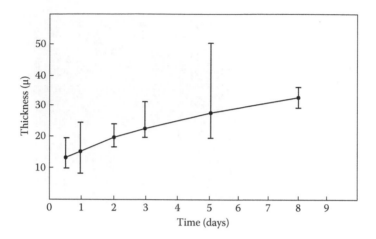

FIGURE 5.7
Transform layer thickness versus corrosion time for ZBLA glass during stagnant test.

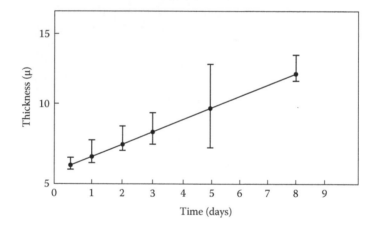

FIGURE 5.8
Transform layer thickness versus corrosion time for ZBLA glass during stirred flow test.
(From D. G. Chen, Graduate Thesis, University of Florida, 1987, p. 101.)

moving corrosive, the precipitate formed is removed as it is formed. Hence, the growth of the transform layer continues proportional to the time. This is shown in Figure 5.8. However, the rate of growth is slow, as the pH does not decrease at the surface as to have a greater corrosion rate.

5.2.3 Glazes and Enamels

Glazes are the ceramic coatings done on ceramics, and enamels are done on metals. Both these coatings perform two functions—corrosion resistance and aesthetics. The substrates for both have a crystalline structure.

The coatings in the form of glazes and enamels possess an amorphous structure, which is more resistant to corrosion than the crystalline one. Also, the surface of an amorphous material is smoother and provides a more pleasing appearance. Corrosion still takes place in amorphous materials. This was shown in the case of glasses.

Glazes and enamels are formed by melting and solidification. The composition of these coatings is similar to that of glass. On cooling, they form amorphous glass structures. When these coatings are brought into contact with an aqueous solution, alkali ions are extracted into the solution in preference to silica and other constituents of the glass [8]. Because of this leaching, an alkali-deficient layer is formed on the surface of the glass. The thickness of this layer depends on the composition of the coating, time, temperature, and the pH of the solution.

When the alkali ion goes into solution, an equivalent amount of hydrogen ion goes into the glass. That means an ion exchange reaction takes place, as shown in Equation 5.6 [27].

$$H^+_{soln.} + [Me^+O - Si]_{glass} = [H^+O^- - Si_{glass}] + Me^+_{soln.} \tag{5.6}$$

As H^+ ions replace Me^+ ions, a surface film is produced, resembling vitreous silica or silica gel, and having properties different from the parent glass [28]. This film swells, acts as a barrier to further reaction, and decreases diffusion rates into and out of the surfaces, thereby inhibiting further attack [8,29,30]. This is the mechanism of corrosion in neutral and acidic media.

The presence of alkali ions in the solution tends to raise the pH to the point at which the silica-rich layer can itself be attacked by hydroxyl ions [27]. The reaction is represented by Equation 5.7. In this reaction, doubly bonded oxygen becomes singly bonded.

$$[Si-O-Si_{glass}] + [Me^+(OH)^-]_{soln.} = [Si-(OH)]_{glass} + [Si-O-Me^+]_{glass} \tag{5.7}$$

This open oxygen interacts with a water molecule, according to Equation 5.8.

$$[Si-O^-]_{glass} + H_2O = [Si-(OH)]_{glass} + (OH)^-_{sln} \tag{5.8}$$

As a result of this reaction, a hydroxyl ion gets regenerated. This liberated hydroxyl ion further reacts with the silica-rich layer, according to Equation 5.7. Thus, it becomes autocatalytic.

For the ion exchange process, the rate of alkali extraction varies linearly with the square root of time at short times and low temperatures [27,31]. The variation is shown in Figure 5.9. In this figure, the amount of sodium oxide extracted per gram of glass is shown on the y-axis. However, when the glass is exposed for longer periods and at higher temperatures, the linearity comes after sometime, as shown in Figure 5.10. This figure also shows

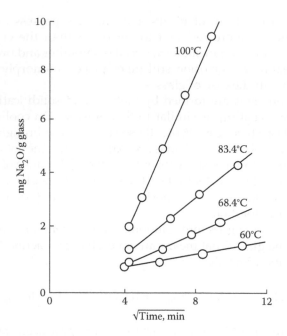

FIGURE 5.9
Amount of Na$_2$O leached as a function of temperature and time.

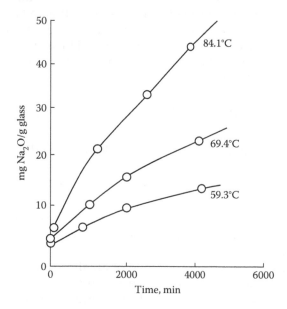

FIGURE 5.10
Long-time water leaching of 15Na$_2$O·85SiO$_2$ glass. (From R. A. Eppler, *Am. Ceram. Soc. Bull.,* 56, 1068–1070, 1977.)

that, as temperature increases, The leaching rate also increases. Higher temperatures normally cause greater kinetics of reactions.

Figure 5.11 shows the effect of pH on the weight loss of enamel. It is seen that the rate of weight loss increases with pH. Thus, time (t), temperature (T), and pH together account for greater than 90% of total corrosion of glass coatings.

The ratio of surface area to volume (S/A) can change the pH of the solution in contact with the surface. Therefore, exposure to water vapor produces a different corrosion pattern than exposure to liquid water [32]. The pH is greater for water deposited on a coating surface than for liquid water. This causes greater corrosion of the coating by the deposited water.

While t, T, and pH are the principal factors governing the attack of liquids on glass, there are some secondary effects [33]. If the solution contains ions, they can form complex ions. Examples for such ions are citrate and acetate. When complex ions are present, the activity of the solution is reduced, with the effect of reduced corrosion. Other ions, if they are present, may chemically interact with hydrous silica ions present on the surface of the coatings. These ions therefore increase the corrosion.

Hydrofluoric acid attacks all silicate glasses. The attack can be represented as follows:

$$SiO_2 + 6HF = SiF_6^{-2} + 2H_2O + 2H^+ \tag{5.9}$$

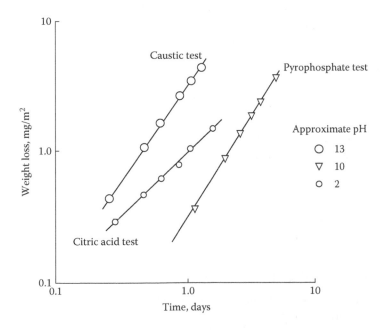

FIGURE 5.11
Weight loss data at 65°C–66°C for enamel. (From T. M. El-Shamy, PhD Thesis, University of Sheffield, 1966.)

This attack by hydrofluoric acid is utilized in the etching of glass surfaces. Another way by which it can attack the glass is by directly attacking its structure. This is given by Equation 5.10.

$$[Si-O-Si]_{glass} + [H^+F^-]_{soln.} = [Si-OH]_{glass} + [Si^+F^-]_{soln.} \qquad (5.10)$$

The rate of extraction of ions from any glass coating is determined by the coating's overall composition, and by the extraction process that is used [34,35]. The rate of release of alkali ions from the coating by the ion exchange process increases with increasing alkali content [36]. K_2O is more soluble than Na_2O, which is more soluble than Li_2O [37,38]. Mixtures of alkalis are usually less soluble than a single alkali [39,40]. Additions of alkaline earth ions decrease coating durability, but much less so than alkalis [36]. Cadmium oxide addition also increases the coating corrosion.

The presence of other ions affects the effect of lead oxide (PbO) on acid resistance. For example, in lead-alkali silicates, the amount of the modifier extracted is directly proportional to the total modifier content (PbO + alkalis) [40]. Additions of alumina (Al_2O_3) are particularly beneficial in increasing acid resistance [36–38,41]. When large amounts of Al_2O_3 are added, the coordination number of aluminum ion changes from 4 to 6. When this happens, Al_2O_3 addition is not beneficial. Acid durability is also improved by silica, titania, and zirconia additions. Up to 12% B_2O_3 improves the durability of alkali silicate glasses [36]. As with fluorine, phosphorus pentoxide also attacks the silicate network. Therefore, their addition reduces the acid durability of the coatings.

A parameter that correlates all these findings regarding ion exchange durability is [34,35]:

$$FM = Good/(Bad)^{1/2} \qquad (5.11)$$

Here, FM is a figure of merit for a glass, expressed in terms of the molar concentrations of its constituents:

$$Good = 2(Al_2O_3) + (SiO_2) + (TiO_2) + (ZrO_2) + (SnO_2) \qquad (5.12)$$

$$Bad = 2[(Li_2O) + (Na_2O) + (K_2O) + (B_2O_3) + (P_2O_5)] + (MgO) \\ + (CaO) + (SrO) + (BaO) + (F) + (ZnO) + (PbO) \qquad (5.13)$$

5.2.4 Corrosion of Glass-Ceramics

Glass-ceramics contain an amorphous phase and a crystalline phase minimum. Their attractive properties are chemical durability, mechanical strength, lower thermal expansion coefficient compared to glasses, dielectric constant, and electromagnetic radiation transmittance. Because of these properties, they find applications in electronics, military, vacuum technology, households, industries, and biology. The corrosion behavior of glass-ceramics must be considered as that of a multiphase material. The glass phase is considered as a homogeneous system. The crystalline phase is constituted by a number of grains. While considering the corrosion behavior of crystalline phases, the grains and boundaries come into the picture. Grain boundaries are more active than grains. Thus, between them, galvanic couples can form, with the more active grain boundaries serving as anodes and the grains as cathodes.

In the corrosion of the glass phase, five mechanisms are to be taken into account. They are ion leaching, dissolution, precipitation, pitting, and weathering. The case of ion leaching has already been dealt with in the corrosion of glasses earlier. In this mechanism, alkali ions in the glass are exchanged for the H^+ and H_3O^+ ions present in the corrosive medium. In dissolution, the silicate structure is broken down, and silica starts dissolving. When the concentration of the dissolved species reaches the saturation limit, they start precipitating on the glass surface. Wherever there are pores in the precipitate layer, attack can be through these pores. The medium can enter through these pores. This attack on the surface of the glass produces pits. Weathering is the result of intermittent exposure of the glass surface to water vapor. Ion exchange takes place between this vapor and the surface.

The basis for a large group of commercial glass-ceramics is the $Li_2O–SiO_2$ system. Corrosion studies on this system have been carried out [42–44], the results of which can be summarized as follows. The corrosion of the lithia–silica system is significantly affected by crystallization. Acidic environment produces the least damage, and the basic one the greatest. In areas where glass is present, the highest and uniform dissolution takes place. When the medium is neutral or basic, interphase and intracrystalline attacks take place.

5.2.5 Construction Materials

Construction materials made of ceramics are used for carrying out chemical processes. There are two groups of these materials—acid proof and specialty materials. Red shale and fireclay are the examples for the first group. The second group consists of silica, silicon carbide, and high alumina. They become useful in chemical environments because of their excellent corrosion resistance. The acid-proof materials are resistant to inorganic acids. Specialty materials are useful for high-temperature applications involving chemical environments. Both these materials are used as structural materials or as a coating.

Published literature [45–52] on the studies of the corrosion behavior of ceramic construction materials yielded the following results:

a. When these ceramics were exposed to acids, ions of Fe, Al, Ca, Mg, Na, K, and Ti got removed. The ion loss versus number of days of exposure gave a parabolic curve.

b. The Fe ion loss was the greatest, followed by the Al ion. The remaining ions were lost to a minor extent.

   c. The weight-loss was proportional to the porosity of the material.

   d. There was no effect on the strength of the materials as a result of corrosion.

   e. The ion loss was a function of the concentration and temperature of the medium. As concentration of the acid was increased, the ion loss decreased. However, the ion loss increased with increase in temperatures.

   f. As moisture was absorbed through the pores present in the ceramics, there was volume expansion. This kind of expansion was divided into two components [53]. The first one is a reversible component. It is due to wetting, and is reversible because the expanded part contracts to the same extent upon drying. The other expansion is irreversible, and continues for many years. The volume expansion depends on porosity. As porosity decreases, this expansion also decreases.

5.2.6 High-Temperature Corrosion

High temperatures are involved in the working of heat engines and heat exchangers. Operating temperatures in these applications can go beyond 1000°C. Also, chemical environments are involved. The environments can contain oxygen, hydrogen, steam, and sulfur oxides as gases and vapors. These environments can lead to condensed-phase deposits. These deposits can be quite corrosive.

One of the harmful deposits has been sodium sulfate. In the case of heat engines, ingestion of sodium takes place through intake air or fuel. Fuel contains sulfur and, on burning, it forms sulfur oxides. The reaction can be represented by Equation 5.14.

$$2NaCl\ (g) + SO_2\ (g) + \tfrac{1}{2}\ O_2\ (g) + H_2O\ (g) = Na_2SO_4\ (l) + 2HCl\ (g) \quad (5.14)$$

Hot corrosion is the term used for sodium-sulfate-induced corrosion. Components made of SiC and Si_3N_4 form a layer of SiO_2 on their surfaces on exposure to oxidizing atmosphere. This thin layer acts as a protection that prevents the corrosion of the substrate. Previous works have shown that, for silicon-based ceramics, the primary mode of attack is dissolution of the silica scale [54–56].

Sodium sulfate forms readily on heat engine parts, according to Equation 5.14. It is a highly stable molecule. Once formed, its corrosion reaction occurs above its melting point. The different reactions leading to corrosion are represented by the following equations:

$$SiC\ (s) + \tfrac{3}{4}\ O_2\ (g) = SiO_2\ (s) + CO\ (g) \quad\quad\quad (5.15)$$

$$Na_2SO_4\ (l) = Na_2O\ (s) + SO_3\ (g) \quad\quad\quad (5.16)$$

$$2SiO_2\ (s) + Na_2O\ (s) = Na_2O{\cdot}2SiO_2\ (l) \quad\quad\quad (5.17)$$

These reactions indicate that, for the corrosion of the SiO_2, Na_2O is required. Hence, any reaction that gives rise to the formation of Na_2O should cause corrosion. This is possible if sodium carbonate and sodium chloride are present. The corresponding reactions are given in Equations 5.18 and 5.19.

$$Na_2CO_3 = Na_2O + CO_2 \tag{5.18}$$

$$2NaCl + H_2O = Na_2O + 2HCl \tag{5.19}$$

Corrosion at high temperature requires that the participating corrosives be in molten condition. Hence, this corrosion is also termed *molten salt corrosion*. In some of the high-temperature processes, there is molten slag formation. An example to cite is the production of pig iron using a blast furnace. These slags can also give rise to corrosion. For example, one type of molten salt corrosion is the formation of silicides as a corrosion product. Equation 5.20 gives this reaction.

$$13SiC + 5Fe \text{ (slag)} = Fe_5Si_{13} + 13C \tag{5.20}$$

The corrosion processes discussed so far involve three steps—deposition, oxidation, and dissolution. The extent of corrosion in the life of a component will be decided by the kinetics of each these steps. In kinetics, one of the steps will be the slowest, and it will decide total rate of the whole process.

5.2.7 Superconductors

Ceramic superconductors were discovered in the 1980s. Ceramic superconductors possess the highest critical temperatures so far achieved. Critical temperatures mark the temperatures below which the resistivity of the material falls to zero. The highest critical temperature achieved so far is 125 K. This has been found in the case of $Tl_2Ba_2Ca_2Cu_3O_{10}$. The use of this compound is limited because of the volatility and toxicity of thallium (Tl). Earlier to the synthesis of this compound, there was another compound synthesized, which is referred to as the *123 compound*, where the numbers represent, respectively, the number of yttrium, barium, and copper atoms present in its molecule. Thus the formula of this compound becomes $YBa_2Cu_3O_7$. This compound has a critical temperature of 93°C, which can be achieved by using liquid nitrogen. Liquid nitrogen's boiling point is 77 K. Only three materials possess critical temperatures greater than 77 K, and all the three are ceramic materials. Those materials possessing critical temperatures greater than 77 K are called high-temperature superconductors. Flux creep is a property that limits the current-carrying capacity of a superconductor. This is lower for the 123 superconductor than the other two. Hence, 123 compound is found to be a more useful superconductor than the other two. Because of this, this material has been characterized the most.

Both the room temperature and low temperature resistivity of the 123 superconductor were found to increase when it was exposed to humid environments [57]. This increase in resistivity is the result of the appearance of an electrochemical electromotive force at the contact area between the electrode and the superconductor, and the formation of insulating barriers at the grain boundaries of the superconductor. The formation of the barriers can be accelerated if carbon dioxide is also present along with moisture.

It was found that barium carbonate had formed on the surface of the superconductor. The mechanism of this phenomenon consists of the following reactions:

$$4YBa_2Cu_3O_7 + 6H_2O = 2Y_2BaCuO_5 + 10CuO + 6Ba(OH)_2 + O_2 \quad (5.21a)$$

$$2Y_2BaCuO_5 = 2BaCuO_2 + 2Y_2O_3 \quad (5.21b)$$

$$Ba(OH)_2 + CO_2 = BaCO_3 + H_2O \quad (5.21c)$$

The particle size of the starting powder had an effect on the degradation of the superconductor. Decreasing particle size accelerated the degradation, because of the larger surface area.

Another factor that affects the deterioration in humid conditions is the temperature. Measurements of the increase in resistivity as a function of temperature were reported to increase with the time required to reduce the conductivity by half. Increase in resistivity is also inversely proportional to the vapor pressure, which rises exponentially with increasing temperature [58].

Liquid water attacks the 123 compound more severely than water vapor. Samples of this material were found to lose one-third of their diamagnetic susceptibility upon exposure to distilled water at 60°C for 5 min [59]. The mechanism of attack followed the same reaction sequence indicated for water vapor, given by Equations 5.21a through 5.21c [60].

Similar to water vapor, the degradation in liquid water is also affected by temperature. The 123 compound is inert upon exposure to cold water for several hours. Heating to 35°C causes significant decomposition [61]. Most of this material was decomposed upon a 30 min soak in water at 65°C, while at 90°C decomposition was completed in less than 30 min [61]. At higher temperatures, aqueous solutions of 10 M NaOH, 1 M NaCl, and 1 M KCl, which are less reactive than water at room temperature, cause decomposition of the 123 compound over short periods of exposure [57].

The 123 compound is a basic oxide. Hence, it dissolves in acidic solutions. The corrosion rate of $YBa_2Cu_3O_7$ is extremely high in strong acidic media, for example, 0.1 M HCl, where all the three constituent oxides (BaO, Y_2O_3, and CuO) are highly soluble, and, accordingly, the dissolution process is congruent [62]. The accelerated rates of corrosion of $YBa_2Cu_3O_7$ in humid atmospheres, which are observed when CO_2 is present, can be attributed to an increase in the acidity of water [63].

Comparative studies of the effects of methanol and aqueous solutions on 123 ceramic powder showed that, at room temperature, sample deterioration took place almost immediately in water, while no reaction was observed in 1 M NaCl over an exposure period of 1 day and in methanol over a period of 5 days [57]. The order of reactivity of this ceramic with various media was

$$HCl > H_2O > NaOH > NaCl > CH_3OH$$

More systematic studies were carried out in order to obtain comparative information on the stability of superconducting oxides in various organic solvents [64,65]. The relative extent of corrosion appears to follow the sequence:

Acetic acid>>2-pentanol>formamide>>ethanol>N-methylformamide>
2-propanol>N,N-dimethyl formamide>acetone>methanol, benzene,
toluene

It was found that the presence of porosity increases the corrosion in superconductors. An increase in density from about 75% to about 90% of the theoretical value results in a decrease by more than a factor of 2 in the rate of attack by liquid water [66]. The exposure of 123 samples to air with 100% relative humidity at 60°C for 20 min gave rise to the growth of needle-shaped crystals at the grain boundaries [58,67]. Stoichiometric polycrystalline specimens of this material usually exhibit precipitation of $BaCuO_2$ at grain boundaries [68,69].

Because of the high sensitivity of sintered specimens to attack by humid environments, encapsulation is required in many applications in order to retard corrosion [70]. Coating of 123 strips with a 0.5-mm-thick, fast-setting epoxy was found to reduce the rate of change of resistance upon exposure to air, at 85°C and 85% relative humidity, by a factor of 3 [63]. Encapsulation of sintered $YBa_2Cu_3O_7$, $Bi_2Sr_{2.2}Ca_{0.8}Cu_2O_8$, and $Tl_2Ba_2Ca_2Cu_3O_{10}$ pellets in an epoxy matrix effectively fills the outer pores and permits the use of such pellets as electrodes for periods of 4, 7, and 14 days, respectively, in water [71,72].

5.3 Corrosion of Refractories

Thermal stability and corrosion resistance are greater in refractories than in other ceramics. This is because refractories operate at higher temperatures. Corrosion, being a chemical reaction, get enhanced by higher temperatures. Corrosion most often determines refractory life, and often enough sets the temperature limit on the contained process.

5.3.1 Thermodynamics of Corrosion

Thermodynamics principles can be applied to corrosion, as with any chemical reaction. Corrosion is related to the free energy of formation, ΔG_f. In the present case, we have to consider it for the refractories.

5.3.1.1 Standard Free Energies of Formation of Compounds

When the reactants forming a compound are in their standard states, the free energy of formation of one mole of this compound is called its *standard free energy of formation*, ΔG_f^0. Standard states are defined in the following manner:

   a. A stable crystalline solid will be below its melting point.

   b. A pure liquid will be above its melting point.

   c. An *ideal* gas will be at 1 atm. partial pressure.

A constant total pressure of 1 atmosphere is also assumed for all the states of matter. The standard free energy of formation is related to the standard driving force for formation in volts (E^0) by the following equation:

$$\Delta G_f^0 = nFE^0 \tag{5.22}$$

Here, n is the number of redox equivalents per mol of compound, and F is Faraday's constant. Equation 5.22 can be rewritten as:

$$E^0 = -\Delta G_f^0 / nF \tag{5.23}$$

Thus, we find that $-\Delta G_f^0 / n$ is proportional to E^0. E^0 is an intrinsic quantity that permits the direct comparison of compounds of different formulae. Figure 5.12 is a plot of ΔG_f^0 vs. T for simple binary oxides. Here, ΔG_f^0 is taken as a positive value so that more stable oxides are represented by upper curves. The *knee* in the curves for MgO and CaO occurs at the metal boiling point, reflecting a change in standard state of the metal. The end of the curves signifies the melting point of the compound.

The rule for computing ΔG_T^0, the standard free energy change for any reaction at a temperature T, from the standard free energies of formation is:

$$\Delta G_T^0 = \sum \Delta G_{f,T}^0 \left(\text{products}\right) - \sum \Delta G_{f,T}^0 \left(\text{reactants}\right) \tag{5.24}$$

When ΔG_T^0 is <O, the reaction is spontaneous.

5.3.1.2 Free Energies of Reactions

Corrosion reactions have to deal with substances that are not in standard states. Examples are (a) substances in solution, and (b) gases at other than 1 atm. partial pressure. In these reactions, therefore, a fundamental thermodynamic relationship is used:

$$\Delta G_T = \Delta G_T^0 + RT \ln Q \tag{5.25}$$

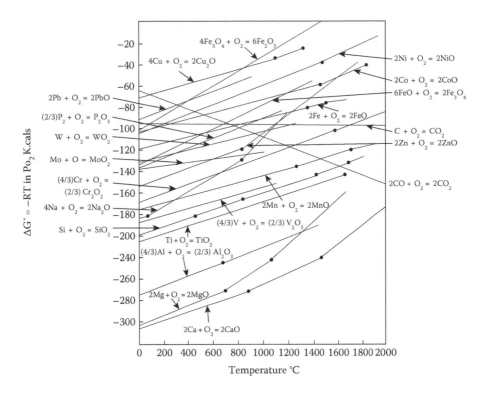

FIGURE 5.12
Plot of ΔG^0 versus T for simple binary oxides.

Here, ΔG_T is the free energy change of a reaction involving at least one substance in a nonstandard state, R is the gas constant, and Q is an activity quotient. Giving the value of R and converting the natural logarithm to a common one, Equation 5.25 becomes:

$$\Delta G_T = \Delta G_T^0 + 0.01914\,T \log Q, \quad \text{for } \Delta G \text{ in kJ} \tag{5.26}$$

The algebraic definition of Q is best given using a generalized chemical equation, written with capital letters representing formulas and lower-case letters representing their coefficients. The equation is given here:

$$uU + vV + \dots = xX + yY + \dots \tag{5.27}$$

Q is a function of the chemical *activities* of the species, given by:

$$Q = [(a_X)^x\,(a_Y)^y\,\dots]/[(a_U)^u\,(a_V)^v\,\dots] \tag{5.28}$$

It remains only to specify the chemical activity of each substance, a_X, a_Y, and so on, by the use of the following guidelines. If a participating substance X is

a. In its standard state, $a_X = 1$

b. A gas (above its boiling point), a_X = its partial pressure p in atm.

c. A solvent, a_X = its mol fraction

d. A solute, a_X = its concentration as a fraction of the saturation value

Equation 5.5 for nonstandard ΔG_T is used when ΔG_T^0 is known and every activity is known or can be estimated.

Example 1

A silica refractory wall has a hot-face temperature of 1800 K and is exposed to a process atmosphere of 90% CO, 10% CO_2 (1 atm. total). The reduction of SiO_2 to SiO takes place according to the following equation:

$$SiO_2 \text{ (s)} + CO \text{ (g)} = SiO \text{ (g)} + CO_2 \text{ (g)} \tag{5.29}$$

A partial pressure of SiO (g) of 10^{-3} in the furnace atmosphere is tolerable. Will this reaction be spontaneous? At 1800 K, $\Delta G_{SiO}^0 = -247.4$ kJ/mol, $\Delta G_{CO_2}^0 = -396.4$ kJ/mol, $\Delta G_{SiO_2}^0 = 551.1$ kJ/mol, and $\Delta G_{CO}^0 = 269.2$ kJ/mol.

Answer

$$\Delta G_{1800}^0 = 551.1 + 269.2 - 247.4 - 396.4 = 176.5 \text{ kJ}$$

$$Q = 10^{-3} \times 0.1/1 \times 0.9 = 1.1 \times 10^{-4}$$

Therefore,

$$\Delta G_{1800} = 176.5 + 0.01914 \times 1800 \times \log (1.1 \times 10^{-4})$$

$$= 176.5 + 0.01914 \times 1800 \times (0.0414 - 4)$$

$$= 176.5 - 136.4 = 40.1 \text{ kJ}$$

This result being >O, the preceding reaction will be not spontaneous.

5.3.1.3 Free Energies and Chemical Equilibrium

By far the most powerful use of thermodynamics in assessing refractory corrosion entails equilibrium calculations. From the preceding section, we have:

$$\Delta G_T = \Delta G_T^0 + 0.01914 \, T \log Q \tag{5.30}$$

Equilibrium occurs at some unique set of nonstandard conditions under which $\Delta G_T = 0$. That is, there is no driving force for the reaction in either direction. If $\Delta G_T = 0$, then:

$$-\Delta G_T^0 = 0.01914 \, T \log K \text{ or } \log K = -52.25 \Delta G_T^0 / T \qquad (5.31)$$

Here, ΔG_T^0 is in kJ. K is the numerical value of Q at *equilibrium*.

Example 2

Calculate the equilibrium constant for the reaction given in *Example 1* at 1800 K. Also, calculate the equilibrium partial pressure of SiO.

Answer
We have

$$\log K = -52.25 \times 176.5/1800 = -5.1234, \text{ or } K = 0.75 \times 10^{-5} = \text{Answer 1}$$

$$K = 0.75 \times 10^{-5} = \left(a_{SiO} \times a_{CO_2}\right)/\left(a_{SiO_2} \times a_{CO}\right) = \left[p(SiO) \times p(CO_2)\right]/\left[1 \times p(CO)\right]$$

Therefore,

$$p(SiO) = 0.75 \times 10^{-5} \times 0.9/0.1 = 6.75 \times 10^{-5} \text{ atm.} = \text{Answer 2}$$

Redox reactions are more common in the corrosion of refractories—for example, there are redox agents in combustion atmospheres.

Reduction–decomposition of MgO and SiO$_2$ is another example. The evaporation of elemental Mg by reduction of MgO has long been of concern as a potential mechanism of recession of basic refractories. Historically, the problem started with a juxtaposition of magnesite bricks with carbon blocks [73]. The reduction reaction is represented by Equation 5.32.

$$\text{MgO (s)} + \text{C (s)} = \text{Mg (g)} + \text{CO (g)} \qquad (5.32)$$

The activity quotient Q for this equation can be written as shown in Equation 5.33.

$$Q = p_{(Mg)} * p_{(CO)}/[1 * 1] \qquad (5.33)$$

The equilibrium constant K was determined for various values of temperature T. Then, using Equation 5.33, the partial pressure of magnesium vapor (p_{Mg}) was calculated at various temperatures by keeping partial pressure of carbon monoxide gas [$p_{(CO)}$] = 1. The calculations showed that p_{Mg} was about 10^{-3} atm. at 1500°C and about 0.1 atm. at 1700°C. Fortunately, for the refractory in service, two factors mitigate the anticipated Mg evaporation rate: (a) mass transport is impeded by slag and other factors affecting diffusion [74], and (b) much of the Mg vapor apparently reoxidizes within the refractory [75] either at lower temperature behind the hot face or upon encountering a less-reducing atmosphere and slag-borne oxidizing agents (e.g., FeO or Fe_2O_3) at the hot face. Nevertheless, the susceptibility of MgO to reduction certainly limits its service temperatures in contact with carbon. They must not be appreciably higher than they are now in steelmaking.

This susceptibility can be diminished by replacing MgO by a stable spinel such as $MgAl_2O_4$. By doing so, the same partial pressure of magnesium will be obtained at about 200°C higher temperature.

Reduction of silica by carbon or CO can produce SiO (g), Si (l), or SiC (s). Below the melting point of SiO_2 (melting point = 1996 K), its reduction to SiO (g) takes place. In *Example 2*, we found that $p_{(SiO)} = 6.75 \times 10^{-5}$ atm. for a $p(CO)/p(CO_2)$ ratio of 9 at 1800 K. Let us consider Equation 5.34. It shows the reduction of SiO_2 by CO.

$$SiO_2 \text{ (s)} + CO \text{ (g)} = SiO \text{ (g)} + CO_2 \text{ (g)} \tag{5.34}$$

Silica is seen to be susceptible to reductive decomposition by carbon at temperatures well below those for MgO. Let us say that, at $p_{SiO} = 10^{-3}$ atm., equilibrium is reached for this reaction. This equilibrium pressure can be reached by many combinations of decreasing $p_{(CO)}/p_{(CO_2)}$ with increasing temperature. In other words, at different combinations of partial pressures of CO and CO_2 with a corresponding change in temperature, the equilibrium partial pressure of SiO can be achieved.

5.3.2 Corrosion in Hot Liquids

Hot liquids exist in processes involving metallurgical slags and fluxes, glasses, hot chemical operations such as combustion, ore roasting, Portland cement manufacture, and smelting. Refractories undergo corrosion in hot liquids by dissolution. In dissolution, mass transport is the dominant factor controlling the process. If it goes on unlimited, then dissolution continues uninhibited. In refractories, the penetration of the liquid takes place through pores and voids. Penetration without dissolution is not corrosion.

5.3.2.1 Liquid Penetration and Dissolution–Corrosion

When a hot face of a refractory is exposed to a liquid, various microstructural features of the refractory interact with the hot liquid. The features that could be present in a refractory are refractory crystal, orientation boundary, phase boundary, segregated impurity, unbonded boundary, matrix, liquid boundary film, connected porosity, and open joint.

The *refractory crystal* is a unit of the one or more principal phases chosen to provide thermal stability and corrosion resistance in the material. *Orientation boundaries* are narrow, but not quite atomically narrow even if free from impurities. *Phase boundaries* are still somewhat thicker, atomically less dense, and crystallographically more disordered because they join crystals of different phases in haphazard orientations. *Segregated impurities* occur in all these boundaries in reality, and thicken them by up to a magnitude [76]. *Unbonded boundaries* are narrow voids between crystallites. *Matrix* is the continuous phase in the refractory. *Liquid films* occur in boundaries near the hot face if their melting temperature is below the service temperature. *Connected porosity* originates in the imperfect packing of particles; in the evaporation/decomposition of additives; and in shrinkages accompanying the reaction, melting, and recrystallization of constituents of a refractory in its manufacture. *Open joints* in a refractory wall or lining can originate in its installation, design, and support, together with its service history.

Liquid intrusion is very rapid through open joints, and it fills these joints fully. In connected porosity, rapid filling takes place progressively. Rapid local liquid filling takes place in unbonded boundary. In the case of liquid film, less rapid local interdiffusion takes place. Moderate progressive debonding takes place in matrix. In segregated impurity, moderate progressive debonding is the result of interaction with the hot liquid. Slow hot face deep grooving results in phase boundary. In orientation boundary, hot face grooving takes place very slowly. Hot face surface recession is the type of interaction with refractory crystal, and it is the slowest of all the processes.

Penetration, which takes place in open joints, connected porosity, unbonded boundary, and liquid film, establishes the overall pace of corrosion, that is, its increasing depth into the refractory with time [77]. Matrix and segregated impurity are the vulnerable ones. They are solid regions; their invasion has to be by dissolution [77]. Phase, orientation boundaries, and refractory crystal are equally exposed to the penetrating liquid, but their resistance to invasive dissolution is the highest among all features present.

Both physical penetration and chemical invasion are favored by the effective liquid–solid wetting and by the low viscosity of the liquid. Wetting is decided by the contact angle. As it decreases, wetting is increased. In the liquid state, silicate slags and glasses are the most viscous, oxidic compounds are less so, and halides and elemental metals are the least viscous.

Two groups of controllable factors influencing refractory dissolution–corrosion by liquids can be identified—penetration and dissolution factors. The penetration factors include freezing of the penetrant, porosity and pore sizing, and wetting and nonwetting. The dissolution factors include resistance of the matrix, extent of intercrystalline bonding, and nature of the grain.

A classical phenomenon associated with *freezing* of a penetrating and aggressive liquid is *slabbing*. Slabbing is also called peeling or chemical spalling of the refractory. Given a negative ΔT across a wall or lining, with time the liquid penetration front approaches the depth at which the liquid finally freezes. This moving front is a surface, roughly parallel to the hot face. Debonding of crystals by dissolution has to follow this penetration in time; in any event, it is a thermally activated process that slows markedly with increasing depth (i.e., decreasing T).

Hence, the weakened zone of a refractory is less deep than the penetrated zone.

Now, this refractory is allowed to cool. Cooling can happen in actual practice by shutting down a furnace or between heats in a converter. Then it is placed on duty again, restoring the steady-state operating temperature and ΔT. As to the intruded liquid, this is freeze–thaw cycling, with a ΔV accompanying each change of phase. A common view is that this ΔV subjects the refractory to a shear strain parallel to the hot face. Failure occurs where the refractory is weakened. Slabs of the refractory are dislodged. Typical depths of penetration at which slabbing occurs are from about 1 cm to well over 5 cm, in a wall or lining backed by a steel shell.

A common way of thermally controlling liquid penetration and invasion is by employing water cooling behind or within the refractory.

A hot process liquid, as it intrudes into the *porosity* and down the temperature ramp in a refractory wall, may crystallize out some of its own highest-melting components progressively as it goes. Upstream dissolution of the refractory augments this process. The resistance of a refractory to penetration by liquids is improved by reducing its apparent porosity.

Wetting is characterized by a wetting angle of <90°, and *nonwetting* is characterized by a wetting angle of >90°. A very low wetting angle signifies that the molecules (or ions, or atoms) of the liquid are attracted more to those of the solid than they are to each other. The meniscus made in a pore channel in that case is concave, and the surface tension of the liquid generates the equivalent pressure driving the liquid into the channel. This pressure increases with decreasing pore size, because the ratio of a channel circumference to its cross-sectional area is proportional to $1/d$. Wetting angles of around 90° provide essentially no driving force for pore penetration, while with still further increasing angle the meniscus becomes increasingly convex, and a resistance to penetration develops. This resistance again increases with decreasing void diameter [77,78].

Massive carbon, graphite, and SiC are resistant to wetting and reaction by molten chlorides, alkalis (M_2O and MOH), and silicates. The only oxidic substance that has been credited with any appreciable nonwetting quality toward oxidic liquids is ZrO_2 [73]. This compound has been found to be remarkably resistant to molten glasses and to their combined or separated alkalis (Na_2O, K_2O). It is the ultimate glass contact refractory, paying its way by durability despite its high cost.

Dissolution of the bond or matrix between crystals or grains of the major phase(s) causes their gradual debonding. A grain or aggregate is the first line of defense against corrosion. The *matrix* is that portion of the final consolidated solid that separates and yet bonds the grain or aggregate particles. It is possible to increase the kinetic resistance of a matrix to dissolution by altering its composition so as to raise its own minimum eutectic melting temperature. That done, the matrix should also be well crystallized, that is, as close as possible to its equilibrium-phase composition. Given a matrix that is not to be eliminated, an obvious approach to its upgrading is via chemical purity. A second way of increasing the melting temperature of the matrix is by chemical addition rather than subtraction. Finely powdered matrix additives are included in the initial refractory mix, with the intention that they shall dissolve when the matrix melts on firing.

If the dissolving power of an invading liquid is finite, then, in the interests of impeding the debonding of refractory crystals, it is logical to provide the largest possible field for the liquid to have to plough. That is, a maximum feasible bonded area of the refractory crystals is desirable.

Large sizing of a dense grain or aggregate impedes dissolution, because only its external or geometric surface can be reached by the corrosive liquid, and that external surface area is inversely proportional to the size of the grain.

5.3.3 Hot Gases and Dusts

Refractory sidewalls, roofs, ducting, heat exchangers, and muffles exemplify exposures to hot gases and dusts. Kiln and steam boiler linings likewise entail corrosive atmospheric exposures without a liquid bath at the same time.

5.3.3.1 Atmospheric Penetration and Condensation

Two phenomena distinguish corrosive gas diffusion into refractories from liquid penetration. One is that the driving force for the former at steady state is the pressure gradient due to consumption of the gas by reaction. On that account, it is reasonable to characterize gas penetration broadly as less rapid in mass terms than that of wetting liquids. The other distinction is that the reaction does not necessarily commence at the hot face. On both accounts, the depth of penetration by gases can be much greater in some circumstances than that by liquids [79–83].

A simple isothermal equation for kinetic order of unity is adapted for use in the case of penetration by gas:

$$p_{o2} = p_{o1} \exp \left[(h_1 - h_2)/k \right] \tag{5.35}$$

Here, $h_2 > h_1$; h_1 is the height at which gas consumption begins; p_{o1} is *initial*; p_{o2} prevails at height h_2; and k is a constant. Corrosion of the refractory heat-exchanger packing or checkerwork is intensely concentrated close to height h_1, and then falls off in quasi-exponential fashion with increasing h.

In *condensation-corrosion* by gases, there are three possible modes of interaction of a gas with the external and internal surfaces of a refractory. First, if the gas is condensable (referred to as a vapor), in the course of penetration of the porosity and moving down the temperature gradient in a wall, the gas may literally condense or liquefy progressively. Second, the gas may condense by virtue of dissolving in the refractory; and third, it may condense by virtue of reacting chemically to form one or more new compounds. These three possible modes, in the order given, are progressively more energetic and rapid, and hence are progressively more concentrated into a narrow band in the porous solid. The resulting corrosion being by virtue of condensation, this kind of gas-phase corrosion is called *condensation-corrosion*.

Once condensation has occurred, however, the basic criteria for liquid corrosion apply. Liquid interaction products are the hallmark of corrosive gases. Rapid slabbing is absent in gas-exposed refractories.

Invasive debonding by liquid interaction products observes the reactivity distinction between refractory matrix and grains. Liquid interaction products tend to protect the underlying solid somewhat. The depth into the refractory at which maximum gas reaction rates occur tends to increase with time. Eventually, the depth of alteration becomes large. However, in the most aggressive cases, blinding of the hot-face pores with liquid occurs early, and the most dramatic alteration is concentrated there. Reaction at depth then becomes paced by diffusion through liquid-filled pores near the hot face.

This concentration of liquid products on macroscopic surfaces causes severe local melting by dissolution, especially in crevices. Mortared joints between bricks are particularly vulnerable in this respect, as mortars tend to be more reactive than well-sintered bricks. In contrast with the blinding of pores, low-viscosity liquids condensed in mortared joints tend to run out of the crevices by gravity, at once carrying out the dissolved mortar and exposing the crevices to further condensation. This gravitational flow is most marked in vertical brick joints in sidewalls, and quite generally in roof joints. Attack is most severe with liquids of lowest viscosity, hence with alkalis and chlorides [84]. It is not uncommon for otherwise-sound bricks to fall out of a roof eventually in such service, requiring shutdown for repair to avoid a massive collapse. Temperature is a major factor, both in reaction kinetics and in its effect on liquid viscosity.

Sulfur dioxide, usually from combustion, gives good examples of most of the condensation–corrosion patterns found in refractory interiors. Unaccompanied by appreciable alkalis, SO_2 reacts with oxidic refractories. Condensation reaction occurs with CaO in refractories at about 1400°C and lesser; with MgO at about 1100°C and lesser; with Al_2O_3 or ZrO_2 at much lower temperatures, about 600°C and lesser; and insignificantly with SiO_2. Given the refractory type and a hot face temperature above the appropriate limit, SO penetrates the porosity but does not condense until a limiting temperature is reached. Refractory alteration in that case occurs well inside a wall instead of at the hot face, while other corrosive gases may attack the latter region. This separation of corrosive reactions has been seen, for example, in oil-fired boiler linings [79,85].

When alkali vapors are present with SO_2 at the same time as, for example, Na_2O, the condensation of Na_2SO_4 starts at still higher temperatures and overlaps with the SO_2 attack on the refractory. Liquid Na_2SO_4 appears in this case at some 1600°C and lesser. Examples are seen in the cement kiln hot zone [82] in coal processing [86], and emphatically in glass furnace checkers [83, 87–90]. Checkers operate over a large $\Delta T/\Delta h$ and dT/dt as well, so that the spectrum of condensation products is ultimately spread out [90–92].

Condensed liquids fill the porosity and engage in debonding of the refractory. They may experience freeze–thaw cycling, and if solidified may create thermal expansion mismatches with the host. The typical consequences are weakening, softening, swelling, and finally even slabbing [79,82,85,90].

There are many acidic oxide gases that react with refractories. They are oxides of nitrogen and sulfur. The nitrates produced by oxides of nitrogen have low melting points. SO_2 and SO_3 form liquid sulfates. Volatile oxides of arsenic are encountered in some nonferrous metal operations, and are corrosive. The volatile basic oxides are the alkalis K_2O and Na_2O, and their corresponding hydroxides KOH and NaOH. These react with the refractories that contain them and yield products that are almost always low-melting.

Even more volatile and more corrosive than the alkalis are the chlorides. The identity of the chlorides boiled out of an inorganic process liquid depends on the cations that are present, and so does the temperature at which the vaporization occurs. Condensation of chlorides in refractories typically produces liquids, but these products are most often oxychlorides: solutions of refractory oxides (including silicates) in chloride hosts. Chloride ion also dissolves substitutionally in refractory oxides, lowering their melting temperatures.

5.3.3.2 Dusts

Hot process dusts originate in feed materials, but also importantly in the grinding action of moving solids within heated vessels. A further source is evaporation–condensation from process chemicals, while the combustion of inexpensive fuels always yields some dust, and may produce soot (carbon dust) in addition. Dusts have two characteristic effects on refractories—deposition and abrasion.

* *Deposition:* Dusts from industrial processes are often accompanied by fog and gases, which condense on the refractories. These accompanying liquids facilitate the gradual accretion of solid particles constituting the dust on the working surfaces. The resulting buildup can be variously a hard scale, or a somewhat porous, lightly sintered cake, or simply a pile of powder. In some less-obvious cases, dust, fog, and condensing gases all fuse together to create an invasive liquid. This gives rise to condensation-corrosion.

Solid buildup on refractory surfaces tends to seal off the porosity and is a deterrent to chemical corrosion. This is often welcomed. In large masonry installations, however, the deposition of hard scales in crevices, together with temperature cycling or mechanical flexing, can eventually fracture and even dislodge bricks.

There are a number of geometrical situations in which solid buildup in inaccessible places actually terminates the life of a refractory installation or component. A dramatic example of this has been in glass checkers [93], whose passageway dimensions have now become much enlarged by redesign to allow for their gradual constriction by caking. The effect of buildup on the transfer of heat from gas to refractory has to be figured in as well, requiring overdesign of the checker height.

- *Abrasion:* Abrasion of the refractory working face by dusts often contributes to wear. Abrasive wear depends on the impinging particle size, shape, hardness, mass, and velocity; on the inertia and viscosity of the entraining fluid; on the presence of intervening films; on the refractory surface geometry and texture; and on the angle of impact. Abrasion is frequently an accompaniment and an aggravator of corrosion. Its severity is hard to predict, but throats, ports, flues, elbows, valves, angular shapes, surface roughness, and similar sites of high gas velocity or turbulence are the most vulnerable.

Problem

5.1 Show that the $\Delta G^0_{1600} = -84.0$ for the reaction:

$3MgAl_2O_4 + 7/2\ SiO_2 = Al_6Si_2O_{13} + 3/2\ Mg_2SiO_4$

Given: $\Delta G^0_{f,1600}$ ($MgAl_2O_4$) = 1619.6 kJ/mol; $\Delta G^0_{f,1600}$ (SiO_2) = 590.6 kJ/mol

$\Delta G^0_{f,1600}$ ($Al_6Si_2O_{13}$) = 4769.4 kJ/mol; $\Delta G^0_{f,1600}$ (Mg_2SiO_4) = 1493.7 kJ/mol.

References

1. Advanced Structural Ceramics/Environment and Engineering of Ceramic Materials—*Advanced Structural Ceramics*—Basu—Wiley, http://as.wiley.com/WileyCDA/WileyTitle/productCd-0470497114.html, pp. 124–126.
2. W. D. Callister and D. G. Rethwisch, *Materials Science and Engineering*, 8th Edn., John Wiley, Hoboken, NJ, 2011, p. 706.
3. H. Schröder, Coatings on glass, *Glastechn. Ber.*, 26: 91–97, 1953.
4. K. Zagar and A. Schimöller, In Uhlig's Corrosion Handbook, 3rd Edn., John Wiley & Sons, Hoboken, New Jersey, p. 415.
5. R. W. Douglas and T. M. M. El-Shamy, Reaction of glass with aqueous solutions, *J. Am. Ceram. Soc.*, 50: 1, 1967.
6. R. H. Doremus, Interdiffusion of hydrogen and alkali ions in a glass surface, *J. Non Crystalline Solids*, 19: 137–144, 1975.
7. R. M. J. Smets and T. P. A. Lommen, In Corrosion of Glass, Ceramics, and Ceramic Superconductors: Principles, Testing, Characterization, and Applications, *Phys. Chem. Glasses*, 8: 140–144, 1967.
8. A. Paul, *Chemistry of Glasses*, Chapman and Hall, New York, 1982, pp. 108–147.

9. A. Paul, Chemical durability of glasses—Thermodynamic approach, *J. Mater. Sci.*, 12: 2246, 1977.
10. R. J. Charles, Static fatigue of glass, *J. Appl. Phys.*, 29: 1549, 1958.
11. R. K. Iler, *Colloid Chemistry of Silica and Silicates*, Cornell University Press, Ithaca, NY, 1955.
12. J. D. Rimstidt and H. Z. Barnes, The kinetics of silica-water reactions, *Geochim. Cosmochim. Acta*, 44: 1683, 1980.
13. M. Pourbaix, *Atlas of Electrochemical Equilibria in Aqueous Solutions*, Eng. Trans. by J. A. Franklin, NACE, Houston, TX, 1974, p. 644.
14. D. B. Levy, K. H. Custis, W. H. Casey, and P. A. Rock, A comparison of metal attenuation in mine residue and overburden material from an abandoned copper mine, *Appl. Geochem.*, 12: 203–211, 1997.
15. R. M. Garrels and C. L. Christ, *Solutions, Minerals, and Equilibria*, Harper and Row, New York, 1965, p. 435.
16. H. C. Helgeson and F. T. Mackenzie, Silicate-sea water equilibria in the ocean system, *Deep Sea Res.*, 17: 877, 1970.
17. J. D. Rimstidt, The kinetics of silica-water reactions, PhD Thesis, Pennsylvania State University, p. 135, 1979.
18. R. G. Newton and A. Paul, A new approach to predicting the durability of glasses from their chemical compositions, *Glass Technol.*, 21: 307, 1980.
19. D. G. Chen, Plant plasma membrane transport systems and osmotic shock signal transduction in Dunaliella salina, Graduate Thesis, University of Florida, 1987, p. 101.
20. D. Ge, Z. Han, Y. Yan, H. Chen, Z. Lou, X. Xu, R. Han, and L. Yang, Corrosion of glass, *J. Non Cryst. Solids*, 80: 341–350, 1986.
21. A. H. Dietzel, On the so-called mixed alkali effect, *Phys. Chem. Glasses*, 24: 172–180, 1983.
22. O. W. Richards, Some fungous contaminants of optical instruments, *J. Bacteriol.*, 58: 453–455, 1949.
23. P. W. Baker, Some fungous contaminants of optical instruments, *Int. Biodetn. Bull.*, 3: 59–64, 1967.
24. Z. G. Razamovskya and L. L. Mitushova, *Optiko Mekan Promysh*, 3: 69–72, 1957.
25. M. Poulain, Halide glasses, *Journal of Non-Crystalline Solids*, 56: 1–14, 1983.
26. D. G. Chen, C. J. Simmons, and J. H. Simmons, Corrosion layer formation of ZrF_4-based fluoride glasses, *Mater. Sci. Forum*, 19–20: 315–320, 1987.
27. R. A. Eppler, Zirconia-based colors for ceramic glazes, *Am. Ceram. Soc. Bull.*, 56: 1068–1070, 1977.
28. J. R. Taylor and A. C. Bull, *Ceramics Glaze Technology*, Pergamon, Elmsford, New York, 1986, pp. 167–183.
29. M. A. Rana and R. W. Douglas, The reaction between glass and water. Part 1. Experimental methods and observations, *Phys. Chem. Glasses*, 2: 179–195, 1961.
30. C. R. Das and R. W. Douglas, Studies on the Reaction between Water and Glass, *Phys. Chem. Glasses*, 8: 178–184, 1967.
31. T. M. El-Shamy, The electrode properties of some cation sensitive glasses, PhD Thesis, University of Sheffield, 1966.
32. D. E. Clark and E. C. Ethridge, Corrosion of glass enamels, *Am. Ceram. Soc. Bull.*, 60: 646–649, 1982.
33. R. A. Eppler, The durability of pottery frits, glazes, glasses and enamels in service, *Am. Ceram. Soc. Bull.*, 61: 989–995, 1982.

34. R. A. Eppler, Formulation of glazes for low Pb release, *Am. Ceram. Soc. Bull.*, 54: 496–499, 1975.
35. R. A. Eppler, Formulation and Processing of Ceramic Glazesfor Low Lead Release, *Proceedings of the International Conference on Ceramic Foodware Safety*, J. F. Smith and M. G. McLaren (Eds.), 1974, pp. 74–76.
36. J. W. Mellor, The durability of pottery frits, glazes, glasses and enamels in service, *Trans. J. Brit. Ceram. Soc.*, 34: 113–190, 1935.
37. J. H. Koenig, Lead frits and fritted glazes, *Ohio State Univ. Eng. Expt. Sta. Bull.*, 95: 116, 1937.
38. International Lead Zinc Research Organization, *Lead Glazes for Dinnerware*, ILZRO Manual Ceramics, No. 1, International Lead Zinc Research Organization, New York, 1970.
39. M. F. Dilmore, D. E. Clark, and L. L. Hench, Chemical durability of Na_2O-K_2O-CaO-SiO_2 glasses, *J. Am. Ceram. Soc.*, 61: 439, 1978.
40. S. C. Yoon, PhD Thesis, Rutgers, London, 1973.
41. F. Singer and W. L. German, *Ceramic Glazes*, Borax Consolidated, London, 1964.
42. V. A. Borgman, V. K. Leko, and V. K. Markargan, *The Structure of Glass*, Vol. 8, Trans. Russian, Consultants Bureau, New York, 1973, pp. 66–68.
43. D. M. Sanders and L. L. Hench, Mechanisms of glass corrosion, *J. Am. Ceram. Soc.*, 56(7): 373–377, 1973.
44. D. M. Sanders, W. B. Person, and L. L. Hench, Quantitative analysis of glass structure with use of infrared reflection spectra, *Appl. Spectrosc.*, 28: 247–255, 1974.
45. I. E. Campbell and E. M. Sherwood, *High-Temperature Materials and Technology*, Wiley, New York, 1967, pp. 141–147.
46. G. V. Samsonov (Ed.), *The Oxide Handbook*, IFI/Plenum, 1973, p. 524.
47. G. V. Samsonov and I. M. Vinitskii, *Handbook of Refractory Compounds*, IFI/Plenum, 1980, pp. 299–341.
48. L. A. Lay, *The Resistance of Ceramics to Chemical Attack*, Report Chem 96, National Physical Laboratory, Middlesex, England, 1979, pp. 1–27.
49. L. A. Lay, *Corrosion Resistance of Technical Ceramics*, National Physical Laboratory, Teddington, Middlesex, England, 1983, pp. 34–44, 50–55, 69–85, 98–107.
50. R. Morrel, *Handbook of Properties of Technical and Engineering Ceramics, Part 1: An Introduction for the Engineer and Designer*, Her Majesty's Stationery Office, 1985, pp. 183–203.
51. D. J. Clinton, L. A. Lay, and R. Morrel, *An Appraisal of the Resistance of Refel Silicon Carbide to Chemical Attack*, Report Chem 113, National Physical Laboratory, Middlesex, England, 1980, pp. 1–6.
52. R. W. Lashway, Sintered alpha-Silicon Carbide: an Advanced Material for CPI Applications, *Chem. Eng.*, 92: 121–122, 1985.
53. J. Lomax and R. W. Ford, Investigations into a method for assessing the long-term moisture expansion of clay bricks, *Trans. J. Br. Ceram. Soc.*, 82: 79–82, 1983.
54. N. S. Jacobson, J. L. Smialek, and D. S. Fox, *Molten Salt Corrosion of SiC and Si_3N_4*, NASA TM-10136, 1988.
55. D. W. McKee and D. Chatterji, Corrosion of silicon carbide in gases and alkaline melts, *J. Am. Ceram. Soc.*, 59: 441–444, 1976.
56. N. S. Jacobson and J. L. Smialek, Hot corrosion of sintered α-Sic at 1000° C. *J. Am. Ceram. Soc.*, 68: 432–439, 1985.
57. R. L. Barns and R. A. Laudise, Stability of superconducting $YBa_2Cu_3O_7$ in the presence of water, *Appl. Phys. Lett.*, 51: 1373–1375, 1987.

58. J. R. Gaier, A. H. Hepp, H. B. Curtis, D. A. Schupp, P. D. Hambourger, and J. W. Blue, in Stability of bulk $Ba_2YCu_3O_{(7-x)}$ in a variety of environments, *High-Temperature Superconductors, Mater. Res. Soc. Symp.*, Boston, MA; NASA TM-101401, NASA Lewis Research Center, Cleveland, OH, 1988.

59. A. Safari, J. B. Watchman, V. Parkhe, R. Coracciolo, D. Jeter, A. S. Rao, and H. G. K. Sunder, in Stability of bulk $Ba_2YCu_3O_{(7-x)}$ in a variety of environments, *High-Temperature Superconductors*, M. B. Brodsky, R. C. Dynes, K. Kitazawa, and H. L. Tuller (Eds.), *Mater. Res. Soc. Symp. Proc.*, Vol. 99, 1988, pp. 269–272.

60. M. F. Yan, R. L. Barns, H. M. O'Bryan, P. K. Gallagher, R. C. Sherwood, and S. Jin, Water interaction with the superconducting $YBa_2Cu_3O_{7-x}$ phase, *Appl. Phys. Lett.*, 51: 532–534, 1987.

61. S. Nasu, H. Kitagava, Y. Oda, T. Kohara, T. Shinjo, K. Asayama, and F. E. Fujita, *Fe Mössbauer effects in high-Tc superconductor: Y-Ba-Cu oxides*, 148(1–3): 484–487, 1987.

62. M. Komori, H. Kozuka, and S. Sakka, Chemical durability of a superconducting oxide $YBa_2Cu_3O_{7-x}$ in aqueous solutions of varying pH values, *J. Mater. Sci.*, 24: 1889–1894, 1989.

63. K. Kitazawa, K. Kishio, T. Hasegawa, A. Ohtomo, S. Yaegashi, S. Kanbe, K. Park, K. Kuwahara, and K. Fueki, in Environmental and Chemical Durability of Oxide Superconductors—Effect of Ambient Water on Superconductivity, *High-Temperature Superconductors*, M. B. Brodsky, R. C. Dynes, K. Kitazawa, and H. L. Tuller (Eds.), *Mater. Res. Soc. Symp. Proc.*, Scientific.Net, Trans Tech Publications Inc. Pfaffikon, Switzerland, Vol. 99, 1988, pp. 33–40.

64. S. E. Trolier, S. D. Atkinson, P. A. Fuierer, J. H. Adair, and R. E. Newnham, Dissolution of $YBa_2Cu_3O_{7-x}$ in various solvents, *Am. Ceram. Soc. Bull.*, 67: 759–762, 1988.

65. T. P. McAndrew, K. G. Frase, and R. R. Shaw, in Thin Film Processing and Characterization of High-Temperature Superconductors, *Proc. Am. Vacuum Soc. Symp.*, J. M. E. Harper, R. J. Colton, and L. C. Feldman (Eds.), 1987, *Am. Inst. Phys. Conf. Proc.*, Vol. 16, 1988.

66. H. Hojaji, A. Barkatt, R. A. Hein, Preparation and properties of highly densified yttrium-barium-copper oxide, *Materials research bulletin*, 23: 869–879, 1988.

67. Z. Dexin, X. Mingshan, Z. Ziqing, Y. Shubin, Z. Huansui, and S. Shuxia, Hydrolytic and deliquescent properties of superconducting Y- Ba-Cu oxide, *Solid State Comm.*, 65: 339, 1988.

68. Y. Wadayama, K. Kudo, A. Nagata, K. Ikeda, S. Hanada, and Izumi O., Crystal growth of high temperature superconductors—Problems, successes, opportunities, *Jpn. J. Appl. Phys.*, 27: L1221–L1224, 1988.

69. H. Hojaji, A. Barkatt, K. A. Michael, S. Hu, A. N. Thorpe, M. F. Ware, I. G. Talmy, D. A. Haught, and S. Alterescu, Yttrium Enrichment and Improved Magnetic Properties in Partially Melted Ya-Ba-Cu-O Materials, *J. Mater. Res.*, 5: 721–730, 1990.

70. R. A. Laudise, L. F. Schneemeyer, and R. L. Barns, Crystal growth of high temperature superconductors—Problems, successes, opportunities, *J. Crys. Growth*, 85: 569–575, 1987.

71. J. T. McDevitt, M. Longmire, R. Golmar, J. C. Jernigan, E. F. Dalton, R. McCarley, and R. W. Murray, Constant Power Functions of Time, *J. Electroanal. Chem.*, 243: 465, 1988.

72. J. T. McDevitt, R. L. McCarley, E. F. Dalton, R. Gollmar, R. W. Murray, J. Collman, G. T. Yee, and W. A. Little, Electrochemistry of high temperature superconductors—Challenges and opportunities, in *Chemistry of High Temperature Superconductors II*, D. L. Nelson and T. F. George (Eds.), *ACS Symp. Ser.*, 377, American Chemical Society, 1988, pp. 207–222.

73. F. H. Norton, *Refractories*, 3rd Edn., 1949, also 4th ed., 1960, McGraw-Hill Book, New York.

74. S. C. Carniglia, Limitations on internal oxidation-reduction reactions in BOF refractories, *Am. Ceram. Soc. Bull.*, 52: 160, 1973.

75. R. J. Leonard and R. H. Herron, Significance of oxidation-reduction reactions within BOF (basic oxygen furnace) refractories, *Am. Ceram. Soc. J.*, 55: 1, 1972.

76. S. C. Carniglia, Grain boundary and surface influence on mechanical behavior of refractory oxides: Experimental and deductive evidence, in *Materials Science Research*, Vol. 3, W. W. Kriegel and H. Palmour, III (Eds.), Plenum Press, New York, 1966, p. 425.

77. D. Clavaud, P. Meunier, and J. P. Radal, Twenty years of experience with high compactness concretes: A link between shaped products and electrocast refractories, in *UNITECR '89 Proceedings*, Vols. 1–2, L. J. Trostel, Jr. (Ed.), American Ceramic Society, Westerville, OH, 1989, p. 1094.

78. J. Mori, N. Watanabe, M. Yoshimura, Y. Oguchi, T. Kawakami, and A. Matsuo, Material design of monolithic refractories for steel ladle, in *UNITECR '89 Proceedings*, Vols. 1–2, L. J. Trostel, Jr. (Ed.), American Ceramic Society, Westerville, OH, 1989, p. 541.

79. G. Mascolo, O. Marina, and V. Sabatelli, Refractory and insulation degradation in an oil-fired power plant, *Sci. Ceram*, 12: 411, 1984.

80. Y. T. Chien and Y. C. Ko, Causes of refractory disintegration in a carbon monoxide environment, *Am. Ceram. Soc. Bull.*, 62: 779, 1983.

81. J. J. Brown, Carbon monoxide disintegration of coal gasifier refractories, *Annu. Conf. Mater. Coal Convers. Util., Proc.*, 6, IV/23, 1981.

82. F. Tanemura and T. Honda, Wear pattern of basic refractories in the burning zone of a rotary cement kiln, *Gypsum Line*, 200: 20, 1986.

83. T. Ishino and A. Yamaguchi, Wear factors of checker bricks in glass-melting furnace regenerators, *Taikabutsu*, 40: 390, 1988.

84. L. Wu, Formation of drops and corrosion cavities on silica crown bricks, *Boli Yu Tangci*, 15: 24, 1987.

85. J. B. Tak and D. J. Young, Sulfur corrosion of calcium aluminate bonded castables, *Am. Ceram. Soc. Bull.*, 61: 725, 1982.

86. R. N. Singh and C. R. Kennedy, Compatibility of refractories with simulated MHD environment consisting of a molten-coal-slag/alkali-seed mixture, *Mater. Energy Syst. J.*, 2: 3, 1982.

87. J. Sauer, L. Hundt, and G. Mlaker, Basic linings in steel-casting and steel-treatment ladles, *Fachber. Huettenprax. Metallweiterverarh.*, 24: 844, 1986.

88. U. Schmalenbach and T. Weichert, Factors influencing wear in areas of soda-lime glass tanks and lining recommendations, *Glass*, 65: 16, 1988.

89. M. L. Van Dreser and R. H. Cook, Deterioration of basic refractories in a glass regenerator, *Am. Ceram. Soc. Bull.*, 40: 68, 1961.

90. U. Schmalenbach and T. Weichert, Checker material of regenerative chambers in soda-lime glass-melting furnaces—Corrosion mechanism and lining recommendations, *Sprechsaal*, 120: 386, 1987.

91. I. Kamidochi and Y. Oishi, Regenerator checker brick for glass-melting furnaces, with emphasis on borosilicate glass-melting furnaces, *Taikabutsu*, 40: 364, 1988.
92. K. Nobuhara, H. Nishiyama, and K. Matsuda, Use of spinel checker bricks for glass melting furnace (regenerators), *Taikabutsu*, 40: 355, 1988.
93. E. Aoki, J. Kishimoto, E. Hosoi, and N. Moritani, Corrosion of castable refractories by alkali (potassium carbonate), *Tatkabutsu*, 35: 260, 1983.

Bibliography

C. Bodsworth, *Physical Chemistry of Iron and Steel Manufacture*, Orient Longmans, Madras, 1963, p. 31.

S. C. Carniglia and G. L. Barna (Eds.), *Handbook of Industrial Refractories Technology*, Reprint Edition, Noyes Publications, Westwood, NJ, 1992, pp. 189–198, 200, 202, 213, 215–219, 226, 228, 230–232, 236, 240–242, 245, 250, 252, 255, 257, 259–261, 266–267.

D. E. Clark and B. K. Zoitos (Eds.), *Corrosion of Glass, Ceramics, and Ceramic Superconductors: Principles, Testing, Characterization, and Applications*, Noyes Publications, Park Ridge, NJ, 1992, pp. 1–6, 124–125, 153–155, 242–243, 269, 273–274, 285–295, 315–328, 372–381, 432–435, 455–479, 514–529, 548–556, 559, 567–568, 571–573.

6

Failure in Ceramics and Refractories

6.1 Introduction

Ceramics and refractories are inherently brittle materials. The reason for this behavior is that the bonding in them is predominantly ionic or predominantly covalent. For plastic deformation, which is required for ductile fracture, there should be dislocation movement. In ionic compounds, formation of dislocation itself is difficult, because, for neutrality of the material, a pair of dislocations should simultaneously form. One should carry negative charge, and the other, positive. This is a difficult thing. If at all a dislocation forms, it requires simultaneous movement of the oppositely charged dislocations. This is still more difficult. In the case of covalent bonds, they are directional and strong. There is no question of any line defect, such as a dislocation, forming. Therefore, any question of dislocation movement does not arise. Ceramic and refractory materials fail by the sudden fracture of their atomic or ionic bonds. Hence, the failure of ceramic and refractory materials can be discussed in terms of the failure of brittle materials. In other words, the theory of brittle materials' fracture can be applied to ceramics and refractories.

6.2 Brittle Fracture

Brittle fracture is also called cleavage fracture. The reason is that the fractured surface looks like a cleaved surface. Fracture is the result of two processes: crack initiation and its propagation. Initially, a microcrack is produced, and it propagates until a time is reached when there is a sudden fracture. To prevent this failure, therefore, there is a need to control either one of the processes, or both.

When the material is isotropic, where the properties of the material are the same in all directions, it fails when a critical uniaxial tensile stress is exceeded. This critical stress represents the maximum tensile strength of the material. Amorphous materials such as glass fail in any direction. In the case of crystalline materials, they fail on planes having maximum density of atoms or ions. These planes are called *cleavage planes*.

6.2.1 Theoretical Strength

When a tensile force is applied to a material containing no defect, at a maximum force, the material fractures along its cleavage plane. The stress corresponding to this force, which is obtained by dividing this force required to fracture by the cross-sectional area, is called the *theoretical strength* of the material. It is represented by σ_{th}. To have a physical picture for finding the σ_{th} of a material, let us consider a cylindrical sample. To this is applied a tensile force, as shown in Figure 6.1.

In this sample, the force of cohesion between two adjacent planes varies according to the distance between them, given by the interatomic distance. On the application of tensile stress, the interatomic distance increases, as shown in Figure 6.2. When the force corresponds to the theoretical strength

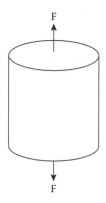

FIGURE 6.1
Conceptualization of theoretical strength.

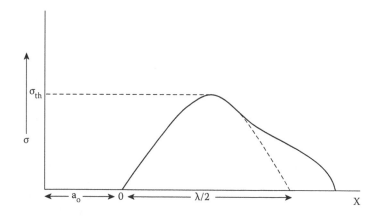

FIGURE 6.2
Stress versus interatomic distance.

of the material, the material fractures. Once it fractures, there is no more force acting on the material, and therefore the force becomes zero. The equation for the variation of the stress with the separation between two adjacent planes, perpendicular to the force, can be written as:

$$\sigma = \sigma_{th} \sin (2\pi X/\lambda) \tag{6.1}$$

Here, X is the distance moved by the force applied. The work per unit area required to separate the two adjacent planes is given by the force multiplied by the distance. This is given by Equation 6.2.

$$\int_0^\lambda [\sigma_{th} \sin (2\pi X/\lambda)]dX = \lambda \, \sigma_{th}/\pi \tag{6.2}$$

This work is utilized to create two new surfaces, and, therefore, it can be equated to the total surface energy of those two surfaces. Hence,

$$\lambda\sigma_{th}/\pi = 2\gamma \quad \text{or} \quad \sigma_{th} = 2\pi\gamma/\lambda \tag{6.3}$$

Here, γ is the surface energy. Until fracture, the ceramics are in an elastic state. Hence, Hooke's law is applicable for their stress–strain behavior. This is expressed as:

$$\Sigma = E \times (X/a_o) \tag{6.4}$$

Here, E is the elastic modulus of the material. The rate of change of the stress with respect to the increase in separation, as a result of the stress σ, can be obtained by differentiating Equation 6.1. That is,

$$d\sigma/dX = (2\pi\sigma_{th}/\lambda) \cos(2\pi X/\lambda) \sim 2\pi\sigma_{th}/\lambda \tag{6.5}$$

Differentiating Equation 6.4 with respect to X, we get Equation 6.6.

$$d\sigma/dX = E/a_o \tag{6.6}$$

Equating the RHSs of Equations 6.5 and 6.6, we get:

$$2\pi\sigma_{th}/\lambda = E/a_o \quad \text{or} \quad 2\pi/\lambda = E/a_o\sigma_{th} \tag{6.7}$$

Substituting the value of $2\pi/\lambda$ in (6.3) and rearranging, we get:

$$\sigma_{th} = (E\gamma/a_o)^{1/2} \tag{6.8}$$

Typically, $E = 3 \times 10^4$ MPa, $\gamma = 1$ J/m^2, and $a_o = 3 \times 10^{-10}$ m. Therefore, the typical value for theoretical strength is given by Equation 6.9.

$$\sigma_{th} = (3 \times 10^{10} \times 1/3 \times 10^{-10})^{1/2} = 10^{10} \text{ Pa} = 10^4 \text{MPa} \tag{6.9}$$

Let us take for granted that λ is similar in magnitude to a_o; then Equation 6.7 becomes:

$$2\pi = E/\sigma_{th} \quad \text{or} \quad \sigma_{th} = E/2\pi \tag{6.10}$$

This means that the theoretical strength is of the order of $E/5$–$E/10$. Such high strengths have been obtained only for thin ceramic fibers and whiskers. For commercial materials, the strengths are in the range of $E/100$–$E/1000$. Therefore, this concludes that, only for defect-free materials, such as thin fibers and whiskers, do the strength correspond to the force required to break the interatomic or interionic bonds.

6.2.1.1 Griffith–Orowan–Irwin Analysis

Griffith suggested that defects in the materials act as stress concentrators. The cracks originate from these defects and propagate through the material causing the final fracture. As the crack propagates, the stored elastic energy is utilized in the formation of the two surfaces. In other words, the stored energy is converted to surface energy. When this energy exceeds the surface energy, the crack propagates without any hindrance until the fracture takes place. Consider a thin plate. Let us assume that this plate contains an elliptical through-and-through crack across its thickness. Let $2c$ be the major axis of the crack. The plate with the crack is shown in Figure 6.3.

In the same figure, the direction of application of force is also shown. Equating the differentials of the two energies, we get:

$$(d/dc)(\pi c^2 \sigma^2 / E) = (d/dc)(4\gamma c)$$

or

$$(\pi \sigma^2 / E) \times 2c = 4\gamma$$

or

$$\sigma = (2E\gamma/\pi c)^{1/2} \approx (E\gamma/c)^{1/2} \tag{6.11}$$

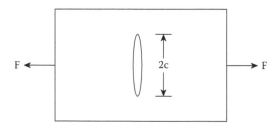

FIGURE 6.3
A thin plate containing an elliptical crack of length 2c.

Fracture will take place when the stress exceeds $(E\gamma/c)^{1/2}$. Hence, fracture strength of the material can be written as:

$$\sigma_f = (E\gamma/c)^{1/2} \qquad (6.12)$$

Inglis had said that the stress concentration takes place at the tip of the crack. This maximum stress is expressed as:

$$\sigma_m = 2\sigma(c/r)^{1/2} \qquad (6.13)$$

Here, r is the radius of the crack tip. When this maximum stress exceeds the theoretical strength of the material, fracture would take place—that is, when, $\sigma_m > \sigma_{th}$. According to Orowan, the minimum value of the radius of a crack tip is an interatomic distance. If the r in Equation 6.13 is replaced by a_o, the condition for failure becomes:

$$\sigma_f = (E\gamma/4c)^{1/2} \qquad (6.14)$$

Comparing Equations 6.12 and 6.14, it is found that the fracture strength obtained by considering stress concentration at the crack tip is only half the one obtained by considering crack propagation due to decrease in stored energy. Equation 6.12 has been found to hold good for brittle oxide glasses. For ductile materials, it was found that their fracture strength is higher. In order to account for this, Orowan introduced a term called γ_p, representing the surface energy of the area created by the work done in the plastic range of the material. The modified equation is then written as:

$$\sigma_f = [E(\gamma + \gamma_p)/c]^{1/2} \qquad (6.15)$$

Irwin considered the crack extension force for explaining the failure of materials. This force is given by:

$$G = (P^2/2)[d(1/m)/dc] \qquad (6.16)$$

Here, P is the load applied to a plate with a crack of length 2c, and m is the slope of the load-extension plot. During the propagation of a crack, the strain energy due to the stress applied is released. The rate of release of this strain energy is also represented as G. In this case, G is given by Equation 6.17.

$$G = \pi c\sigma^2/E \qquad (6.17)$$

As the crack propagates, and when the value of G exceeds a critical value, the crack becomes unstable, and it propagates suddenly, resulting in fracture.

This value of the strain energy release rate is denoted as G_c. Hence, fracture stress can now be written as:

$$\sigma_f = (EG_c/\pi c)^{1/2} \tag{6.18}$$

A factor called stress intensity factor is defined as:

$$K_I = \sigma \, (Yc)^{1/2} \tag{6.19}$$

Here, Y is a geometry factor. This factor depends on specimen and crack geometry. If the crack is located at the center of a thin infinite plate, then Y is equal to 1, σ is the normal stress acting at the crack tip, and 2c is the length of the crack if it is inside the plate. In the case of a thin plate, it will be under plane stress condition. Under this condition, we can write:

$$K_I^2 = GE \tag{6.20}$$

When G assumes critical value, the value of K_I is called the critical stress intensity factor. It is denoted as K_{Ic} and is called the fracture toughness of the material.

Following D. S. Dugdale [1], a plastic zone was found to be formed at the crack tip when the material yielded at the crack tip. This is shown schematically in Figure 6.4. The radius of this zone is given by Equation 6.21.

$$R = (\pi/8) \, (K_{Ic}/\sigma_y)^2 \tag{6.21}$$

Here, σ_y is the yield strength of the material.

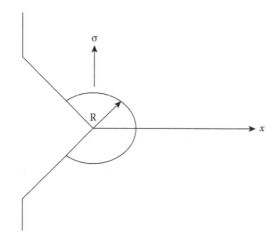

FIGURE 6.4
Plastic zone at the crack tip.

6.2.1.2 Statistical Nature of Strength

Griffith theory is based on probability. It requires the existence of a flaw of sufficient length. The length should be such that it is able to propagate. This length of the crack is called the *critical length*. This length should be acquired by means of the applied stress, or a crack of this length should be present initially in the material. When there is probability of finding such a crack, the material will fail by fracture. This explains the statistical nature of fracture strength. When we compare tensile strength and flexural strength, it will be seen flexural strength is greater. This is because the probability of finding a flaw of critical size is greater during tensile loading compared to bending load application. When we apply a tensile load, the whole of the cross-section is subjected to the same load. This is shown in Figure 6.5.

Hence, the probability of finding a crack of sufficient length is greater than that in a bending test. In a bending test, half the cross-section will be under tension, and the remaining half under compression. This is shown in Figure 6.6. Also, at the neutral axis, there is no force acting.

Hence, the probability of finding a crack that can be propagated is very much less.

Weibull [2] developed a statistical theory for the strength of brittle materials. His assumption was that the risk of rupture is proportional to the stress and volume of the body. Expressing this in the form of an equation, we get:

$$R = \int_V f(\sigma) \, dv \qquad (6.22)$$

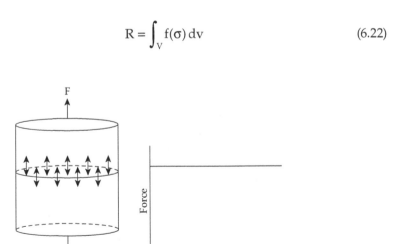

(a)
(b)

FIGURE 6.5

Tensile loading of a cylindrical specimen: (a) uniform distribution of the force on the cross-sectional area; (b) variation of force along any diameter of the cross-sectional area.

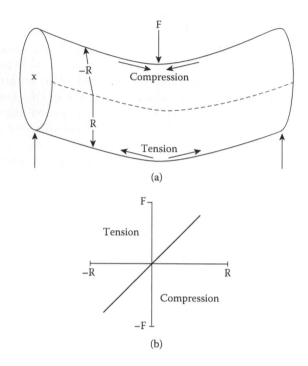

FIGURE 6.6
Flexural loading of a cylindrical specimen: (a) three-point bend test; (b) variation force along a diameter in the line of the force applied.

Weibull used the following equation for $f(\sigma)$:

$$f(\sigma) = (\sigma/\sigma_0)^m \tag{6.23}$$

Here, σ_0 is a characteristic strength, and m is a constant related to material homogeneity. As the homogeneity of the material increases, the value of m also increases. When m becomes zero, Equation 6.23 says that $f(\sigma)$ equals one. This amounts to telling us that the probability of failure is equal for all values of the stress (σ). As m tends to ∞, $f(\sigma)$ tends to be zero for all values of $\sigma < \sigma_0$. When $\sigma = \sigma_0$, the probability of fracture becomes one. This means that the fracture will take place only when the applied stress equals the characteristic strength. The average strength of the material then can be written as:

$$\sigma_{R = 1/2} = (1/2)^{1/m} \, \sigma_0 \tag{6.24}$$

This equation suggests two things:

1. Larger samples are weaker than smaller ones.
2. Dispersion in values of fracture stress increases as the median strength increases.

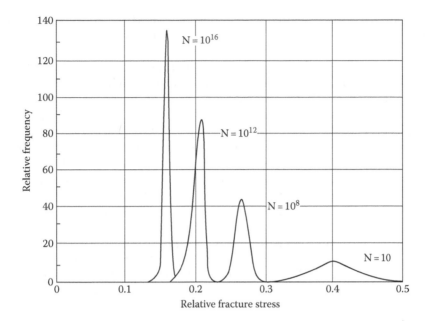

FIGURE 6.7
Calculated frequency distribution.

The first suggestion has already been explained earlier when the fracture under tensile and flexural loads was compared. The second suggestion is illustrated by Figure 6.7.

When the strengths are of modest values, statistical predictions agree with the experimental values. The mean strength decreases with increasing sample size. When the mean strength is larger, there is greater variation for breaking strength. For strengths above about 689 MPa, it is found that the strength is independent of sample size, and also that the dispersion of the fracture stress is independent of the mean strength. These were found to be the case with two works [3,4]. In these works, glass fibers in the range of 20–60-micron diameter were studied, and it was found that the tensile strength was independent of the diameter. The mean strengths quoted by these workers were 2756 MPa and 3652 MPa, respectively. In both works, the dispersion of the fracture stresses was also small.

6.2.1.3 Strength versus Fracture

Depending on the type of material, the strength of ceramics ranges from 689 Pa to 6895 MPa. The strength of highly porous firebrick corresponds to the lower limit, and that of silica fibers to the upper limit.

It is found that surface condition greatly decides the fracture strength of ceramics. The surface would contain flaws if it is not prepared carefully.

If the surface is not prepared with a compressive stress on it, even touching with hand can reduce the strength values of ceramics by an order of magnitude.

Bonded particles on glass surfaces are detrimental to their strength [5,6]. Cracks are seen to propagate from these particles. The other factors that contribute to the reduction in strength are the differences in moduli and thermal expansion coefficient of the materials of the glass and the particles. Also, local chemical attacks can also take place at the site of the particles.

In polycrystalline ceramics, microcracks can develop at the grain boundaries. The reason for this is the difference in the orientation of two adjacent grains. When the material is heated to a higher temperature, the grains will expand differentially along their different axes. On cooling from this higher temperature, the contraction also will be differential. Because of this differential expansion and contraction, thermal stresses will be built up in the material.

These stresses can then produce microcracks. An instance where heating and cooling take place is the firing process in the manufacture of bricks. Thermal stresses can also set up during the usage of the various products. Microcracks will be seen more on the surfaces of the parts because surfaces are heated to a greater extent than interiors. Most of the normal ceramics being insulators, the heat on the surface gets concentrated there. Another source of thermal stress on the surface is thermal etching. Thermal etching can produce notches on the surface, and notches are the sites at which stresses can get concentrated.

Crystalline materials are generally ductile or semibrittle. However, these can exhibit brittle fracture under certain conditions. Tw of such conditions are low temperature impact loading and the presence of notches. Notches are impediments to plastic deformation. In case there is plastic deformation under the application of a load, then dislocation movement takes place. During their movement, there is a chance for coalescence of dislocations, and then microcracks can develop.

When the dislocations move on a slip plane in large numbers, they pile up at the barriers. The barriers can be in the form of slip bands, grain boundaries, and surfaces. Due to pileup of dislocations, locally, high stresses are produced. When the dislocations are released from these pileups by these stresses, cracks get nucleated at this site.

An example for a pileup at the intersection of two slip bands is shown in Figure 6.8. This pileup is the result of the application of a tensile load. Dislocations move at an angle of 45° to the tensile stress produced.

Once a microcrack is produced, there will be stress concentration at the tip of this crack. This stress gets relieved by plastic deformation at this tip. This formation of a plastic zone where the deformation has taken place slows down the crack propagation. In the case of semibrittle materials, where the slip bands are limited, the stress builds up at the grain boundaries.

When the stress thus accumulated exceeds the strength of the material, fracture takes place. Dislocations move along the slip bands within the grains.

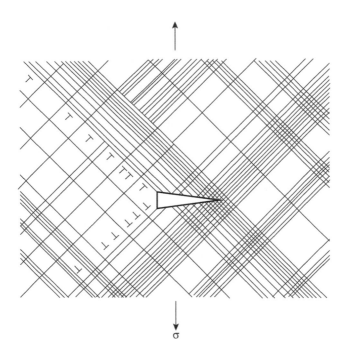

FIGURE 6.8
Dislocation pileup at the intersection of two slip bands resulting in the nucleation of a crack.

Let us say that the length of the slip band is proportional to the grain size. Then, the fracture stress can be written as:

$$\sigma_f = \sigma_0 + k_1 d^{-1/2} \tag{6.25}$$

where

$$k_1 = [3\pi\gamma E/(1-v^2)]^{1/2} \tag{6.26}$$

Equation 6.25 has the form of the Hall–Petch relation. Here, d represents the grain size. It says that the fracture stress should increase as the grain size decreases. In the case of brittle materials, if the initial flaw size is limited by the grain size, the relation between the fracture strength and grain size is given by the Orowan relation. This relation is expressed by Equation 6.27.

$$\sigma_f = k_1 d^{-1/2} \tag{6.27}$$

In polycrystalline ceramics, another cause for fracture is the internal stress. In such ceramics, anisotropic contraction takes place on cooling. Grain boundaries contain defects such as pores. Cracks are assumed to initiate at these pores and propagate along the grain boundaries [7]. It is easy for cracks to move along the grain boundaries because grain boundaries possess higher energy than grains. Also, the mobility of atoms is more in grain boundaries as they are not as tightly bound as inside the grains. When the pores are much smaller than the grain size, spontaneous fracture is expected when:

$$\varepsilon = (48\gamma_b/Ed)^{1/2} \qquad (6.28)$$

In this equation, ε is the strain in the grain boundary and γ_b is the grain boundary energy.

When the grain size is large, the fracture strength is found to follow the Orowan relation. The failure depends on the severest flaw. The severest flaws are associated with porosity, surface damage, abnormal grains, or foreign inclusions.

In single crystals and glasses, cracks can propagate easily, because there is no hindrance for the propagation, such as grain boundaries. Hence, crack initiation is the critical stage in them. In the case of polycrystalline materials, grain boundaries provide hindrance to the crack propagation. The direction of crack propagation will be changed. The increase in stress for a crack to change at a grain boundary has been analyzed [8], and has been found to be 2–4 times.

In the case of materials containing a ductile phase such as cemented carbides, the crack propagating in the brittle phase will get blunted when it encounters the ductile phase, which is metallic. Once it is blunted, the stress concentration at the crack tip is reduced because of the increase in the radius of the crack tip. The stress intensity factor reduces as the radius of the crack tip increases. The reduction in the stress will be such that it is not sufficient any more for the propagation of a blunt crack. Hence, a new crack will have to get nucleated for the progress of the fracture process under the action of the applied stress. This kind of crack nucleation will have to take place at all heterogeneous-phase boundaries. The heterogeneous-phase boundary in the present case is the boundary between the ductile and brittle phases. The fracture stress further increases as compared to a single-phase polycrystalline material.

In single crystals, fractures proceed along those planes having the least surface energy. In the case of polycrystalline ceramics, as mentioned, the crack propagates along the grain boundaries. This gives rise to greater $(\gamma + \gamma_p)$ values. The irregular crack path is shown in Figure 6.9. Because of the greater area created due to the irregular path, fracture stress also correspondingly increases. The crack has to meet many obstacles in its longer path.

6.2.1.4 Static Fatigue

In the case of silicate glasses, which break in a brittle manner, the fracture is time dependent. That is, for a given load, the fracture takes place after a period of time. This phenomenon is called *static fatigue*. The time to failure is related to the fracture stress by the following equation:

$$\log t = (A/\sigma_f + B) \qquad (6.29)$$

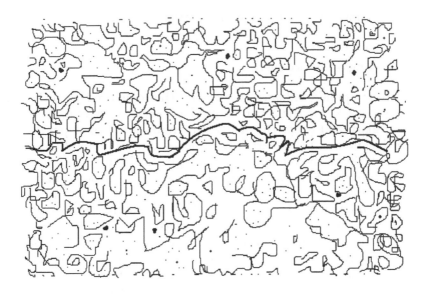

FIGURE 6.9
Fracture path in a sample of thermally cracked polycrystalline material.

Static fatigue depends on the environment. If moisture is present, the time of fracture decreases. The phenomenon of static fatigue is also temperature dependent. At room temperature, the strength increases. This is attributed to the decreasing atomic mobility as temperature is lowered. It is also found that, at temperatures above 450 K, the strength increases with temperature. The reasons given for this occurrence are the following:

a. Decreasing surface adsorption of water. At temperatures above 373 K, water is changed in the form of steam.

b. At these elevated temperatures, the effect of the occluded particles also decreases.

c. The viscous work at the crack tip increases.

In the temperature range between −323 K and 423 K, it is found that the time to fracture and temperature follow an Arrhenius relation. The activation energy for soda-lime-silica glass was found to be equal to 18.1 kcal/mole. This kind of behavior shows that the static fatigue is a temperature-dependent activated process, similar to any chemical reaction or diffusion.

In room temperature tests carried out on soda-lime-silica glasses that were given different abrasion treatments, it was found that each treatment gave a different static fatigue curve [9]. Static fatigue strength is about 20% of the low-temperature strength.

Two phenomena happen during static fatigue:

1. Stress corrosion
2. Lowering of surface energy

The stress applied enhances corrosion. This enhanced corrosion happens at the crack tip. The corrosion product being brittle makes the crack tip more sharp and deep. This increased length of the crack penetrates to greater thickness of the material. This causes more concentration of stress at the crack tip, as the remaining area perpendicular to the applied stress decreases. This decrease in the cross-sectional area is evident in Figure 6.4. More stress gives rise to more corrosion, and then more crack propagation, and so on, until the final fracture results. Lowering of the surface energy is a result of adsorption of active species. As surface energy decreases, the stress needed for the creation of new surface as a result of the crack propagation decreases. Thus, crack propagates more easily, and fracture results finally.

Etching removes surface cracks, and crack bottoms are made rounded. Rounding of crack bottoms reduces stress concentration at the crack tips as the radius of the cracks is increased. Stress concentration is inversely related to the stress intensity factor. The small radius of curvature of the crack tip increases its chemical potential and solubility. Thus, etching ultimately increases the fracture strength.

The factor causing more rapid dissolution at the crack tip than at the sides is that the high tensile stress at the tip leads to an expansion of the glass network or crystal lattice at that point, and an increase in the rate of corrosion [10]. Dissolution at the crack tip extends the crack length and thereby increases the stress concentration. Quenched glass is more vulnerable to corrosion than annealed glass. This is because quenched glass has a more expanded structure. The tensile stress makes the glass elastically expanded. As the expansion is more at the crack tip, sodium ions tend to move toward the crack tip. There, these ions get dissolved.

The plot of crack velocity versus applied force in soda-lime glass showed three regions [11]. The plot is shown in Figure 6.10. The stress intensity factor is proportional to the applied force.

The three regions are represented by Roman numerals in Figure 6.10. The uppermost linear line is for the test in water. The region I part of the curves below this line are parallel to it. These curves are for different humidity values. The humidity values are written on the right-hand side of each curve in terms of relative humidity. The plot shows that the crack velocity increases exponentially as the stress intensity factor increases in the region I. As humidity increases, so does the velocity. In region II, the crack velocity is independent of the stress intensity factor. In this region, the velocity is affected only by the water-vapor pressure. In region III, all the curves lie on the same straight line. It shows that crack velocity increases exponentially with the stress intensity factor, independent of the environment.

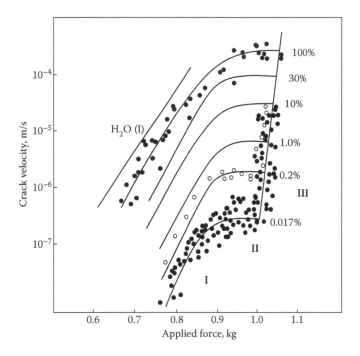

FIGURE 6.10
Dependence of crack velocity on applied force for soda-lime-silicate glass for various humidity values. Each curve shows three regions.

There is an empirical relation between crack velocity and the stress intensity factor. It is given by Equation 6.30. In this equation, A and n are constants. The value of n is in the range of 30–40.

$$V = AK_I^n \qquad (6.30)$$

Slow crack is also observed for systems free of alkali. Results similar to those found for silicate glasses at room temperature are also found for a variety of other ceramic materials, such as porcelains, glassy carbon, Portland cement, high-alumina ceramics, silicon nitride, lead zirconite, and barium titanate [12].

Strength also depends on the loading rate. As the loading rate is decreased, strength also decreases. This relation can be explained as follows. As the loading rate is decreased, more time is allowed for the crack to propagate. As crack velocity is decreased, the stress intensity factor is also decreased, as can be obtained from Equation 6.30. That means the critical stress intensity factor (K_{Ic}), at which the fracture takes place, is lowered. This lower K_{Ic} is reached at lower applied stress, which is same as the fracture strength of the material.

In the case of cyclic loading, crack propagation takes place during the tensile part of the cycle. Hence, the effective time taken for the fracture will be the sum of the times during which the sample was under tensile stress. This total time can be used to calculate the fracture strength of the material.

When the stress is varied sinusoidally with time, the relation between stress and time is given by:

$$\sigma_t = \sigma_o + \sigma_1 \sin 2\pi v t \qquad (6.31)$$

Here, σ_t is the stress at any time t; σ_o and σ_1 are constants; and v is the frequency of cyclic loading—that is, the number of cycles per unit time. Following Equation 6.30, the average stress intensity factor is related to the average crack velocity per cycle, according to Equation 6.32.

$$V_{av} = gAK_{Iav}^n \qquad (6.32)$$

where

$$g = v \int_0^{1/v} (\sigma_t/\sigma_o)^n \, dt \qquad (6.33)$$

Taking K_I as K_{Iav} in Equation 6.30, and writing the resulting RHS of this equation as equal to V_{static}, we get:

$$V_{av} = gV_{static} \qquad (6.34)$$

Values of g for different values of n are available [13]. Reasonable agreement was found between the predicted and experimental values for the crack velocity under sinusoidal stress condition in the case of ceramics. Hence, crack propagation under this condition involves the same mechanism as under static conditions.

6.2.1.5 Creep Failure

Creep is the phenomenon of high-temperature deformation under a constant load. When the deformation results in the failure of the material, the failure is termed as the *creep failure*. In the case of polycrystalline materials, the deformation is mainly due to grain boundary sliding. Tensile stresses are built up at the boundaries. These stresses develop pores and cracks at the boundaries. As deformation increases, the pores grow.

As the pores increase in size, the cross-sectional area decreases. This reduction in cross-sectional area increases the stress. Increase in stress results in the nucleation of cracks and their subsequent propagation. Finally, fracture results. The development of cracks is shown in Figure 6.11. The material was deformed in bending. Therefore, the cracks that had nucleated from the top surface (which was under tension) had propagated to greater depths. The time required for creep failure is more for low stresses and low temperatures.

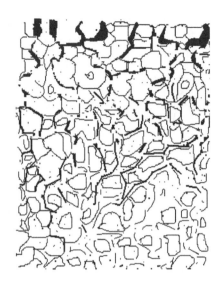

FIGURE 6.11
Development of cracks in alumina during creep deformation.

Creep failure is represented by creep–rupture curves. In this plot, applied stress is plotted against time to failure in the logarithmic scale.

6.2.1.6 Effect of Microstructure on Fracture

Porosity has a major effect on the fracture of materials. There are two ways by which pores influence fracture strength:

1. Presence of porosity decreases the cross-sectional area on which the load is applied.
2. Pores act as stress concentrators.

The first effect has already been discussed in connection with Figure 6.4. If there is a spherical pore present, the stress at the pore is increased by a factor of 2. An empirical relation between the strength and porosity is given in Equation 6.35 [14].

$$\sigma = \sigma_0 \exp(-nP) \tag{6.35}$$

Here, σ is the strength of the material having a porosity value of P in terms of volume fraction; σ_0 corresponds to the strength without any pores; and n is a constant. The value of n ranges from 4 to 7. This equation suggests that, if there is 10% porosity, the strength is reduced by half. Earthenware contains 10–15% porosity, and has less strength than hard porcelain, which has a porosity of about 3%. In Figure 6.12, the ratio of σ/σ_0 is plotted against the volume fraction of pores following the Equation 6.35. Along with this curve,

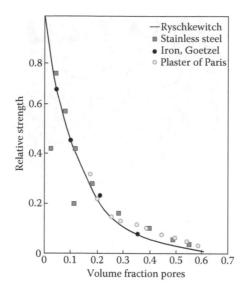

FIGURE 6.12
Effect of porosity on the fracture strength of materials.

the plot also contains the data for iron, stainless steel, and plaster of Paris. This plot proves that porosity has the same effect on the fracture strength of both metals and ceramics.

In two-phase systems such as glazed ceramics, it is better to have a glaze having less thermal expansion coefficient than the body. This will give rise to a compressive stress in the glaze when the material is cooled from a higher temperature. Compressive stress tends to close any crack present, thereby preventing fracture. Cracks normally originate from the surface. In this case, the surface is that of the glaze.

When the ceramic contains more than one phase, each phase will have different thermal expansion coefficients. On heating and cooling, this differ-ence can give rise to thermal stresses at the phase boundaries. These stresses can cause internal cracks. Such a crack is seen in the microstructure of a triaxial whiteware composition, shown in Figure 6.13.

Because of the difference in the thermal expansion coefficients of the quartz and cristobalite phases, thermal stresses develop during cooling. Obviously, cristobalite has greater thermal coefficient than quartz. As the material cools, cristobalite exerts tensile stress on quartz. This tensile stress causes the crack-ing of the quartz grain. The fracture strength of a material is also affected by particle size. As the particle size decreases, the fracture strength increases. It has been found that fine-grained quartz gives higher strength for porcelains.

A ductile phase in a multiphase ceramic is found to increase the fracture strength. An example to cite is the case of cemented carbide. This composite contains cobalt as the ductile phase. In this material, cracks tend to originate

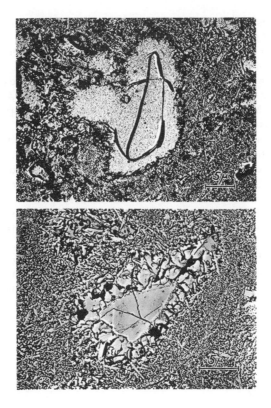

FIGURE 6.13
Microstructure of a triaxial whiteware composition. (From W. D. Kingery et al., *Introduction to Ceramics*, 2nd Edn., John Wiley, 1976, pp. 783–812.)

in the brittle carbide phase. When the crack reaches the phase boundary, it meets the cobalt phase. Further propagation of the crack is hindered because the applied stress now is utilized in the plastic deformation at the crack tip, which gives rise to strain hardening. Also, the crack gets blunted in the plastic zone. Correspondingly, the stress intensity factor is decreased. These mechanisms increase the fracture strength.

References

1. D. S. Dugdale, Yielding of steel sheets containing slits, *J. Mech. Phys. Solids*, 8: 100, 1960.
2. S. B. Batdorf, Fracture statistics of brittle materials with intergranular cracks, *Nuclear Engineering and Design*, 35(3): 349–360, 1975.
3. W. H. Otto, Relationship of tensile strength of glass fibers to diameter, *J. Am. Ceram. Soc.*, 38: 122, 1955.

4. W. F. Thomas, An investigation of the factors likely to affect the strength and properties of glass fibres, *Phys. Chem. Glasses*, 1: 4, 1960.
5. A. Woldemar A. Weyl, Structure of subsurface layers and their role in glass technology, *Journal of Non-Crystalline Solids*, 19: 1–25, 1975.
6. N. M. Cameron, The effect of environment and temperature on the strength of E-glass fibres. Part 2. Heating and ageing, *Glass Technol.*, 9: 14, 21, 1968.
7. A. G. Evans, Microfracture from thermal expansion anisotropy—I. Single phase systems, *Acta Metallurgica*, 26(12): 1845–1853, 1978.
8. E. Smith, Cleavage crack formation and the effect of the structure of the nucleating deformation process, *Acta Metallurgica*, 16(3): 313–320, 1968.
9. R. E. Mould and R. D. Southwick, Strength and static fatigue of abraded glass under controlled ambient conditions: II, Effect of various abrasions and the universal fatigue curve, *J. Am. Ceram. Soc.*, 42: 582, 1959.
10. R. J. Charles, Static fatigue of glass, *J. Appl. Phys.*, 29: 1549, 1958.
11. S. M. Wiederhorn, Influence of Water Vapor on Crack Propagation in Soda-Lime Glass, *J. Am. Ceram. Soc.*, 50: 407, 1967.
12. R. C. Bradt, D. P. H. Hasselman, and F. F. Lange (Eds.), *Fracture Mechanics of Ceramics*, Vols. 1 and 2, Plenum Publishing Corporation, New York, 1974.
13. A. G. Evans and E. R. Fuller, Crack propagation in ceramic materials under cyclic loading conditions, *Met. Trans.*, 5: 27, 1974.
14. E. Ryschkewitsch, Compressive Strength of Porous Sintered Alumina and Zirconia, *J. Am. Ceram. Soc.*, 36: 65, 1953.

Bibliography

W. D. Kingery, H. K. Bowen, and D. R. Uhlmann, *Introduction to Ceramics*, 2nd Edn., John Wiley, New York, 1976, pp. 783–812.

7

Design Aspects

7.1 Introduction

The defining of the requirements of an application is the first step in the design of the material for that application. For example, the requirement for wear-resistant application is hardness. If there is more than one requirement, then the designer has to prioritize the different requirements. Redesign may sometimes be necessary. The second step is to compare the properties of different materials with the requirements of the application. Next is to consider the fabrication method to fabricate the part with the required properties. Last, the reliability of the part also should be specified.

7.2 Design Approaches

There are five approaches to design:

1. Empirical
2. Deterministic
3. Probabilistic
4. Linear elastic fracture mechanics
5. Combined approach

7.2.1 Empirical Approach

In this approach, an initially designed part is fabricated and tested for its suitability in the application. If it does not serve the purpose fully, it is redesigned, refabricated, and retested. This kind of cycle is repeated until it becomes 100% suitable for the purpose. This approach becomes effective in the following cases:

a. The part is already in use but requires modification.
b. Available property data are limited.
c. The survival of the part is strongly affected by the environment.

7.2.2 Deterministic Approach

In this approach, the maximum stress in a component is calculated by finite element analysis or closed-form mathematical equations. The material for the component is then selected that has a strength with a reasonable margin of safety over the calculated peak stress. The safety factor is decided based on previous experience.

Figure 7.1 shows the strength distribution for a high-strength ceramic material. It is evident from Figure 7.1 that the curve is broad. It is not symmetrical. If the average strength value is calculated, it will be seen that it is not the same as the maximum strength value. Because of these characteristics of the curve, it will not be possible to fix a factor of safety for this ceramic. The reason is that the design stress and the strength curve will overlap.

The average strength can be taken for the typical strength of the material. It is got in the following manner:

1. Conduct the test to determine the strength of the material using a number of specimens.
2. Add all the strength values obtained.
3. Divide the total value obtained by the number of tests conducted.

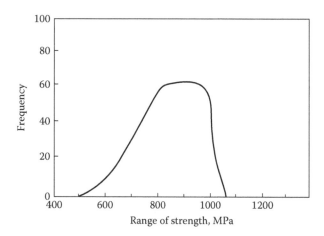

FIGURE 7.1
The strength distribution for a high-strength ceramic material.

Mathematically, the calculation can be represented by the following equation:

$$\overline{\sigma} = \sum_{i=1}^{N} \sigma_i \tag{7.1}$$

Here, $\overline{\sigma}$ is the average strength, σ_i is the strength value got for the i-th test, and N is the total number of tests conducted.

7.2.3 Probabilistic Approach

The probabilistic approach considers the flaw distribution and stress distribution in the material. This approach will be useful when high stresses and their complex distributions are present. In these situations, both empirical and deterministic designs have limitations. The flaw distribution can be characterized by the Weibull approach [1]. This approach is based on the weakest link theory. This theory says that a given volume of a ceramic under a uniform stress will fail at the most severe flaw. Thus, the probability of failure F is given by the following equation:

$$F = f(\sigma, V, A) \tag{7.2}$$

Here, F is shown as a function f of the applied stress σ, volume V of the material, and area A under the stress.

The applied stress is related by a threshold stress σ_μ, a characteristic stress value σ_0, and the Weibull modulus m. The characteristic stress value is taken as the stress at which the probability of failure is 0.632. The Weibull modulus describes the flaw size distribution. The relationship of the applied stress with σ_μ, σ_0, and m is expressed by the following relation:

$$f(\sigma) = [(\sigma - \sigma_\mu)/\sigma_0]^m \tag{7.3}$$

The probability of failure is a function of the volume. It is given by the following equation:

$$F = 1 - \exp - \left\{ \int_V \left[(\sigma - \sigma_\mu)/\sigma_0 \right]^m \right\} dV \tag{7.4}$$

The curve for this equation is shown in Figure 7.2. The probability of failure can be calculated for each value of the strength obtained from the tests. The strength values are arranged in the increasing order. From the rank obtained for each strength value, F is calculated as $n/(N + 1)$. In this, n represents the rank in the order, and N is the total number of tests conducted. Table 7.1 gives an idea about how F is calculated.

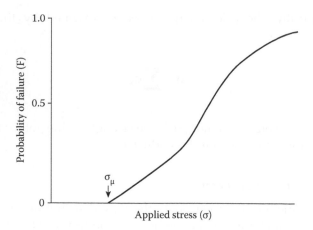

FIGURE 7.2
Probability of failure versus applied stress.

TABLE 7.1

Calculation of Probability of Failure for N + 1 = 10

Strength Values in Increasing Order	Rank, n	Probability of Failure, F = n/(N + 1)
S1	1	0.1
S2	2	0.2
S3	3	0.3
S4	4	0.4
S5	5	0.5
S6	6	0.6
S7	7	0.7
S8	8	0.8
S9	9	0.9

Figure 7.3 gives the schematic of the plot of the probability of failure versus strength. This plot does not give the flaw size distribution. In order to get it, we rearrange Equation 7.4 as follows:

$$\exp - \int_V \left[(\sigma - \sigma_\mu)/\sigma_0 \right]^m dV = 1 - F \tag{7.5}$$

or

$$\exp \int_V \left[(\sigma - \sigma_\mu)/\sigma_0 \right]^m dV = 1/(1 - F) \tag{7.6}$$

$$\ln \int_V \left[(\sigma - \sigma_\mu)/\sigma_0 \right]^m dV = \ln \left[1/(1 - F) \right] \tag{7.7}$$

or

$$m \ln \sigma - K = \ln \left[1/(1 - F) \right] \tag{7.8}$$

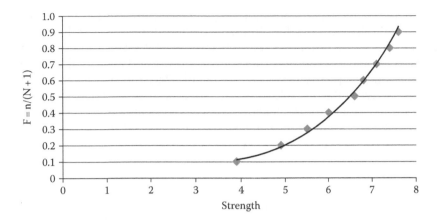

FIGURE 7.3
Probability of failure versus strength.

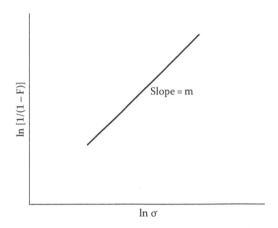

FIGURE 7.4
Schematic plot of ln [1/(1 − F)] against ln σ.

Now, a plot of ln [1/(1 − F)] against ln σ will be a straight line. This plot is drawn schematically in Figure 7.4. The slope of this straight line gives the value for m. The larger the value of m, the less the scatter for the flaw size.

Equations 7.3 and 7.4 are called the three-parameter Weibull functions. The three parameters are σ_μ, σ_0, and m. For ceramics, a two-parameter form is used. In this, the σ_μ is considered to be zero. Then the function can be written as:

$$F = 1 - \exp - \left\{ \int v [\sigma/\alpha_0]^m \right\} dv \tag{7.9}$$

To obtain 90% confidence in the value of m, at least 60 samples should be tested. The more the value of m, the more reliable the material.

Sometimes the data do not fit a single straight line, and may give rise to two straight lines. Such a distribution is called a *bimodal one*. If that happens, we can think of eliminating the distribution that gives a low value of m. What we should do is observe the fracture surface and identify the two types of cracks and the phases in which they occur. Once the identification is done, then we arrive at the possible causes for their occurrence. Once we know the causes, we can modify the fabrication steps to eliminate the cracks giving rise to the lower value of m. Thus, Weibull curves and fracture analysis can be used to improve a ceramic material.

7.2.3.1 Design Using Weibull Distribution

Weibull distribution can also be used to find the failure probability of a component, apart from its use to determine the failure probability against applied stress. This is done by integrating the Weibull distribution for the material with the stress distribution for the component. Finite element analysis is useful to do this. Comparison is made between the material strength data in the form of a Weibull probability curve and a finite element in the component. This is repeated for all the elements in the component. Thereafter, the probabilities of all the elements are added. This sum gives the probability of failure of the component.

Probabilistic design allows designing closer to the properties of the material. That means a material can be used even at a high localized stress. In this approach, peak stresses and stress distributions cannot be adequately defined. Also, the true strength–flaw size distribution cannot be defined adequately. This is because local heat transfer conditions, the effects of geometry, precise loads, load application angles, and so on, are not accurately defined for the calculation of mechanical stresses.

7.2.4 Approach Based on Linear Elastic Fracture Mechanics

This approach is useful for brittle materials such as ceramics. In it, it is assumed that the material contains flaws, which is the case for any real material. It also assumes that fracture is depended on three factors, viz., the fracture toughness, stress intensity factor, and flaw size.

Fracture takes place after crack initiation and its propagation. If the criteria and the characteristics of these two steps are available, then fracture mechanics can be based on them. These are obtained from the following:

a. Analyzing the stress and temperature distribution.
b. Analyzing the time for which each region in the component is exposed to a specific stress and temperature.
c. Identifying the number of cycles the component will have to undergo.

d. Using the fast fracture date and determining whether the material will survive the steady-state and peak stresses in application.

e. Using data from stress rupture, creep, oxidation, fatigue, and so on, to slow crack growth and formation of new defects. This will help in the prediction of the life of the component.

This approach is still evolving. A combination of all four approaches will be needed in the selection of a material, deciding its configuration, prediction of life, and determining the reliability. This is called the *combined approach*.

Reference

1. W. A. Weibull, A statistical distribution function of wide applicability, *J. Appl. Mech.*, 18(3): 293–297, 1951.

Bibliography

D. W. Richerson, *Modern Ceramic Engineering—Properties, Processing, and Use in Design*, CRC Press, Boca Raton, 2006.

Section II

Ceramics

8

Bonding in Ceramics

8.1 Introduction

The properties and the way its atoms are arranged are determined primarily by the nature and directionality of interatomic bonds. If the bond is strong, those are the *primary bonds*. The primary bonds may be ionic, covalent, or metallic. Van der Waals and hydrogen bonds are *secondary bonds*, and they are weaker than the primary bonds. Ceramics contain two types of primary bonds—ionic and covalent. There are a few ceramics that contain the third type of primary bond (i.e., the metallic bond). For example, in Fe_3C, the bond is partly ionic and partly metallic.

Ionic compounds generally form between very active metallic elements and active nonmetals. Thus, the metals mostly involved in the formation of ceramics by ionic bonding are those from groups IA, IIA, and part of IIIA, as well as some of the transition metals. Moreover, the most active nonmetals of groups VII A and VIA form ionic bonding.

For covalent bonding to occur, ionic bonding should be unfavorable to form. This unfavorability can be semi-quantified using a scale of relative *electronegativity*. This scale was proposed by Pauling. In this scale, Cs is given a value of 0.79 and F, 3.98. If two elements forming a bond have similar electronegativities (electronegativity difference Δx becoming very small), they tend to share electrons. Homo-polar bonds are purely covalent. All other bonds in ceramics possess both ionic and covalent character. In such ceramics, which bond predominates can be predicted from the value of Δx. A bond will be predominantly ionic when $\Delta x > 1.7$ and predominantly covalent if $\Delta x < 1.7$.

Coming back to the case of Cs and F, we have the Δx value between these two elements equal to 3.19, which is much above 1.7. Obviously, therefore, CsF is a predominantly ionic solid.

8.2 Ionic Bonding

Ionic bonding gives rise to ionically bonded solids.

8.2.1 Ionically Bonded Solids

The stability of ionically bonded solids depends on the lattice energy. The greater the lattice energy, the more stable the solid will be. The stability in turn determines the melting temperature, thermal expansion, and stiffness of the solid. More stable solids will possess greater melting points, less thermal expansion, and greater stiffness.

8.2.1.1 Lattice Energy Calculations

To calculate the lattice energy, let us consider a common ionic solid NaCl. To confirm that NaCl is a predominantly ionic solid, we can find its Δx value. Putting the values for the electronegativity of Cl and Na in the equation for calculating Δx, we get $\Delta x = 3.16 - 0.93 = 2.13$, which is very high, showing that NaCl is predominantly ionic bonded. Figure 8.1a shows the model of a NaCl unit cell.

It shows that the centrally placed Cl ion is surrounded by six Na ions. That means the central anion is attracted to six Na^+ cations at a distance r_0 (Figure 8.1b). It is repelled by 12 Cl^- anions at a distance $\sqrt{2}r_0$ (Figure 8.1c), and attracted to eight Na^+ cations at $\sqrt{3}r_0$ (Figure 8.1d), and so on.

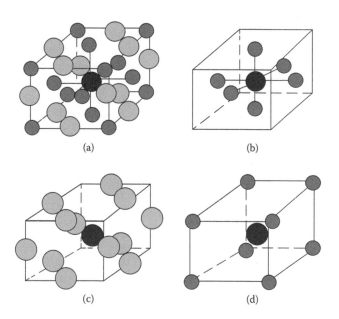

(a)   (b)

(c)   (d)

FIGURE 8.1
(a) Schematic of the NaCl structure; (b) the first six nearest neighbors at a distance of r_0 are attracted to the central anion; (c) the second 12 nearest neighbors at a distance of $\sqrt{2}r_0$ are repelled; (d) the third eight nearest neighbors are attracted.

Summing up electrostatic interactions, we can write that:

$$E_{sum} = \left(z_1 z_2 e^2 / 4\pi\varepsilon_0 r_0\right)\left(1 - 1/n\right)\left[\left(6/1\right) - \left(12/\sqrt{2}\right)\right.$$
$$\left. + \left(8/\sqrt{3}\right) - \left(6/\sqrt{4}\right) + \left(24/\sqrt{5}\right) - \cdots\right]$$

(8.1)

where

z_1 and z_2 are the net charges on the ions

n is the *Born exponent* that lies between 6 and 12

r_0 is the equilibrium interatomic distance

The third term in parentheses is an alternating series that converges to some value α, called the *Madelung constant*.

The total electrostatic attraction for 1 mol of NaCl in which there are twice the Avogadro's number N_{AV} of ions but only N_{AV} bonds is:

$$E_{latt} = N_{AV}(z_1 z_2 e^2 \alpha / 4\pi\varepsilon_0 r_0) \; (1 - 1/n)$$

(8.2)

Here, E_{latt} is the lattice energy of an ionic solid. For NaCl, it can be calculated to be equal to –750 kJ/mol.

8.2.1.2 Born–Haber Cycle

In the derivation of the expression for calculating the lattice energy, we have assumed that the ionic solid is made up of ions attracted to each other by coulombic attractions. To test this model, we can use the Born–Haber cycle. This cycle is based on the first law of thermodynamics. As we know, this law states that energy can neither be created nor destroyed.

The cycle devised for NaCl is shown in Figure 8.2. For this cycle, the following equation is valid, following the first law of thermodynamics:

$$\Delta H_{form}(exo) = E_{vap}(endo) + E_{ion}(endo) + E_{diss}(endo) + E_{EA}(exo) + E_{latt}(exo) \quad (8.3)$$

Here, ΔH_{form} is the enthalpy of formation for NaCl; E_{vap} and E_{ion} are the sublimation and ionization energies, respectively, for 1 gram atom of Na atoms; and E_{diss} and E_{EA} are the dissociation energy and electron affinity, respectively, for 1 gram atom of Cl atoms. The exo/endo indicate whether the energy involved is exothermic/endothermic, respectively. Exothermic energy is indicated by the – sign and endothermic by the + sign. Let us now consider each of these energies.

8.2.1.2.1 Enthalpy of Formation

When the reaction:

$$Na(s) + 1/2 \; Cl_2(g) = NaCl(s)$$

(8.4)

occurs, ΔH_{form} amount of thermal energy gets liberated. For NaCl at 298 K, $\Delta H_{form} = -411$ kJ/mol.

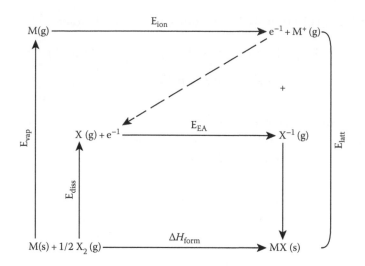

FIGURE 8.2
The Born–Haber cycle.

8.2.1.2.2 Heat of Vaporization

Latent heat of vaporization E_{vap} for the reaction

$$Na(s) = Na(g) \tag{8.5}$$

equals 107.3 kJ/mol.

8.2.1.2.3 Ionization Energy

Energy required to completely remove an electron from an isolated atom in its gas phase is called its *ionization energy*.

For Na, $E_{ion} = 495.8$ kJ/mol.

8.2.1.2.4 Dissociation Energy

E_{diss} for

$$1/2\, Cl_2 = Cl(g) \tag{8.6}$$

equals 121 kJ/mol.

8.2.1.2.5 Electron Affinity

It is the energy change that occurs when an electron is added to the valence shell. Electron affinity of Cl is −348.7 kJ/mol. If we substitute the values, we get:

$$\Delta H_{form}(exo) = 107.3 + 495.8 + 121 - 348.7 - 750 = -374.6 \text{ kJ/mol} \tag{8.7}$$

This compares favorably with the experimentally determined value of −411 kJ/mol.

This result becomes important for two reasons: (1) the model of an ionic solid constructed by assuming the interaction between the ions present is correct for the most part, and (2) it supports the notion that NaCl is ionically bonded.

8.2.1.3 Pauling's Rules

Pauling's rules are used to interpret the majority of the ionic structures. The *first rule* states that a coordination polyhedron of anions formed is formed about each cation. The cation–anion distance is equal to the sum of their radii. The coordination number is determined by the ratio of the radii of the two ions. This relation between the coordination number and radii ratio is based on the stability of the resulting structure. This is illustrated in Figure 8.3.

In a two-dimensional representation, Figure 8.3a is a stable structure, because the repulsive force due to similarly charged ions is lesser than the attractive force between the oppositely charged ions. This is obvious because the similarly charged ions are not in contact, whereas the oppositely charged ions are touching each other. Figure 8.3b is less stable because the repulsive force has increased owing to the contact between the similarly charged ions. Figure 8.3c is unstable because of the greater repulsive force compared to the attractive force by virtue of the noncontact between the oppositely charged ions. The relation between the coordination number and the range of cation-to-anion radius ratio is shown in Table 8.1. Figure 8.4 shows the resulting structures for various coordination numbers.

The *second rule* describes a basis for evaluating local electrical neutrality. It states that, in a stable structure, the total strength of the bonds reaching an anion from the surrounding cations should be equal to the charge of the anion. The strength of an ionic bond donated from a cation to an anion is the formal charge on the cation divided by its coordination number.

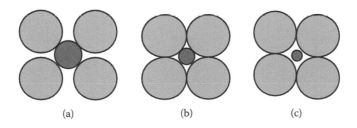

(a)   (b)   (c)

FIGURE 8.3

Stable and unstable coordination configuration (a) Stable, (b) Stable, and (c) Unstable.

TABLE 8.1

Relation between Coordination Number (CN)
and Range of Cation-to-Anion Radius Ratio

Coordination Number (CN)	Range of Cation-to-Anion Radius Ratio
8	≥ 0.732
6	≥ 0.414
4	≥ 0.225
3	≥ 0.155
2	≥ 0

Coordination number	Disposition of ions about central ion	Cation radius ratio / anion radius	
8	Corners of cube	≥ 0.732	
6	Corners of octahedron	≥ 0.414	
4	Corners of tetrahedron	≥ 0.225	
3	Corners of triangle	≥ 0.155	
2	Linear	≥ 0	

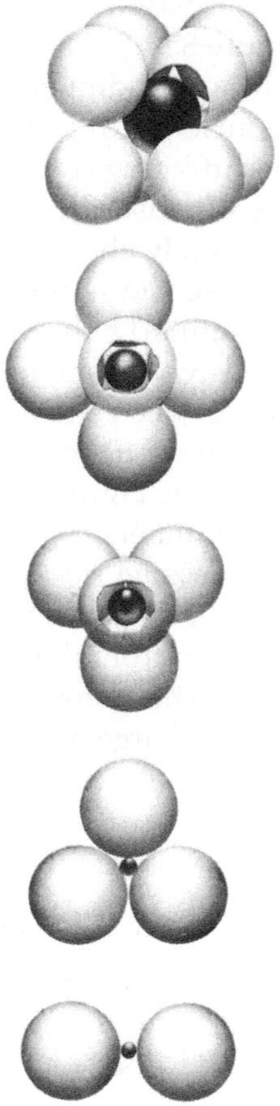

FIGURE 8.4
Polyhedra for different coordination numbers. (From W. D. Kingery, H. K. Bowen, and D. R. Uhlmann, *Introduction to Ceramics*, 2nd Edn., John Wiley, 1976, pp. 783–812.)

The *third rule* further concerns the linkage of the cation coordination polyhedral. In a stable structure, the corners, rather than the edges and especially the faces of coordination polyhedra, tend to be shared. If an edge is shared, it tends to be shortened. The *fourth rule* says that polyhedra formed about the cations of low coordination number and high charge tend especially to be linked by corner sharing. The *fifth rule* states that the number of different constituents in a structure tends to be small.

8.3 Covalently Bonded Solids

Covalent bonds arise by electron sharing. Many predominantly covalently bonded ceramics, especially Si-based ones, such as SiC, Si_3N_4, and silicates, are composed of Si atoms simultaneously bonded to four other atoms in a tetrahedral arrangement. Ground state configuration of Si has the valence electron in $3s^2 3p^2$ configuration, as shown in Figure 8.5a. It shows that Si can have two primary bonds when it combines with another element atom, because an Si atom has two unpaired electrons.

In reality, there are four primary bonds associated with each Si atom. This can happen by *hybridization* between s and p wave functions. The s orbital hybridizes with all the three p orbitals to form *sp^3 hybrid orbitals* (Figure 8.5b). Hybrid orbitals possess both s and p character, and directionally reaches out in space as lobes in a tetrahedral arrangement with a bond angle of 109° (Figure 8.5c). Each orbital is populated by one electron (Figure 8.5b). Now, each Si can bond to four other Si atoms or any other *four* atoms to give three-dimensional structures.

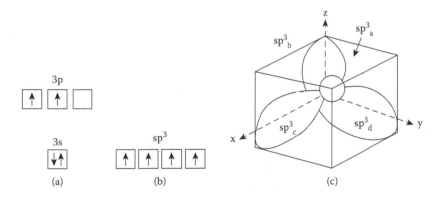

FIGURE 8.5
(a) Ground state of the Si atom; (b) electronic configuration after hybridization; (c) directionality of sp^3 bonds.

Problems

8.1 The radius of K^+ is 1.38 Å and of O^{2-} is 1.40 Å. If they form an ionic solid, find the structure of the polyhedron.

8.2 The structure of sodium chloride has anions in cubic close packing, with Na^+ ions occupying all octahedral positions.

   8.2.1 Compute the value of the lattice constant.

   8.2.2 Calculate the density of NaCl.

   8.2.3 What is the maximum radius of a cation that can be accommodated in the vacant interstice of the anion array?

   8.2.4 Calculate the density of 0.01 mole% KCl solid solution in NaCl.

8.3 Sketch the atomic plan of (110) and ($\bar{1}\,\bar{1}0$) planes of CaO. Show the direction of closest packing. Point out the tetrahedral and octahedral sites.

8.4 In a cubic close packing of O^{2-} ions, what is the ratio of octahedral sites to O^{2-} ions? What is the ratio of tetrahedral sites to O^{2-} ions?

8.5 Explain, on the basis of Pauling's rules in cubic close packing of O^{2-} ions, what valency ions are required to have stable structures, in which:

   8.5.1 All octahedral sites are filled

   8.5.2 All tetrahedral sites are filled

   8.5.3 Half the octahedral sites are filled

   8.5.4 Half the tetrahedral sites are filled

   Give an example for each.

8.6 Barium titanate has a perovskite structure. What are the coordination numbers of each ion?

8.7 Obtain the ionic ratio for a structure in which the anions are at the corners of a triangle.

8.8 The anions in boron trioxide are at the corners of triangles. If the radius of B^{3+} is 0.20 Å, and that of O^{2-} is 1.40 Å, then find out the coordination number of B^{3+}.

8.9 Calculate the O/Si ratio for tremolite, $(OH)_2Ca_2Mg_5(Si_4O_{11})_2$, and talc, $(OH)_2Mg_3(Si_2O_5)_2$.

8.10 Given 1 mol of K^+ and 1 mol of Cl^-, calculate its lattice energy when the coulombic interaction energy is restricted to:

   8.10.1 The first nearest neighbors

   8.10.2 The second nearest neighbors

   8.10.3 The third nearest neighbors

   NOTE: Take Born exponent as infinity.

8.11 Plot the attractive, repulsive, and net energy between Ca^{2+} and O^{2-} from 0.2 nm in increments of 0.01 nm to 0.3 nm. Given: $n = 9$, $B = 0.4 \times 10^{-105}$ J·m^9.

8.12 Calculate the Madelung constant for CsCl if the coulombic interaction is restricted to the first three terms.

8.13

8.13.1 To what inert gases do the ions of Ba^{2+} and O^{2-} correspond?

8.13.2 What is the interionic distance if these two ions are in contact in the BaO solid?

8.13.3 What is the force of attraction between a Ba^{2+} and an O^{2-} ion if the ion centers are separated by 1 nm? State all assumptions.

8.14 The fraction ionic character of a bond between elements A and B can be approximated by:

$$\text{Fraction ionic character} = 1 - \exp. \, -(X_A - X_B)^2/4$$

Here, X_A and X_B are the electronegativities of the respective elements.

Using this expression, compute the fractional ionic character for the following compounds:
KCl, CaO, NiO, PbO, and NaF.

Bibliography

M. W. Barsoum, *Fundamentals of Ceramics*, Taylor & Francis Group, LLC, New York, 2003, pp. 13, 23–33, 49–50.

W. D. Kingery, H. K. Bowen, and D. R. Uhlmann, *Introduction to Ceramics*, 2nd Edn., John Wiley, New York, 1976, pp. 88–89.

9

Structures of Ceramics

9.1 Introduction

The structure of ceramics can be considered by dividing them into three groups—oxides, silicates, and glasses. Both oxides and silicates can form crystalline and amorphous structures. The crystalline form will be considered separately for oxides and silicates. The amorphous forms of both will be considered under glasses.

9.2 Oxide Structures

Oxide ceramics are those in which the nonmetal is oxygen. These ceramics possess different structures. These structures are discussed in the following text.

9.2.1 Rock Salt Structure

In this structure, the large anions are arranged in cubic close packing, and all the octahedral interstitial positions are filled with cations. The structure is shown in Figure 9.1.

Oxides having this structure are MgO (Mg^{2+}/O^{2-} = 0.51), CaO, SrO, BaO, CdO, MnO, FeO, CoO, and NiO. For stability, the radius ratio should be between 0.732 and 0.414, and the anion and cation valencies should be the same.

9.2.2 Zinc Blende Structure

The structure of zinc blende has tetrahedral coordination. A BeO polymorph at high temperature has this structure. The structure is shown in Figure 9.2.

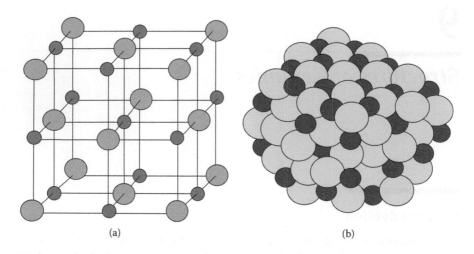

(a)   (b)

FIGURE 9.1
Crystal structure of sodium chloride. Smaller spheres represent Na$^+$, and larger spheres Cl$^-$ ions.
(a) Shows locations of the ions in the FCC structure. (b) Schematic of the actual structure.

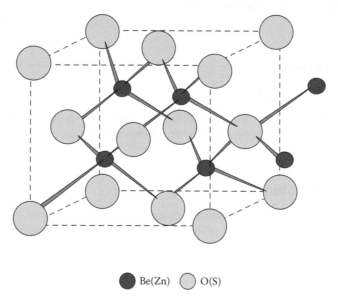

● Be(Zn)   ○ O(S)

FIGURE 9.2
Zinc blende (ZnS) structure.

9.2.3 Cesium Chloride Structure

The radius ratio between cesium and chloride ions requires eightfold coordination. From this, we can find that the bond strength is 1/8. These two criteria give rise to simple cubic (SC) array for Cl$^-$ ions. All the body center positions are filled with Cs$^+$ ions. The structure is shown in Figure 9.3.

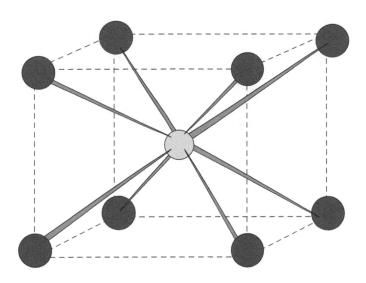

FIGURE 9.3
Cesium chloride structure.

9.2.4 Fluorite and Antifluorite

The example for an oxide-possessing fluorite (CaF_2) structure is thoria (ThO_2). The charge on Th^{4+} is large. This gives rise to the largest coordination number (CN) of 8. Therefore, the bond strength comes to 1/2. The number of valence bonds becomes four to satisfy the valency of 4 of Th^{4+} to each O^{2-} ion. Thus, there will be simple cubic packing for O^{2-} with Th^{4+} in half the sites with eightfold coordination. This is the structure of fluorite (CaF_2), shown in Figure 9.4. The cation lattice is face-centered cubic (FCC), with all the tetrahedral interstices filled with anions. Other examples of oxides possessing the structure are TeO_2 and UO_2.

Antifluorite structure is the reverse of a fluorite structure having a cubic close-packed array of O^{2-} with cations in the tetrahedral sites. Examples of oxides with this structure are Li_2O, Na_2O, and K_2O.

9.2.5 Perovskite Structure

The chemical formula of perovskite is $CaTiO_3$. The structure is shown in Figure 9.5. Let us apply Pauling's rule to this structure. The strength of Ti–O bond = 2/3, and that of Ca–O bond = 1/6.

Since each O^{2-} is coordinated with 2 Ti^{4+} and 4 Ca^{2+}, the total bond strength becomes (4/3) + (4/6) = 2, which is the same as the anion oxide valency. Hence, Pauling's rule is satisfied.

Other examples for compounds having perovskite structure are $BaTiO_3$, $SrTiO_3$, $SrSnO_3$, and $CaZrO_3$.

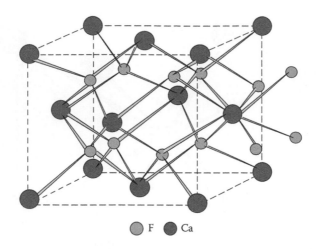

◯ F ● Ca

FIGURE 9.4
Fluorite structure.

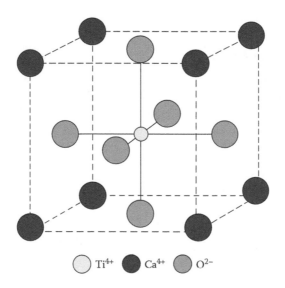

◯ Ti^{4+} ● Ca^{4+} ◯ O^{2-}

FIGURE 9.5
Perovskite structure (idealized).

9.2.6 Spinel

The general formula for spinel structure is AB_2O_4. An example for a ceramic possessing this structure is magnesium aluminate ($MgAl_2O_4$). Thus, A represents divalent cations, and B, trivalent ones. The structure is cubic, having rock salt and zinc blende structures combined. A subcell of this structure is formed by O^{2-} ions in FCC packing (Figure 9.6). The figure shows four

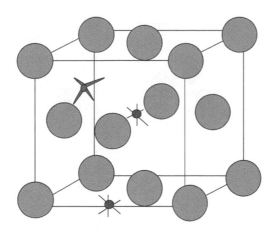

FIGURE 9.6
Subcell of spinel structure.

octahedral interstices and eight tetrahedral interstices. Out of the four octahedral sites, only two will be filled in the spinel structure, and out of the eight tetrahedral sites, only one will be filled. Eight elementary cells are arranged to form a unit cell.

There are two types of spinel: *normal* and *inverse*. In normal spinel, the A^{2+} ions are in tetrahedral sites and B^{3+} ones in octahedral sites. The examples for normal spinel are $ZnFe_2O_4$, $CdFe_2O_4$, $MgAl_2O_4$, $CoAl_2O_4$, $MnAl_2O_4$, and $ZnAl_2O_4$.

In *inverse* spinel, A^{2+} and half B^{3+} will be in octahedral sites and the other half in tetrahedral sites. The arrangement can be represented as B (AB) O_4. These compounds are more common than the normal one. Examples are Fe (MgFe) O_4, Fe (TiFe) O_4, Fe^{3+} ($Fe^{2+}Fe^{3+}$) O_4, Fe (NiFe) O_4, and many other magnetic ferrites.

9.2.7 Alumina Structure

Al_2O_3 has a common structural arrangement shared by many oxides of the transition metals with the formula A_2O_3 [1]. In Al_2O_3, the preferred CN for Al^{3+} is 6, so bond strength becomes equal to 1/2; hence, there should be four Al^{3+} adjacent to each O^{2-} to give the total bond strength equal to the valency of oxide anion. This is possible when the unit cell is hexagonally closed packed (HCP) of O^{2-}, with Al^{3+} filling two-thirds of octahedral sites. The structure is shown in Figure 9.7.

9.2.8 Ilmenite

Ilmenite is iron titanate ($FeTiO_3$). The $FeTiO_3$ crystal structure is an ordered derivative of the Al_2O_3 structure. Here, the two Al^{3+} are replaced by a bivalent Fe^{2+} and a tetravalent Ti^{4+} in the formula. The structure then becomes an

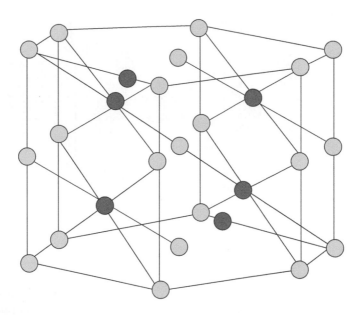

FIGURE 9.7
Unit cell of alumina crystal.

HCP of O^{2-}, with cations filling two-thirds of the octahedral sites. Obviously, half the cation sites are occupied by Fe^{2+}, and the other half by Ti^{4+} ions. Other examples having ilmenite structure are $MgTiO_3$, $NiTiO_3$, $CoTiO_3$, and $MnTiO_3$.

9.2.9 Rutile Structure

Rutile's chemical formula is TiO_2. In its structure, the CN for Ti is 6, its valence is +4, and therefore the bond strength becomes 2/3. This gives rise to threefold coordination for O^{2-}. Cations fill only half the available octahedral sites, and the closer packing of O^{2-} around filled cation sites leads to distortion of anion lattice, which is FCC. In Figure 9.8, the cationic lattice is shown. Other examples for compounds possessing rutile structures are GeO_2, PbO_2, SnO_2, and MnO_2.

9.3 Structure of Silicates

Silicates are also oxides, but they contain two or more oxides, out of which one is essentially silica. Earth's crust contains 48% by weight O_2, 26% Si, 8% Al, 5% Fe, and 11% Ca, Na, K, and Mg combined. Thus, the earth's crust and also mantle consist mainly of silicates.

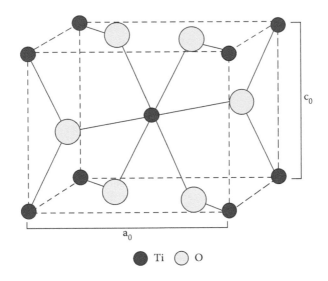

FIGURE 9.8
The cationic lattice of rutile.

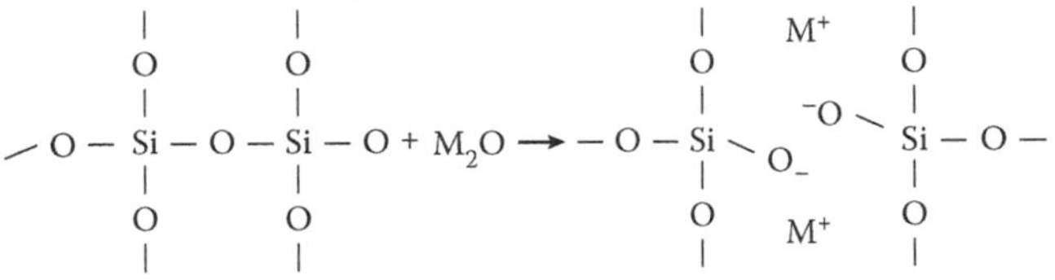

$$\begin{array}{ccccc} | & | & & | & M^+ \quad | \\ O & O & & O & O \\ | & | & & | & \end{array}$$

$-O-Si-O-Si-O + M_2O \longrightarrow -O-Si \underset{O_-}{\overset{^-O}{\diagdown}} \underset{M^+}{\overset{}{Si}}-O-$

FIGURE 9.9
Formation of nonbridging oxygen by the addition of an alkali oxide to silica.

In silicates, two types of oxygens can be seen. They are *bridging* and *non-bridging oxygens* (NBOs). Oxygen atoms are bonded to two Si atoms in the case of bridging oxygens, whereas they are bonded to only one in the case of nonbridging oxygens.

NBOs are formed by the addition of either alkali or alkaline earth metal oxides to silica, as shown in Figure 9.9. In this, O^- denotes an NBO. NBOs are negatively charged. Local neutrality is maintained by having cations end up adjacent to the NBOs. The following are the salient points in the formation of NBOs:

1. The number of NBOs are proportional to the number of moles of alkali or alkaline earth metal oxides added.

2. Addition of alkali or alkaline earth metal oxides to silica increases the overall *O/Si ratio* of the silicate.

3. Increasing the number of NBOs results in the progressive breakdown of the silicate structure into smaller units.

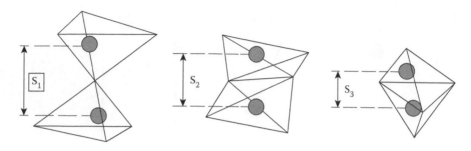

FIGURE 9.10
Effect of corner edge and face sharing on cation–cation separation. The distances S_1:S_2:S_3 are in the ratio 1:0.58:0.33; that is, cation–cation repulsion increases on going from left to right, which tends to destabilize the structure.

A critical parameter that determines the structure of silicate is the number of NBOs per tetrahedron, which, in turn, is determined by the O/Si ratio.

The principles involved in the formation of silicate structures are the following:

1. The basic building block is the SiO_4 tetrahedron. The Si–O bond is partly covalent, and tetrahedron satisfies both the bonding requirements of covalent directionality and the relative size ratio for ionic structures.

2. Because of the high charge on the Si^{4+} ion, the tetrahedral units are rarely joined edge to edge, and never face to face, but almost always share the corners. The reason was first stated by Pauling, and it is illustrated by Figure 9.10. The reason is that the cation separation distance decreases in going from the corner to the edge to the face sharing. This in turn results in cation–cation repulsions and a decrease in stability of structure.

The relationship between the O/Si ratio, which can vary between 2 and 4, and the structure of silicate, is given in Table 9.1.

9.3.1 Silica

As already stated, the basic building block of silicates are $(SiO_4)^{4-}$ tetrahedra. This building block is possessed by the silica structure. When this block is joined to the neighboring tetrahedra, the four negative charges on each block is balanced, and the structure becomes neutral. This gives rise to the O/Si ratio of 2 and the chemical formula of silica as SiO_2. In this structure, each O is linked to two Si, and each Si is linked to four O. Since the tetrahedron is a three-dimensional structure, the network formed by the combination

TABLE 9.1

Relationship between Silicate Structure and O/Si Ratio

Structure	O/Si Ratio	Number of Oxygens per Si		Structure and Examples
		Bridge	Nonbridge	
	2.00	4.0	0.0	Three-dimensional network: quartz, tridymite, cristabolite—are all polymorphs of silica
Repeat unit $(Si_4O_{10})^{4-}$	2.50	3.0	1.0	Infinite sheets: $Na_2Si_2O_5$, clays (kaolinite)
Repeat unit $(Si_4O_{11})^{6-}$	2.75	2.5	1.5	Double chains: asbestos
Repeat unit $(SiO_3)^{2-}$	3.00	2.0	2.0	Chains: $(SiO_3)_n^{2n-}$, Na_2SiO_3, $MgSiO_3$
Repeat unit $(SiO_4)^{4-}$	4.00	0.0	4.0	Isolated SiO_4^{4-} tetrahedra: Mg_2SiO_4, olivine, Li_4SiO_4

of the tetrahedra also becomes a three-dimensional network. Examples for this kind of structure are the different allotropes of silica that, depending on the exact arrangement of the tetrahedra, include quartz, tridymite, and cristobalite. When the long-range order is lacking, the resulting solid is called *amorphous silica* or *fused quartz*.

9.3.2 Sheet Silicates

When three out of four oxygens are shared, with an O/Si ratio of 2.5, the sheet structure results. Examples for this structure are clays, talc $[Mg_3(OH)_2(Si_2O_5)_2]$, and mica $[KAl_2(OH)_2(AlSi_3O_{10})]$. Kaolinite $[Al_2(OH)_4(Si_2O_5)]$ is a clay (Figure 9.11a). In this, $(Si_2O_5)^{2-}$ sheets are held together by the positively charged sheets of $Al_2^{6+} - O_2^{4-},(OH)_4^{4-}$ octahedral (Figure 9.11b). This structure

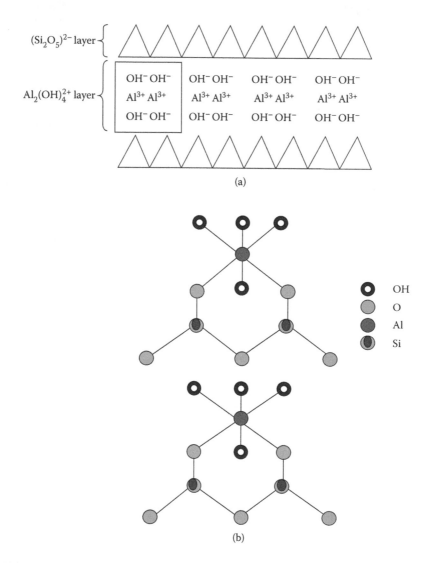

(a)

(b)

FIGURE 9.11
Structure of kaolinite (a) shows the different layers in the structure; (b) shows the composition of the two layers.

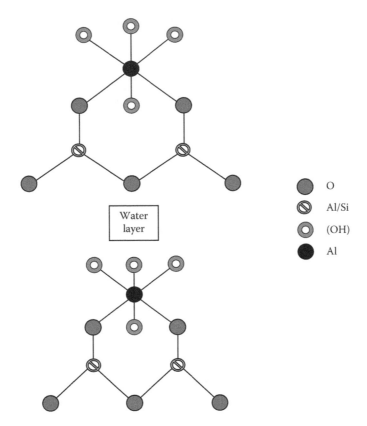

Water
layer

○ O
◉ Al/Si
◎ (OH)
● Al

FIGURE 9.12
Structure of hydrated clay.

helps explain why clays absorb water so readily; the polar water molecule is easily absorbed between the top of positive sheets and the bottom of silicate sheets. This is shown in Figure 9.12. In mica (Figure 9.13), Al ions substitute for one-fourth of Si atoms in sheets, requiring an alkali ion such as K^+ in order for structure to remain electrically neutral. Alkali ions fit in *holes* of silicate sheets and bond sheets together with an ionic bond that is stronger than that in clays. The resulting structure is shown in Figure 9.14. Mica does not absorb water as readily as clays do. Little effort is required to flake off a very thin chip of the material.

9.3.3 Chain Silicates

For an O/Si ratio of 3, infinite chains or ring structures result. An example for a silicate material having this structure is asbestos. Chains are held together by weak electrostatic forces.

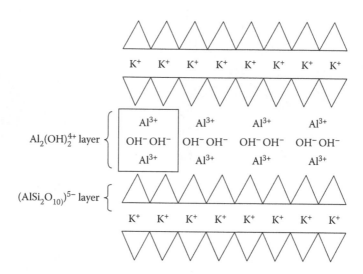

FIGURE 9.13
Structure of mica.

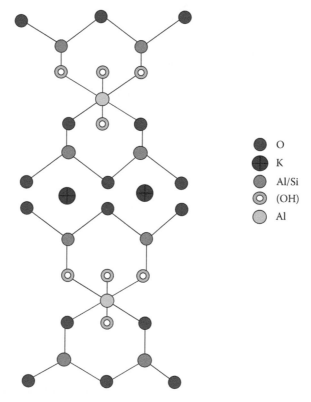

FIGURE 9.14
Structure of mica showing the position of alkali ions.

9.3.4 Island Silicates

When the O/Si ratio is 4, the structural units are isolated $(SiO_4)^{4-}$ tetrahedra. These units are connected by positive ions. The resulting structure gives rise to what is called an *island silicate*. In garnets $(Mg,Fe^{2+},Mn,Ca)_3(Cr,Al, Fe^{3+})_2(SiO_4)_3$ and olivines $(Mg,Fe^{2+})_2(SiO_4)$, $(SiO_4)^{4-}$ becomes the anion. These anions are bonded ionically by the metallic cations.

9.3.5 Aluminosilicates

Aluminosilicates are formed by the combination of Al_2O_3 and SiO_2. We have seen in mica that Al ions substitute for Si^{4+} ions. For each of this substitution, the resulting negative charge has to be compensated for by an additional cation. In clays, Al ions occupy the holes in the silicate network. When Al substitutes for Si in the network, the appropriate ratio for determining the structure is the O/(Al + Si) ratio. For albite $(NaAlSi_3O_8)$, anorthite $(CaAl_2Si_2O_8)$, eucryptite $(LiAlSiO_4)$, orthoclase $(KAlSi_3O_8)$, and spodune $(LiAlSi_2O_6)$, the ratio becomes 2. The structure is three dimensional, corresponding to the three-dimensional tetrahedral units. As the bonds are both covalent and ionic primary bonds, the melting points of these aluminosilicates are the highest among all the silicates. In aluminosilicates in general, bonding can be mixed within the silicate network. Si–O–Si bonds are different from those bonds holding the units together, which can be ionic or weak secondary bonds.

Example 1

    a. Derive a generalized expression relating the number of non-bridging oxygens per Si atom present in a silicate structure to the mole fraction of metal oxide added.

    b. Calculate the number of bridging and nonbridging oxygens per Si atom for $CaO \cdot 2SiO_2$. What is the most likely structure for this compound?

Answer

    a. The number of NBOs is equal to the total cationic charge. Let us say that there are y mols of SiO_2. The addition of η moles of $M_\zeta O$ results in $z(\zeta \eta)$ NBOs, where z is the charge on the modifying cation. Therefore,

$$NBO = z(\zeta \eta)/y$$

For $CaO \cdot 2SiO_2$, z = 2, ζ = 1, η = 1, and y = 2.

Therefore, $NBO = 2 \times (1 \times 1)/2 = 1$.
Therefore, the number of bridging oxygens per Si atom is
    $4 - 1 = 3$.
Therefore, the most likely structure is the sheet structure.

9.4 Structures of Glasses

Glasses are the most important group of inorganic noncrystalline solids. The structure of glasses may be considered on three scales:

1. The scale of 2–10 angstrom units (AU), or that of local atomic arrangements
2. The scale of 30 to a few thousand angstroms, or that of a submicrostructure
3. The scale of microns to millimeters or more, or that of microstructure and macrostructure

On the scale of atomic structure, the distinguishing structural characteristic of glasses, such as the liquids from which many are derived, is the absence of atomic periodicity or long-range order.

9.4.1 Models of Glass Structure

The glass structure can be considered in terms of two models: the crystallite and the random network models.

9.4.1.1 Crystallite Model

X-ray diffraction patterns from glasses generally exhibit broad peaks centered in the range in which strong peaks are seen in the diffraction patterns of the corresponding crystals. These patterns are shown in Figure 9.15. This figure shows that, for the diffraction patterns of three different forms of silica, a prominent peak is seen at the sin θ/λ value of 0.12. In these forms, cristobalite is crystalline, and silica gel consists of fine particles. In crystalline form, there will sharp peaks, due to the different sets of parallel crystal planes diffracting, corresponding to their Bragg angles. In a particular solid, the particles broaden the peaks, depending on their size. The smaller the size, the greater will be the broadening. Since broadening also has taken place for diffraction from glass, it led to the suggestion that glasses are composed of assemblages of very small crystals, termed *crystallites*, having the size of the cristobalite unit cell. This is the crystallite model for glass. This model is not supported by further investigations. Hence, the next model was proposed.

9.4.1.2 Random-Network Model

According to this model, glasses are viewed as three-dimensional networks or arrays, lacking symmetry and periodicity. The broadening seen in the x-ray diffraction is seen as those from an amorphous solid. This model is the accepted model for glass.

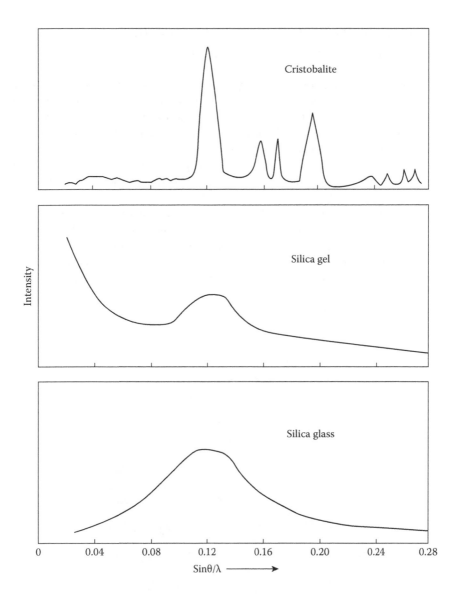

FIGURE 9.15
X-ray diffraction patterns of cristobalite, silica gel, and vitreous silica.

9.4.2 Zachariasen Rules

W. H. Zachariasen suggested four rules for the formation of an oxide glass:

1. Each oxygen ion should be linked to not more than two cations.
2. The CN of oxygen ions about the central cation must be small, 4 or less.

3. Oxygen polyhedra share corners, not edges or faces.

4. At least three corners of each polyhedron should be shared.

In practice, the glass-forming oxygen polyhedra are triangles, and tetrahedra and cations have been termed the *network formers*. The whole network is formed by the joining of these polyhedra formed around the network formers. In alkali silicate glasses, alkali ions occupy random positions distributed throughout the structure, and they are called *network modifiers*. They modify the structure by creating nonbridging oxygens. Their major function is viewed as providing additional oxygen ions. Figure 9.16 shows the schematic representation of sodium silicate glass. Cations of higher valence and lower coordination number than the alkalis and alkaline earths may contribute in part to the network structure, and are referred to as *intermediates*. They substitute some of the network formers. Thus, the role of cations depends on their valence and coordination numbers. An example for intermediates is Pb^{2+}, which has a higher valence than alkali ions, and its CN is 2, which is lower than that of Si^{4+}.

Now, let us consider the structure of some groups of glasses.

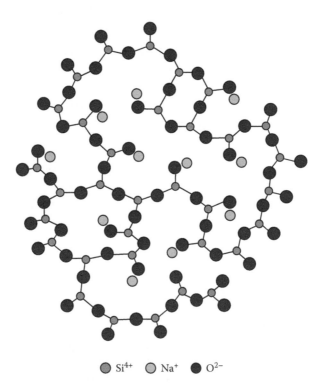

⊙ Si^{4+} ◯ Na^+ ● O^{2-}

FIGURE 9.16
Schematic representation of the structure of a sodium silicate glass.

9.4.3 Structure of Oxide Glasses

9.4.3.1 Silica

Assuming that silica glass obeys the crystallite model, from the width of the main broad diffusion peak in the glass diffraction pattern, the crystallite size in the case of SiO_2 was estimated at about 7–8 AU. Since the size of a unit cell of cristobalite is also about 8 AU, any crystallites would be only a single unit cell in extent. Further, in contrast to silica gel, there is no marked small-angle scattering from a sample of fused silica. This indicates that the structure of silica is continuous, and not composed of discrete particles as with the gel. Hence, if crystallites of reasonable size are present, there must be a continuous spatial network connecting them that has a density similar to that of the crystallites. Later, another x-ray diffraction study of fused silica was carried out. From this, the distribution of Si–O–Si bond angles was determined. These angles are distributed over a broad range, from about 120° to about 180°. This is shown in Figure 9.17. The figure

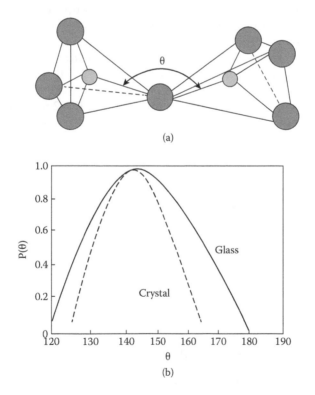

FIGURE 9.17
(a) Schematic representation of adjacent SiO_4 tetrahedra showing Si–O–Si bond angle: smaller spheres = Si, bigger spheres = O; (b) distribution of Si–O–Si bond angles in fused silica and crystalline cristobalite. (From R. L. Mozzi, Mit, R. L. Mozzi, x-Ray Diffraction Study of SiO_2 and B_2O_3 Glasses, Sc. D. Thesis, MIT, Cambridge, MA, 1967.)

shows that, for crystal, the range is narrower. Thus, the x-ray diffraction work provides strong evidence for a random distribution of rotation angles of one tetrahedron with respect to another. This study also shows that there appears to be no pronounced preference in fused silica for edge-to-face sharing of tetrahedra. The structure of fused silica is well described then by a random network of SiO_4 tetrahedra with significant variability occurring in the Si–O–Si bond angles.

9.4.3.2 B_2O_3

X-ray diffraction and nuclear magnetic resonance studies of glassy B_2O_3 indicate that the structure is composed of BO_3 triangles. The triangles are linked in a boroxyl configuration, shown schematically in Figure 9.18.

9.4.3.3 Silicate Glasses

The addition of alkali or alkaline earth oxides to SiO_2 increases the ratio of oxygen to silicon to a value greater than 2, and breaks up the three-dimensional network with the formation of singly bonded oxygen, which do not participate in the network. The structure also changes with the formation of singly bonded oxygen. The progressive change in the structure as the number of singly bonded oxygen increases is shown in Figure 9.19. It is sometimes convenient to describe the network character of silicate glasses in terms of the average number R of oxygen per network-forming ion (e.g., R = 2 for SiO_2). For a glass containing 12 g mole% Na_2O, 10 g mole% CaO, and 78 g mole% SiO_2:

$$R = (12 + 10 + 156)/78 = 2.28 \qquad (9.1)$$

For glasses containing only one type of network-forming cation surrounded by Z oxygen (Z = 3 or 4), with X nonbridging (i.e., singly bonded) and Y bridging oxygen per polyhedron, one may write:

$$X + Y = Z \quad \text{and} \quad X + 0.5Y = R \qquad (9.2)$$

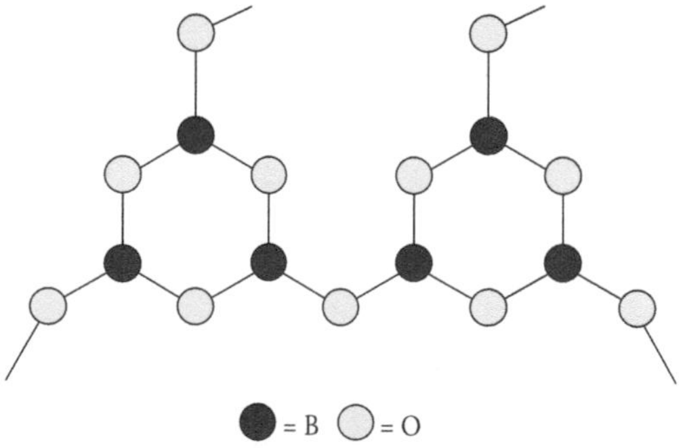

$\bullet$ = B $\bigcirc$ = O

FIGURE 9.18
Schematic representation of boroxyl configuration.

$\dfrac{O}{Si}$		Structure
2	Network, $(SiO_2)_n$	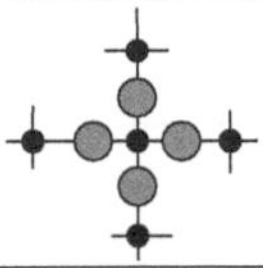
2–2.5	Network	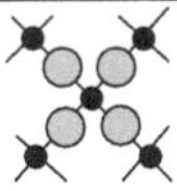 and 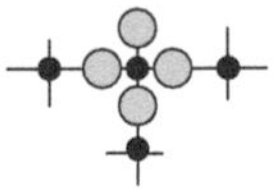
2.5	Network	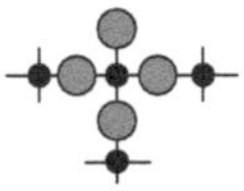
2.5–3.0	Network and chains or rings	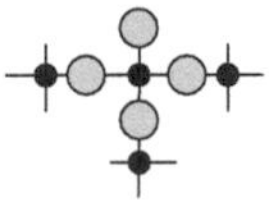 and 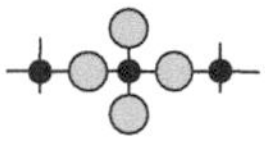
3.0	Chains and rings	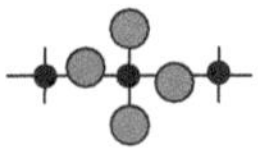 , 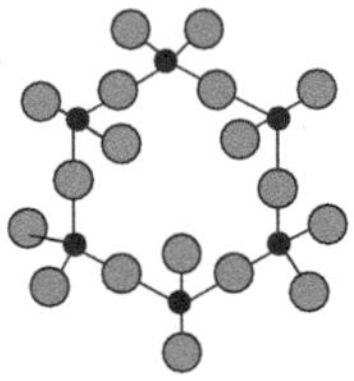 , etc
3.0–3.5	Chains, rings, and pyrosilicate ions	
3.5	Pyrosilicate ions	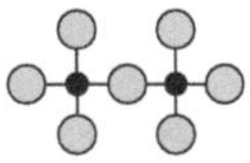
3.5–4.0	Pyrosilicate and orthosilicate ions	
4.0	Orthosilicate ions	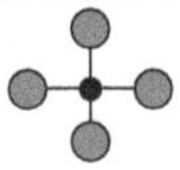

FIGURE 9.19
Effect of oxygen–silicon ratio on silicate network structures.

For silicate glasses, when the oxygen polyhedra are SiO_4 tetrahedra, $Z = 4$, and the preceding equation becomes:

$$X = 2R - 4 \quad \text{and} \quad Y = 8 - 2R \tag{9.3}$$

In the case of silicate glasses containing more alkali and alkaline earth oxides than Al_2O_3, the Al^{3+} is believed to occupy the centers of AlO_4 tetrahedra.

TABLE 9.2

Values of the Network Parameters X, Y, and R for Representative Glasses

Composition	R	X	Y
SiO_2	2	0	4
$Na_2O \cdot 2SiO_2$	2.5	1	3
$Na_2O \cdot (1/2)Al_2O_3 \cdot 2SiO_2$	2.25	0.5	3.5
$Na_2O \cdot Al_2O_3 \cdot 2SiO_2$	2	0	4
$Na_2O \cdot SiO_2$	3	2	2
P_2O_5	2.5	1	3

Hence, the addition of Al_2O_3 in such cases introduces only 1.5 oxygen ions per networking cation, and nonbridging oxygen ions of the structure are used up and converted to bridging oxygen ions. Table 9.2 gives the values of the network.

9.4.3.4 Borate Glasses

The addition of alkali and alkaline earth oxides to B_2O_3 results in the formation of BO_4 tetrahedra. Up to the alkali oxide concentration of about 30 mole%, nearly all the modifier oxides have the effect of converting BO_3 triangles to BO_4 tetrahedra. Beyond this, singly bonded oxygen ions are produced in appreciable numbers. These singly bonded oxygen ions are associated with BO_3 triangles rather than with BO_4 tetrahedra.

9.4.3.5 Germanate and Phosphate Glasses

Glassy GeO_2 is composed of GeO_4 tetrahedra, with a mean Ge–O–Ge bond angle of about 138°. In contrast to fused silica, however, the distribution of inter-tetrahedral angles (Ge–O–Ge) for germanium is quite sharp.

As with silicate and most germanate glasses, phosphate glasses are composed of oxygen tetrahedra, but unlike the silicate and germanate analogs, a PO_4 tetrahedron can be bonded to utmost three other similar tetrahedra. The most familiar structural units in phosphate glasses are rings or chains of PO_4 tetrahedra—that is, each tetrahedron most commonly bonds to the neighboring ones at two corners.

Problems

9.1 Show that the minimum cation to anion radius ratio for a coordination number of 6 is 0.414.

9.2 Calculate the ratio of the sizes of the tetrahedral and octahedral sites.

9.3 When oxygen ions are in a hexagonal close-packed arrangement:

    9.3.1 What is the ratio of the octahedral sites to oxygen ions?

    9.3.2 What is the ratio of the tetrahedral sites to oxygen ions?

9.4 Starting with the cubic close packing of oxygen ions:

    9.4.1 How many tetrahedral and how many octahedral sites are there per unit cell?

    9.4.2 What is the ratio of octahedral sites to oxygen ions? What is the ratio of tetrahedral sites to oxygen ions?

    9.4.3 What oxide would you get:

        a. If one-half of the octahedral sites are filled?

        b. If two-thirds of the octahedral sites are filled?

        c. If all the octahedral sites are filled?

    9.4.4 Locate all the tetrahedral sites, and fill them with cations. What structure do you obtain? If the anions are oxygen, what must be the charge on the cation for charge neutrality to be maintained?

    9.4.5 Locate all the octahedral sites, fill them with cations, and repeat 9.4.4.

9.5 Draw the zinc blende structure. What, if anything, does this structure have in common with diamond cubic structure? Explain.

9.6 The structure of Li_2O has anions in cubic close packing, with Li ions occupying all tetrahedral positions.

    9.6.1 Draw the structure and calculate the density of Li_2O. In the structure, oxygen ions do not touch, but Li ions touch oxygen ions.

    9.6.2 What is the maximum radius of a cation that can be accommodated in the vacant interstice of the anion array in Li_2O without strain?

9.7 Making use of Pauling's size criterion, among Ti^{4+}, Ba^{2+}, and O^{2-}:

    9.7.1 Choose the most suitable cage for each cation.

    9.7.2 Based on your results, choose the appropriate composite crystal structure and draw the unit cell of $BaTiO_3$.

    9.7.3 How many ions of each element are there in each unit cell?

9.8 BeO forms a structure in which the oxygen ions are in FCC arrangement.

    9.8.1 Which type of interstitial site will Be^{2+} occupy?

    9.8.2 What fraction of available interstitial sites will be occupied?

9.9 Cadmium sulfide has a density of 4.82 g/cm³. Using the radii of the ions, show that a cubic unit cell is not possible.

9.10 The compound MX has a density of 2.1 g/cm³, and a cubic unit cell with a lattice parameter of 0.57 nm. The atomic weights of M and X are 28.5 and 30 g/mol, respectively. Based on this information, which of the following structures are possible: NaCl, CsCl, and ZnS?

9.11 What type of structures (sheets, chains, islands, etc.) are expected of the following silicates:

9.11.1 Tremolite, $Ca_2Mg_5(OH)_2Si_8O_{22}$

9.11.2 Mica, $CaAl_2(OH)_2(Si_2Al_2)O_{10}$

9.11.3 Kaolinite, $Al_2(OH)_4Si_2O_5$

9.12 Determine the expected crystal structure including the ion positions of the hypothetical salt AB_2, where the radius of A is 154 pm and that of B is 49 pm. Assume that A has a charge of +2.

9.13 What would be the formulas of the silicate units shown in the following figures?

9.14 Derive an expression relating the mole fractions of alkali earth oxides to the number of nonbridging oxygens per Si atom present in a silicate structure.

9.15 Calculate the number of bridging and nonbridging oxygens per Si atom for $Na_2O\cdot(1/2)CaO\cdot2SiO_2$. What is the most likely structure for this compound?

9.16 Show that chains of infinite length would occur at a mole fraction of Na_2O of 0.5, and balance SiO_2.

9.17 Show that, for any silicate structure, the number of nonbridging oxygens per Si is given by NBO = 2R – 4, and the number of bridging oxygens is 8 – 2R, where R is the O/Si ratio.

Reference

1. R. L. Mozzi and B. E. Warren, The structure of vitreous silica, *J. Appl. Cryst.*, 2: 164, 1969.

Bibliography

M. W. Barsoum, *Fundamentals of Ceramics*, Taylor & Francis, New York, 2003, pp. 70–75, 84–87.

W. D. Kingery, H. K. Bowen, and D. R. Uhlmann, *Introduction to Ceramics*, 2nd Edn., John Wiley, New York, 1976, pp. 61–69, 91–110.

10

Defects in Ceramics

10.1 Introduction

Real crystals are not perfect, but contain imperfections that are classified according to their geometry and shape into point, line, and planar defects. An example for a *point defect* is the lattice point not occupied by the proper ion or atom needed to preserve long-range periodicity.

Dislocations in crystals that cause lattice distortions centered on a line are examples for *linear defects*.

Planar defects are surface imperfections. Examples are the boundaries of separate grains or domains of different orientations—that is, grain and twin boundaries.

Study of defects is important because many properties of the materials are affected by them. For example, there is one-to-one correlation between concentration of point defects and diffusion. As the concentration increases, diffusion also increases. There is also a correlation between grain size and mechanical strength. This relation is given by the Hall–Petch equation:

$$\sigma_y = \sigma_0 + Kd^{-1/2} \tag{10.1}$$

Here, σ_y is the yield strength, σ_0 and K are constants, and d is the grain size. This relation shows that, as the grain size decreases, the yield strength increases.

Let us now consider each of the three groups of defects.

10.2 Point Defects

One overriding constraint operative during the formation of ceramic defects is the preservation of electrical neutrality. All the point defects fall in three categories. They are discussed in the following text.

10.2.1 Stoichiometric Defects

When stoichiometric defects form, the crystal chemistry, that is, the ratio of cations to anions, does not change. The examples for such defects are the Schottky and Frenkel defects. These are shown in Figure 10.1. In a Schottky defect, a pair of cation and anion is missing from their sites in the lattice, whereas in a Frenkel defect, one of the ions goes from its lattice position to an interstitial position. Normally, the ion going will be the cation, as the interstice is small in size and the cations are smaller than the anions.

10.2.2 Nonstoichiometric Defects

Nonstoichiometric defects form by the selective addition or loss of one (or more) of the constituents of the crystal chemistry, and consequently lead to

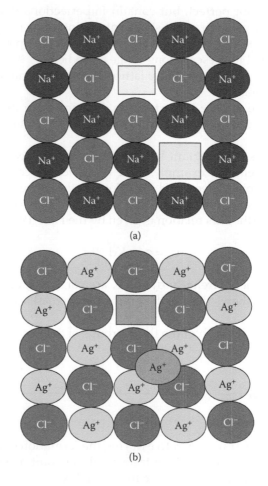

(a)

(b)

FIGURE 10.1
(a) Schottky defect in NaCl; (b) Frenkel defect in AgCl.

a change in crystal chemistry. The material accommodates those changes in composition by selectively losing one of its constituents to its environment by the creation or elimination of defects. For example, if an oxide were annealed in a high oxygen partial pressure, the number of oxygen ions becomes greater than the number of cations. If the oxygen partial pressure were very low, the cation concentration becomes higher.

Many physical properties such as color, diffusivity, electrical conductivity, photoconductivity, and magnetic susceptibility vary markedly with small changes in composition.

10.2.3 Extrinsic Defects

These are the third category of point defects, and are created by the presence of impurities in the host crystal. When we dope a crystal with a different material, it creates extrinsic defects proportional to the amount of the dopant.

10.3 Point Defects and Their Notation

We can find a number of defects in a *pure* binary compound, which is a compound made up of two different elements. Let us review these defects and discuss their notation. The different types of point defects are shown in Figure 10.2.

Vacancies are the sites in the lattice from where atoms are missing. In Figure 10.2, a cation and an anion vacancy are shown.

10.3.1 Interstitial Atoms

Atoms are normally found in lattice sites. When they are found in interstices, those atoms are called *interstitial atoms*. When a cation is found in an interstice, it is called cation interstitial (marked M in Figure 10.2).

10.3.2 Misplaced Atoms

Atoms of one type are sometimes found at a site normally occupied by other types of atoms. These are called *misplaced atoms*. This type of point defect is possible only in covalent ceramics, where the atoms are not charged. It cannot happen in ionic solids because, if it happens, like charges will become adjacent, and the structure will not be stable. In Figure 10.2, it is shown that the lattice atoms M and O have interchanged their positions. Thus, they have formed a pair of misplaced atoms.

Apart from the preceding list of point defects, a pure binary compound could also contain the following electronic defects.

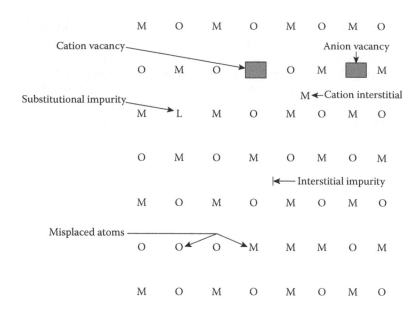

FIGURE 10.2
Various types of defects typically found in ceramics.

10.3.3 Free Electrons

Free electrons are those found in the conduction band. In ceramics, there will not normally be any free electron in the conduction band at ordinary temperatures. That is why their electrical and thermal conductivities are very low at these temperatures. Hence, if by any chance there are free electrons in a ceramic solid, it amounts to the formation of electronic defects.

10.3.4 Electron Holes

Electron holes are positive mobile electronic carriers. As in the case of free electrons, electron holes are also not normally present in ceramics. They can be present when free electrons are generated. Free electrons get generated when electrons in the valence band jump to the conduction band.

Coming now to an impure crystal, in addition to the preceding point defects, it will also contain *interstitial* and *substitutional impurities*. These defects are marked I and L, respectively, in Figure 10.2.

10.3.5 Kroger–Vink Notation

Kroger–Vink notation is the universally accepted scheme to denote defects. This notation involves three symbols. The *main symbol* denotes either the species involved, by the chemical symbol of the element, or the vacancy by V. The *subscript* for the main symbol represents either the crystallographic position occupied by the species involved, or the letter I for the interstitial. The *super-script* of the main symbol denotes the effective electric charge on the defect. This effective electric charge is defined as the difference between the real charge of the defect species and that of the species that would have occupied that site in the perfect crystal. The symbol prime is used for each negative charge, a dot is used for every positive charge, and x is used for zero effective charge.

Examples

1. Vacancy on Na$^+$ sublattice: V'_{Na}
2. Vacancy on Cl$^-$ sublattice: $V^{\bullet}_{Cl}$
3. Interstitial position on Na sublattice: Na_i
4. Consider the addition of $CaCl_2$ to NaCl
   Ca can substitute for Na ion or go as an interstitial
   In the first case, the defect notation is Ca_{Na}
   The interstitial Ca ion can be denoted as Ca_i
5. Now consider the addition of KCl to NaCl
   If the K ion substitutes for Na ion, the notation is K^x_{Na}
   If the K ion goes interstitial, the notation is K_i
6. Now let us dope NaCl with Na_2S
   Here, anions substitute anions, or they go interstitial
   These two possibilities are represented as S'_{Cl} and S''_i, respectively.

10.3.6 Defect Reactions

The formation of the various point defects can be described by chemical reactions for which the following rules have to be followed:

- *Mass balance*: Mass can neither be created nor destroyed. Vacancies have zero mass.

- *Electro-neutrality or charge balance*: Charges can neither be created nor destroyed.

- *Preservation of regular site ratio*: The ratio between the numbers of regular cation and anion sites must remain constant and equal to the ratio of the parent lattice.

To generalize, for an M_aX_b compound, the following relationship has to be maintained:

$$a(X_x + V_x) = b(M_M + V_M)$$

This does not imply that the ratio has to be maintained for the *number of atoms* or *ions*, but only for the *number of sites*.

10.3.6.1 Stoichiometric Defect Reactions

A stoichiometric defect reaction by definition is the one where the chemistry of the crystal does not change as a result of the reaction. The three most common stoichiometric defects are Schottky defects, Frenkel defects, and antistructure disorder or misplaced atoms.

Schottky defects: When these defects are formed, electric-charge-equivalent numbers of vacancies are formed on each sublattice.

For an M_aO_b oxide, the reaction can be written as shown in Equation 10.2.

$$\text{Null (or perfect crystal)} \Rightarrow aV_M^{b-} + bV_O^{a+} \tag{10.2}$$

Frenkel defects: When these defects are formed, a vacancy is created by having an ion in a regular lattice site migrating into an interstitial site.

The Frenkel defect reaction for a trivalent cation is given in Equation 10.3.

$$M_M^X \Rightarrow V_M''' + M_i^{\cdot\cdot\cdot} \tag{10.3}$$

The same reaction on the oxygen sublattice is given in Equation 10.4.

$$O_O^X \Rightarrow O_i'' + V_O^{\cdot\cdot} \tag{10.4}$$

Since there is a restriction for the formation of a Frenkel defect, it is found to form only in a few oxides, such as FeO, NiO, CoO, and Cu_2O.

Antistructure disorder or misplaced atoms: Misplaced atoms are also called *antistructure disorder*. As illustrated earlier, these are the sites in the crystal lattice where one type of atom is found at a site normally occupied by another. It was also stated that this kind of defect does not occur in ionic ceramics. However, it can occur in covalent ceramics such as SiC. The defect reaction in such case can be written as given in Equation 10.5.

$$Si_C^X \Rightarrow C_{Si}^X \tag{10.5}$$

10.3.6.2 Nonstoichiometric Defects

In nonstoichiometric defect reactions, the composition changes. In other words, mass is transferred across the boundaries of the crystal. One common nonstoichiometric reaction is found to occur at low oxygen partial pressures. This is shown in Figure 10.3. In this case, oxygen leaves the crystal, and the reaction is given in Equation 10.6.

$$O_O^X \Rightarrow 1/2\, O_2\,(g) + V_O^X \tag{10.6}$$

Given that oxygen has to leave as a neutral species, it has to leave two electrons behind (Figure 10.3a).

These electrons are weakly bound to the defect site and are easily excited into the conduction band: that is, V_O^X acts as a donor. The ionization reaction occurs in two stages, and are given in Equations 10.7 and 10.8.

$$M^{2+} \; O^{2-} \; M^{2+} \; O^{2-} \; M^{2+} \; \bigg| \; M^{2+} \; O^{2-} \; M^{2+} \; O^{2-} \; M^{2+} \; \bigg| \; M^{2+} \; O^{2-} \; M^{2+} \; O^{2-} \; M^{2+}$$

$$O^{2-} \; M^{2+} \qquad M^{2+} \; O^{2-} \; \bigg| \; O^{2-} \; M^{2+} \qquad M^{2+} \; O^{2-} \; \bigg| \; O^{2-} \; M^{2+} \qquad M^{2+} \; O^{2-}$$

$$M^{2+} \; O^{2-} \; M^{2+} \; O^{2-} \; M^{2+} \; \bigg| \; M^{2+} \; O^{2-} \; M^{2+} \; O^{2-} \; M^{2+} \; \bigg| \; M^{+} \quad O^{2-} \; M^{2+} \; O^{2-} \; M^{2+}$$

$$V_O^x \qquad\qquad\qquad V_O^{\cdot} \qquad\qquad\qquad V_O^{\cdot\cdot}$$

(a) (b) (c)

FIGURE 10.3
(a) The formation of an oxygen vacancy by the loss of an oxygen atom to the gas phase; (b) a $V_O^{\cdot}$ site is formed when one of these electrons is excited into the conduction band; (c) the escape of the second electron creates a $V_O^{\cdot\cdot}$.

$$V_O^x \Rightarrow V_O^{\cdot} + e' \tag{10.7}$$

$$V_O^{\cdot} \Rightarrow V_O^{\cdot\cdot} + e' \tag{10.8}$$

The net reaction reads as shown in Equation 10.9.

$$O_O^x \Rightarrow 1/2 \; O_2 \, (g) + V_O^{\cdot\cdot} + 2e' \tag{10.9}$$

Thus, the oxygen vacancy becomes doubly ionized.

Another nonstoichiometric defect reaction can be when oxygen is incorporated into the crystal interstitially. The reaction is represented by Equation 10.10.

$$1/2 \; O_2 \, (g) \Rightarrow O_i^x \tag{10.10}$$

When oxygen enters an ionic crystal, it becomes ionized. When it gets ionized by absorbing electrons from the valence band, it creates holes in the valence band. In this case, the defect, oxygen interstitial, acts as an acceptor. Again, this ionization reaction can be represented in two stages, as shown in Equations 10.11 and 10.12.

$$O_i^x \Rightarrow O_i' + \overset{\cdot}{h} \tag{10.11}$$

$$O_i' \Rightarrow O_i'' + \overset{\cdot}{h} \tag{10.12}$$

The net reaction becomes as shown in Equation 10.13.

$$1/2 \; O_2 \, (g) \Rightarrow O_i'' + 2\overset{\cdot}{h} \tag{10.13}$$

The reaction in Equation 10.9 will result in an oxygen-deficient oxide, whereas the reaction in Equation 10.13 will result in an oxygen-rich oxide.

10.3.6.3 Extrinsic Defects

The discussion so far has been with respect to pure crystals. In reality, most crystals are not pure. Impurities can substitute for host ions of electronegativity nearest their own. For example, in NaCl, Ca and O can occupy the cation and anion sites, respectively. In writing a defect incorporation reaction, the following simple bookkeeping operation can be of help:

a. Sketch a unit or multiple units of the host (solvent) crystal.
b. Place a unit or multiple units of the dopant (solute) crystal on top of the sketch drawn in step 1, such that the cations are placed on top of the cations and anions on top of the anions.
c. Whatever is left over is the defect that arises.

Example 1

Incorporate $CaCl_2$ into NaCl.

From Figure 10.4a, it is immediately obvious that one possible incorporation reaction is as shown in Equation 10.14.

$$CaCl_2 \Rightarrow 2NaCl\ Ca_{Na}^{\cdot} + V_{Na}' + 2Cl_{Cl}^{X} \tag{10.14}$$

A second incorporation is shown in Figure 10.4b. The reaction is given in Equation 10.15.

$$CaCl_2 \Rightarrow NaCl\ Ca_{Na}^{\cdot} + Cl_i' + Cl_{Cl}^{X} \tag{10.15}$$

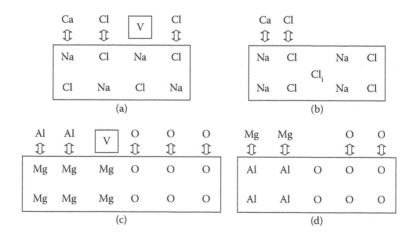

FIGURE 10.4

Bookkeeping technique for impurity incorporation reactions; (a) $CaCl_2$ in NaCl leaves a vacancy on cation sublattice; (b) an alternate reaction is for the extra Cl ion to go interstitial; (c) Al_2O_3 in MgO creates a vacancy on the cation sublattice; (d) MgO in Al_2O_3 creates a vacancy on the anion sublattice.

Example 2

When we dope MgO with Al_2O_3 (Figure 10.4c), the reaction becomes that shown in Equation 10.16.

$$Al_2O_3 \Rightarrow 2Al_{Mg}^{\cdot} + V_{Mg}'' + 3O_O^X \tag{10.16}$$

Example 3

For doping Al_2O_3 with MgO (Figure 10.4d), the incorporation reaction becomes that shown in Equation 10.17.

$$2MgO \Rightarrow 2Mg_{Al}' + V_{\ddot{O}} + 2O_O^X \tag{10.17}$$

Now, let us consider *oxides with multiple substitution of ions.*
Consider the clay structure shown in Figure 9.12.
The substitution of divalent cations for the trivalent Al ions between the sheets occurs readily as long as, for every Al^{3+} substituted, the additional incorporation of a singly charged cation:

$$Al_2(OH)_4(Si_2O_5) \Rightarrow (Al_{2-x}Na_xMg_x)(OH)_4(Si_2O_5) \tag{10.18}$$

In the preceding reaction, when the Al_{2-x} ion is substituted by the divalent Mg_x ions, to compensate for the extra-negative charge, Na_x monovalent ions are incorporated.

10.3.6.4 Electronic Defects

In a perfect semiconductor or insulating crystal at 0 K, all the electrons are localized and are firmly in the grasp of the nuclei. At finite temperatures, some electrons are knocked loose as a result of lattice vibrations, and end up in the conduction band. For an *intrinsic* semiconductor, the liberation of an electron also results in the formation of an electron hole, such that an intrinsic electronic defect can be written as:

$$\text{Null} \Leftrightarrow \acute{e} + \dot{h} \tag{10.19}$$

10.4 Linear Defects

In ionic solids, the structure of dislocations can be quite complex because of the need to maintain charge neutrality. For example, for an edge dislocation to form in an NaCl crystal, it is not possible to insert one row of ions. Two half planes are to be inserted (Figure 10.5).

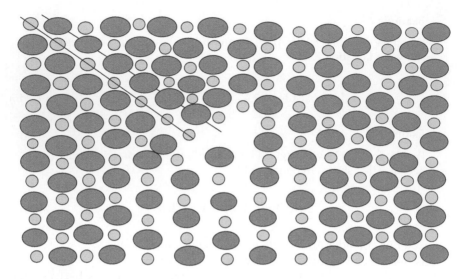

FIGURE 10.5
Existence of a dislocation in an ionic crystal.

10.5 Planar Defects

Grain boundaries are the examples for planar defects.

10.5.1 Grain Boundary Structure

Grain boundary is the interface between two grains. When these two grains are of the same material, the boundary is called a *homophase boundary*; when they are of different materials, then it is a *heterophase boundary*. Often, there are other phases that are only a few nanometers thick and can be present between the grains of two different materials; in such case, the grain boundary represents three phases. These phases may be crystalline or amorphous. The presence or absence of a third phase has important ramifications on processing, electrical properties, and creep of the material concerned.

Grain boundaries are distinguished according to their structure as *low-angle* (<15°), *special*, or *random* ones. Low-angle grain boundary (Figure 10.6) consists of arrays of dislocations separated by areas of strained lattice. The angle of the grain boundary can be determined from the dislocation spacing λ_d and its Burgers vector **b**. From Figure 10.6, the angle of tilt:

$$\sin\theta = b/\lambda_d \qquad (10.20)$$

Special grain boundaries are those in which a special orientational relationship exists between the two grains on either side of the grain boundary.

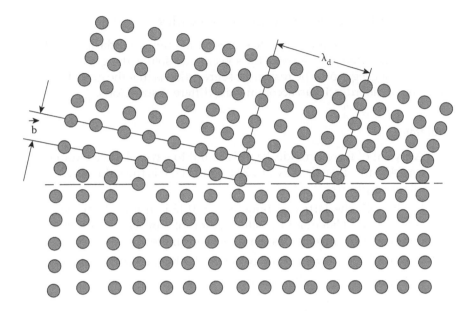

FIGURE 10.6
Schematic of a low-angle boundary.

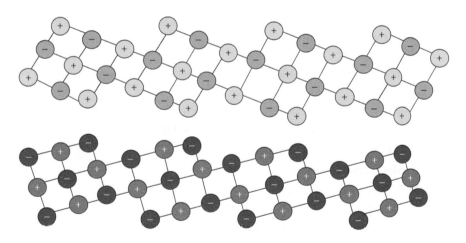

FIGURE 10.7
Schematic of a special grain boundary.

An example of such a boundary is shown in Figure 10.7. Here, the special orientational relationship is that a fraction of the total lattice sites between the two grains coincides periodically.

Random grain boundaries are the ordinary high-angle (>15°) grain boundaries.

10.5.2 Impurity Segregation at Grain Boundaries

There is a tendency for impurities to segregate at grain boundaries. The larger the ionic radii difference between the solvent and solute, the greater the driving force for segregation. The reason for segregation is that the grain boundary is a region of disorder. It can accommodate different-sized ions.

Problems

10.1 A crystal of ferrous oxide, Fe_yO, is found to have a lattice parameter $a = 0.43$ nm, and a density of 5.72 g/cm³. What is the value of y? State all assumptions.

10.2 For $Fe_{0.98}O$, the lattice parameter a = 0.4352 nm and the density is 5.7 g/cm³. Calculate the site fraction of iron vacancies and the number of iron vacancies per cubic centimeter.

10.3 Indicate the following types of defects by Kroger–Vink notation:

　a. Vacancy in the Ca^{2+} sublattice of $CaCl_2$

　b. Vacancy in the Cl^- sublattice of $CaCl_2$

　c. Ca^{2+} in the interstitial position of $CaCl_2$

　d. Cl^- in the interstitial position of $CaCl_2$

10.4 Write two possible defect reactions that would lead to the formation of a metal-deficient compound. Cite an example of an oxide that you think would likely form each of the defect reactions you have chosen.

10.5 Write down two possible defect reactions for the dissolution of CaO in ZrO_2.

10.6 Write the defect reactions for the following:

　a. Oxygen from atmosphere going to the lattice

　b. Schottky defect in alumina

　c. Metal loss from FeO

　d. Frenkel defect in Cr_2O_3

　e. Dissolution of CaO in Fe_2O_3

　f. Dissolution of K_2O in CoO

　g. Oxygen from atmosphere going interstitial

　h. Metal loss from ZnO

　i. Frenkel defect in Al_2O_3

　j. Dissolution of MgO in Al_2O_3

　k. Dissolution of Li_2O in NiO

　l. Addition of $BaCl_2$ to KCl

　m. Addition of $BaCl_2$ to $BaBr_2$

    n. Addition of KCl to $BaCl_2$

    o. Formation of Schottky defect in CaO

    p. Formation of Frenkel defect in Cu_2O

    q. Formation of antistructure disorder in PbSn

    r. Defect formation in Al_2O_3, when it is heated in the presence of O_2

Bibliography

M. W. Barsoum, *Fundamentals of Ceramics*, Taylor & Francis, New York, 2003, pp. 137–173.

11

Ceramic Microstructures

11.1 Introduction

The properties of ceramic products are determined not only by the composition and structure of the phases but also by the arrangement of the phases. This arrangement of the phases can be determined by studying the microstructures. The arrangement of the phases or their distribution depends on the initial fabrication techniques, raw materials, phase–equilibrium relations, and kinetics of phase changes, grain growth, and sintering.

11.2 Characteristics of Microstructure

To characterize a microstructure, the following are to be determined:

1. The number and identification of phases present, including porosity
2. The relative amounts of each phase present
3. Characteristics of each phase, such as size, shape, and orientation

11.2.1 Techniques of Studying Microstructure

There are optical methods and electronic methods to study the microstructure of ceramics. The two most widely used optical methods are (1) observations of thin sections with transmitted light and (2) observations of polished sections with reflected light. The thin section should be of 0.015 mm–0.030 mm thickness in order for ordinary light to pass through it. It is prepared by cutting a thin slice, polishing one side of it, cementing this side to a transparent microscopic slide, and grinding and polishing the other side to obtain a section of the required uniform thickness. The advantage of this method is that the optical properties of each phase can be determined, and, from this, the phases identified.

The following are the disadvantages of this method:

1. The specimens are difficult to prepare.
2. Individual grains in many fine ceramic materials are smaller in size than the section thickness, which leads to confusion, as the images in those cases will be superimposed ones.

The polished sections are prepared by mounting the cut specimen in Bakelite or Lucite plastic and then grinding and polishing one face smooth, using a series of abrasive papers, followed by abrasive powders suspended in water on cloth wheels.

The polished surfaces are observed directly with reflected light in a metallurgical microscope to distinguish the differences in relief or reflectivity between the different phases. The various phases are best distinguished after chemical etching, when the differences will be revealed more clearly.

The polished sections are advantageous because they are relatively easy to prepare and simpler to interpret than the image obtained by transmitted light. The phases are identified by the etching characteristics of the phases. There will be differential attack by the etchant, which is a reactive chemical, on the phases. The reflectivity of a transparent material depends on its index of refraction; for a silicate with refractive index of 1.5, only 4% of the incident light is reflected. In order to increase the intensity of the light reflected for these materials, a more intense xenon or arc light source is preferred; as an alternative, thin layer of highly reflective gold is evaporated from a tungsten filament onto the sample surface. The resolution obtainable is limited by the wavelength, and because the optical light has higher wavelength (of the order of microns) than the other radiations, the optical microscope has lower resolution (again of the order of microns). In addition, because there is a limitation for the physical size of the microscope, and since the focal length of the optical lens is small, optical microscopy is limited to a magnification of about 2000x. In optical microscopy, the use of polarized light aids in the phase identification

By using an electron beam, which has its wavelength in angstrom units (AU), the resolution can be increased to some tens of AU. In this case, the magnification is the ratio of the image monitor size to the area on the sample on which the electron beam falls. Thus, a magnification of 100,000x can be achieved. Thin sections are viewed with transmitted electrons. The sample thickness required in this case is <1 μ, which is an order of magnitude lesser than that required in optical microscopes. This is because of the lesser transmittance value for electrons than that for photons.

This much thickness is obtained by chemical thinning or thinning by argon ion bombardment in addition to grinding. An advantage here is the selected area electron diffraction (SAD) patterns. This facility can be used to identify and characterize individual phases.

Scanning electron microscopy (SEM) uses the electrons obtained after the interaction of the primary electrons from the electron source with the sample surface. Here, the primary electron beam scans the surface, causing emission of secondary electrons suitable for viewing. The range of magnification for SEM is 20x–100,000x, and a smooth surface is not required. In the case of ceramics, which are generally nonconductors, electrostatic charging may take place. This will lead to charring of the surface, which obliterates the real surface features. To prevent such electrostatic charging, the surface is coated with a thin layer of gold evaporated from a tungsten filament onto the sample surface. SEM is particularly useful for observing the surface features and fracture surfaces. One special advantage of this microscope is that, by the analysis of the electron emission energy spectrum from a point on the sample, the approximate chemical composition of the surface can be determined.

11.2.2 Porosity

Porosity is characterized by the volume fraction of pores, and their size, shape, and distribution. In ceramics, porosity can vary from 0 to >90% of the total volume of the material. Figure 11.1 shows how the electrical and thermal conductivities change with porosity. If the spherical pores are isolated, the conductivities decrease linearly with the volume fraction of the pores. On the other hand, in the case of porosity being continuous, there is almost a sudden decrease of conductivities to a very low value, and for larger values of porosity, the decrease becomes gradual.

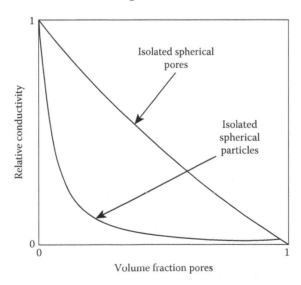

FIGURE 11.1
Effect of porosity on direct electrical conductivity or on thermal conductivity: upper curve, isolated spherical pores in a continuous solid matrix; lower curve, isolated spherical particles in a continuous pore matrix.

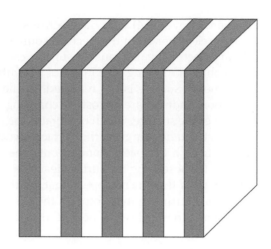

FIGURE 11.2
Parallel slabs of pore space and solid material.

These two limits in the distribution of porosity can be visualized by considering parallel slabs (Figure 11.2).

For electrical or heat flow normal to the slabs, akin to the isolated pores:

$$(1/k_t) = (f_s/k_s) + (f_p/k_p) \tag{11.1}$$

Here, k_p and k_s are the conductivities of the pore and solid phases, respectively, and f is the volume fraction. If $k_p \ll k_s$, which is the actual case, $k_t \sim k_p/f_p$; in other words, the conductivity becomes inversely proportional to the porosity, as shown by the upper curve of Figure 11.1.

For flow parallel to the slabs,

$$k_t = f_s k_s + f_p k_p \tag{11.2}$$

If $k_p \ll k_s$, $k_t \sim f_s k_s$, in which case the conductivity is dependent on that of the solid. When the solid becomes the isolated phase, as shown by the lower curve of Figure 11.1, the conductivity decreases as a function of its volume fraction. The case of parallel slabs, where the solid is also continuous, does not exactly match this situation of isolated spherical solid particles. This discrepancy leads to the inconsistent equation derived.

Samples with low porosity approach the continuous solid phase, and samples with high porosity tend to approach the continuous pore phase. This does not always happen. Flat cracks along grain boundaries approach continuous pore phase, but in this case, fraction porosity is considerably low. Similarly, foam glass structure is essentially a continuous solid phase with isolated pore structure, but here the porosity is much higher.

Porosity is characterized by relations to other phases. For example, the volume percent of pores is obtained as the difference between the theoretical

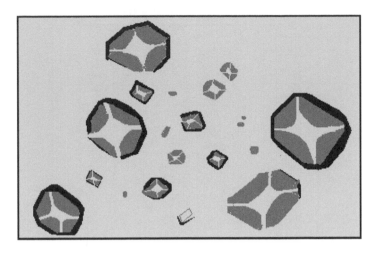

FIGURE 11.3
Pores that have formed *negative* crystals in UO_2; (100) planes are parallel to the surface (18,000x).

and actual density. Pores can be present inside or outside the grains. There can be same-size pores. Porosity is sometimes seen with crystallographic orientation (Figure 11.3). The pores are termed as *negative crystals*, as they contain no mass but have a crystallographic orientation, just like the solid crystals.

One of the common methods of characterizing porosity is finding the *apparent porosity*. This corresponds to the amount of *open pores*. The total porosity includes both open and *closed pores*. Open pores affect permeability, vacuum tightness, and the surface available for catalytic reactions and chemical attacks. Permeability increases with porosity, vacuum tightness decreases with it, and the surface available for catalytic reactions and chemical attacks increases. Before the firing process starts, almost the entire pores are open.

During firing, the volume fraction of porosity decreases. Some open pores get eliminated directly; many get transformed into closed pores. As a result, volume fraction of closed pores increases initially and decreases only toward the end of the firing process. Open pores get eliminated when the total porosity has decreased to 5%, as shown in Figure 11.4. Permeability is a measure of open porosity.

11.2.3 Single-Phase Polycrystalline Ceramics

In the case of single-phase polycrystalline ceramics, in addition to porosity, it is necessary to determine the amount, size, shape, and distribution of the other constituents to characterize the microstructure completely. The microstructure of polycrystalline ceramics develops as grains that meet at faces whose intersections form angles of 120°. In some materials,

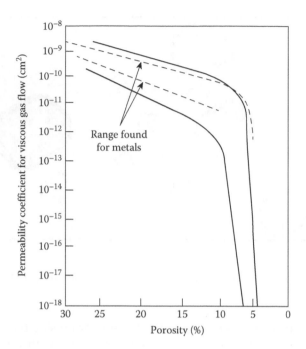

FIGURE 11.4
Permeability coefficient for viscous gas flow in beryllia ware of differing porosities.

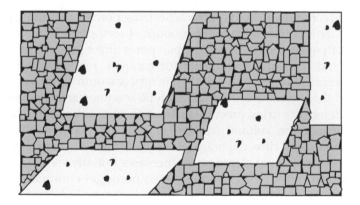

FIGURE 11.5
Idiomorphic grains in a polycrystalline spinel. The large grain edges appear as straight edges; shape of small grains is controlled by surface tension (350x).

a duplex structure is developed, where some large grains are found in a fine-grained matrix (Figure 11.5). Grains are columnar, prismatic, cubic, spheroidal, or acicular in shape. Each of these shapes gives rise to special properties. In single-phase ceramics, the nature and composition of dislocations, sub-grain boundaries, and grain boundaries are important.

FIGURE 11.6
Crystal–liquid structure of a forsterite composition (150x).

These have definite structural characteristics such as specific orientation, concentration, and impurity segregations. In addition, the properties change with these characteristics.

11.2.4 Multiphase Ceramics

In multiphase compositions, we have to consider the relationships among amount, distribution, and orientation of separate phases. The most common structure is one or more phases dispersed in a continuous matrix. These can be prismatic crystals in a glass, such as forsterite glass ceramic (Figure 11.6), or there can be crystals precipitated in a crystalline matrix.

11.3 Quantitative Analysis

The following sections deal with the quantitative analysis of the differential microstructural characteristics.

11.3.1 Relative Amounts of Phases Present

For a random sample, the volume fraction (V_V) of a phase is equal to the cross-sectional area fraction (A_A) of the phase in a random plane through the sample. It is also equal to the linear fraction (L_L) of the phase intersecting a random line drawn through the sample, as well as the fraction of points (P_P) randomly distributed along a line over a cross-sectional area or throughout a volume, which falls within the phase. Therefore, we can write:

$$V_V = A_A = L_L = P_P \tag{11.3}$$

Experimental measurements on a ceramic sample can be done using a thin section (relative to particle size) of it or a polished plane prepared on it. Following Equation 11.3, the measurements are done by area, lineal, or point analysis.

For point analysis, *point counting* is done by randomly distributing a grid of points over the sample in the microscope or on a photomicrograph, or by randomly moving a microscope crosshair and counting the fraction of the total point count that falls in a given phase (P_P).

In lineal analysis, a line in the microscope objective or one drawn across a micrograph is used, and the fractional length intercepting each phase, L_L, is measured. If the line inscribed on the objective lens is divided into equally spaced divisions, then the number of divisions falling in each phase is manually counted. This number divided by the total number of divisions gives the linear fraction for one trial. A number of trials is to be carried out. Then the average value is taken as L_L. The measurements can also be with an integrating stage micrometer.

Areal analysis is done to determine the relative area of each phase A_A in the planar section. This is less commonly done, because it is slower than point counting or lineal analysis. Relative areas can be determined by three ways:

1. From tracings with planimeter
2. By cutting out separate phases from a micrograph and weighing them
3. By counting squares on a grid placed over the micrograph

In addition to determining the volume fraction of phases, relatively simple measurements and statistical analyses give a range of additional information about microstructure characteristics. This additional information is important for special purposes. There are two such basic measurements. These are with a known length of test line L. How much is the length magnified during the measurements should be known. The measurements are as follows:

1. The number of intersections of a particular feature such as phase boundaries along the line, P_L
2. The number of intersections of the objects such as particles along the line, N_L

Some related measurements can also be made. These are the number of points, such as the three-grain intersections observed in a known area A, which is P_A, or the number of an object, such as grains, in a known area, N_A.

From these measurements, one can calculate directly the surface area per unit volume S_V, the length of linear elements per unit volume L_V, and the number of point elements per unit volume P_V. The equations for the calculations are as follows:

$$S_V = 2P_L \tag{11.4}$$

$$L_V = 2P_A \tag{11.5}$$

$$P_V = (1/2)L_V S_V = 2P_A P_L \tag{11.6}$$

11.3.2 Sizes and Spacings of Structure Constituents

In observing microstructures, it is frequently found that there are two or more levels of structure:

1. A structure associated with the distribution of relatively large pores, grog (prefired clay) particles, or other grains in a matrix phase
2. A structure of phase distribution within the large grains and within the bond phase
3. A structure that shows the distribution of dislocations and the separation between various structure constituents

It is often necessary to use different techniques for a study of each of these characteristics. For example, we may use a thin or polished section to study the large-scale structure coming under level 1. We will have to use scanning electron microscopy to study a fine clay matrix of level 2 structures. Determining which level of structure is the most important depends on the particular ceramic and the particular property of interest.

As a particle parameter, the mean intercept length $\bar{L} = L_L/N_L$ is a simple and convenient measurement to characterize particle size. For spherical particles or rods of uniform size, the mean intercept length gives a measure of the particle radius r:

$$\bar{L} = (4/3)r \text{ sphere} \tag{11.7}$$

$$\bar{L} = 2r \text{ rod} \tag{11.8}$$

And the thickness of a plate-shaped phase:

$$\bar{L} = 2t \tag{11.9}$$

In addition, for space-filling grains, since $\bar{L} = 1/N_L$ and $P_L = N_L$, the surface area per unit volume is given by:

$$S_V = 2P_L = 2N_L = 2/\bar{L} \tag{11.10}$$

For separated particles, $P_L = 2N_L$, and

$$S_V = 2P_L = 4N_L = 4V_V/\bar{L} \tag{11.11}$$

Another important parameter in determining spatial distributions is the mean free distance between the particles, λ. This distance is the same as the mean edge-to-edge distance along a straight line between the particles or phases.

$$\lambda = (1 - V_V)/N_L \tag{11.12}$$

This value is related to the mean intercept length:

$$\lambda = \bar{L}(1 - V_V)/V_V \tag{11.13}$$

In addition, if the mean spacing between the particle center is S,

$$S = 1/N_L \tag{11.14}$$

and

$$L = S - \lambda \tag{11.15}$$

11.3.3 Porosity

The best method to characterize porosity is by the use of polished sections with lineal or area analysis. Difficulties in the characterization of porosity by this method arise in the following instances:

1. When the specimens are relatively soft, it is difficult to polish.
2. In some samples, grains tend to pull out during polishing. If it happens, since the areas from where the grains have pulled out also look like pores, the total porosity measured will be high.

The difficulty associated with the polishing of soft specimens can be eliminated by impregnating them with a resin before polishing. All the pores will be filled up with resin, and their sizes will not change during polishing. The pull out is caused by differences in constituent hardness, or microfissures or microstresses already present. Hardness differences causes high relief that encourages pullout. This is corrected by using a suitable hard abrasive for polishing, such as diamond powder, and a hard flat polishing surface. Microstresses can cause flat cracks or microfissures along grain boundaries. These flat cracks or microfissures allow the grains to pop out with the very small added stresses arising from polishing.

11.3.4 Density

Total porosity can be calculated by determining the bulk density ρ_b (total weight/total volume, including pores) and comparing this with true density ρ_t (total weight/volume of solids):

$$f_P = (\rho_t - \rho_b)/\rho_t = 1 - (\rho_b/\rho_t) \tag{11.16}$$

It is sometimes convenient to express bulk density as the fraction of the theoretical density achieved:

$$\rho_b/\rho_t = 1 - f_p \tag{11.17}$$

For a crystalline solid, the density can be calculated from the determination of the crystal structure and lattice constant, since the atomic weight for each constituent is known. True density can also be determined by comparing pore-free samples with a liquid having known density. For glasses and single crystals, which are examples for pore-free samples, this is done by weighing in air and then after suspending in liquid. The volume is then calculated following Archimedes principle. The suspension in the liquid medium is more precisely achieved by adjusting the composition or temperature of a liquid column just to balance the density of the solid.

For complex mixtures and porous solids, the sample is pulverized until no residual closed pores are present. The density is then determined by the pyknometer method. In this method, a known weight of the sample is put in the pyknometer bottle, of which the volume is inscribed on it and then weighed. Then a liquid of known density is added to fill the remaining volume of the pyknometer. To ensure the penetration of the liquid among all particles, the sample and the liquid should be boiled or heated under vacuum and then cooled, and then weighed again. The difference between the second and the first weights gives the weight of the liquid. From the knowledge of the liquid's density, its volume is calculated. The difference between the pyknometer volume and the liquid volume is the volume of the sample. From the weight and volume of the sample, its true density is calculated.

Bulk density of porous bodies requires the determination of its total volume of solid plus pores. For samples such as bricks, this is done by measuring the dimensions and calculating the volume. For smaller samples, the bulk density is determined by measuring the weight of mercury (or of any non-wetting liquid that does not penetrate pores) displaced by the sample with a mercury volumeter or the force required to submerge the sample. According to Archimedes principle, the weight of the mercury displaced by the sample is the same as the loss of weight of the sample in mercury. This weight is also proportional to the force required to submerge the sample in mercury, where the sample floats in mercury.

For small samples, bulk density can also be determined by coating the sample with an impermeable film such as that of paraffin. The weight of the film is the difference in the weights of the coated and uncoated samples. Knowing the density of the film material, its volume is calculated from its weight and density. The volume of the sample plus film is determined by Archimedes' method, and the sample volume is calculated by finding the difference between the two volumes obtained.

Open-pore volume is measured by first weighing the sample in air (W_a) and then heating in boiling water for 2 h to fill all the open pores completely with water. After cooling, the weight of the saturated piece is determined after (1) suspending in water (W_{sus}) and (2) in air (W_{sat}). The difference between these last two values gives the sample volume and allows calculation of bulk density. The difference between saturated and dry weights gives open pore volume.

11.3.5 Open Porosity

There are various methods devised for characterizing open pores for their size by regarding them as capillaries and determining their equivalent diameter from the rate of fluid flow through them or the extent to which liquid mercury can be forced into them.

In the mercury method, for example, the sample is placed in a container and evacuated. Mercury is then admitted by applying pressure. The pressure necessary to force mercury into the capillary depends on the contact angle and surface tension, and is given by:

$$P = 4\gamma\cos\theta/d \approx 14/d \tag{11.18}$$

Here, P is in kg/cm^2, d is the pore diameter (μ), γ is the surface tension of mercury, and θ is the contact angle (140° for most oxides). As the pressure is increased, smaller-sized pores are permeated at the increased pressure. Their amount is measured by the decrease in the apparent volume of mercury plus sample. Thus, the results obtained can be used to determine the distribution of open pore sizes.

The following sections deal with the evolution of microstructure in different ceramic compositions.

11.4 Triaxial Whiteware Compositions

Ceramic compositions forming the basis for the whiteware industry are the mixtures of clay, feldspar, and flint. These compositions include hard porcelain for tableware, sanitary ware, electrical porcelain, and so on.

The clays serve the following dual purpose:

1. Providing fine particle sizes and good plasticity for forming
2. Forming fine pores and a more or less viscous liquid essential to the firing process

The feldspar acts as a flux, forming a viscous liquid at firing temperature and aids in vitrification. Flint is an inexpensive filler material that, during firing, remains unreactive at low temperatures, and at high temperatures forms a high-viscosity liquid.

11.4.1 Changes during Firing

The initial mix of the constituents used to form whiteware compositions is composed of relatively large quartz and feldspar grains in a fine-grained clay matrix.

During firing, the feldspar grains melt at about 1140°C, but, because of their high viscosity, there is no change in the shape of the product until the temperature reaches above 1200°C. At around 1250°C, all the feldspar grains smaller than about 10 μ disappear by reaction with the surrounding clay, and the larger grains interact with the clay. This interaction causes alkali ions to diffuse out of the feldspar, and mullite crystals to form in a glass.

The clay phase initially shrinks, and fissures frequently appear. Fine mullite needles appear at about 1000°C but cannot be resolved with an optical microscope until temperatures of at least 1250°C are reached. With further increases of temperature, the mullite crystals continue to grow. After firing at temperatures above 1400°C, mullite is present as prismatic crystals up to about 0.01 mm in length.

No change is observed in the quartz phase until temperatures of about 1250°C are reached. The rounding of edges can then be noticed in small particles. The solution rim of high-silica quartz around each quartz grain increases in amount at higher temperatures. By 1350°C, grains smaller than 20 μ are completely dissolved. At above 1400°C, little quartz remains.

Finally, the porcelain consists almost entirely of mullite and glass. The heterogeneous nature of the microstructure is shown in Figure 11.7. Quartz grains are surrounded by a solution rim of high-silica glass. The outlines of the glass–mullite areas correspond to the original feldspar grains, and

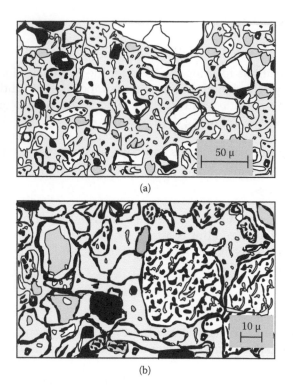

(a)

(b)

FIGURE 11.7
Photomicrographs of electrical insulator porcelain (etched 10 s, 0°C, 4% HF) showing liquid quartz grains with solution rim, feldspar relics with indistinct mullite, unresolved clay matrix, and dark pores. (a) at lower magnification, (b) at higher magnification.

the unresolved matrix corresponding to the original clay can be clearly distinguished. Pores are also seen to be present.

Although mullite is the crystalline phase in both the original feldspar grains and in the clay matrix, the crystal size and development are quite different (Figure 11.8).

Larger mullite needles grow into the feldspar relicts from the surface as the composition changes by alkali diffusion. The changes taking place during firing occur at a rate depending on the time, temperature, and particle size. The slowest process is quartz solution. Under normal firing conditions, equilibrium at the firing temperature is only achieved at temperatures above 1400°C, and the structure consists of a mixture of siliceous liquid and mullite. In all cases, the liquid at firing temperatures cools to form a glass, so that the resulting phases at room temperature are normally glass, mullite, and quartz in amounts depending on the initial composition and conditions of firing treatment. Compositions with larger feldspar content form larger amounts of siliceous liquid at lower temperatures and correspondingly vitrify at lower temperature than compositions with larger clay content.

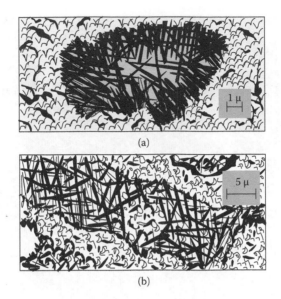

(a)

(b)

FIGURE 11.8
Mullite needles growing into a feldspar relict (etched 10 s, 0°C, 40% HF). (a) at lower magnification, (b) at higher magnification.

11.5 Refractories

Refractories are generally composed of large grog or refractory grain particles held together with a fine-grained bond. Both the bond material and refractory grains have a fine structure, and both are multiphase.

11.5.1 Fireclay Refractories

The largest group of fireclay refractories is based on mixtures of plastic fireclay, flint clay, and fireclay grog. All these materials tend to form mullite on heating. In addition, quartz is often present as an impurity in plastic fireclay, and is sometimes added to reduce firing and drying shrinkage.

The fine structure in the grog (prefired clay) or flint clay particles is difficult to resolve with an optical microscope, but consists of fine mullite crystals in a siliceous matrix (Figure 11.9).

Alkali, alkaline earth, iron, and similar impurities that are present largely combine with the siliceous material to form a low-melting-point glass and decrease the refractoriness of the brick.

11.5.2 Silica Refractories

Silica refractory bricks are manufactured from ground ganister rock (quartzite) containing 98% SiO_2, to which 2% CaO is added as milk of lime.

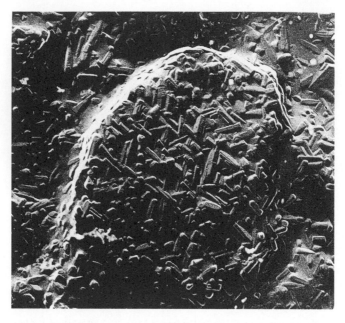

FIGURE 11.9
Mullite crystals in silica matrix formed by heating kaolinite (37,000x). (From W. D. Kingery et al., *Introduction to Ceramics,* 2nd Edn., John Wiley, New York, 1976, pp. 783–812.)

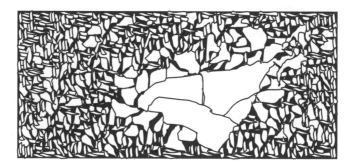

FIGURE 11.10
Quartz grain with cristobalite formed at surface (etched 20 min, 100°C, 50% NaOH, silica replica).

The added lime serves as a mineralizer (dissolving cristobalite and precipitating tridymite) during firing. Fired brick consists of shattered quartz grains that have been almost completely transformed into cristobalite (starting at the edges of the grain, as is evident in Figure 11.10) in a matrix of fine tridymite, cristobalite, and glass. Small amounts of unconverted quartz (about 10%) normally remain, with equal amounts of cristobalite and tridymite formed.

11.5.3 Basic Refractories

In the class of basic refractories are included bricks manufactured from chrome ore [(MgFe)(AlCr)$_2$O$_4$], periclase (MgO), calcined dolomite (CaO, MgO), olivine [(MgFe)$_2$SiO$_4$], and mixtures of these materials.

Chrome ore contains serpentine and other silicates as impurities, which are low melting and deleterious. If magnesia is added, it reacts with this material to form forsterite, which is refractory. Brick consists of large chromite grains that usually contain a precipitate of (Fe, Al, Cr)$_2$O$_3$ resulting from iron oxidation during firing. The bond phase consists of fine chrome ore, magnesioferrite, and forsterite. A typical structure is illustrated in Figure 11.11.

When more than a small amount of magnesite is added, we enter a range of chrome ore–periclase compositions in which there are usually large grains of chromite in a matrix of fine MgO, MgFe$_2$O$_4$, and Mg$_2$SiO$_4$.

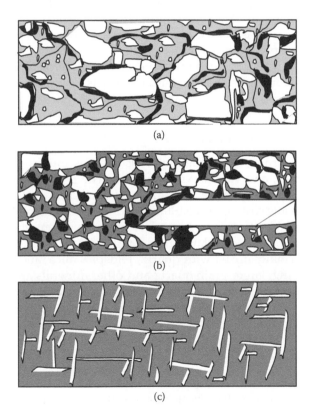

(a)

(b)

(c)

FIGURE 11.11
Basic brick: (a) chrome ore–magnesite brick at low magnification; light grains are chrome ore, gray phase is periclase, and dark gray are porosity (51/4x); (b) different chrome ore–magnesite brick at higher magnification; angular chrome ore grain and magnesite grain (rounded periclase particles in silicate matrix) in fine-grained bond (150x); (c) higher magnification of chromite grain showing lamellar (CrFe)$_2$O$_3$ precipitate parallel to the (111) plane in spinel (1400x).

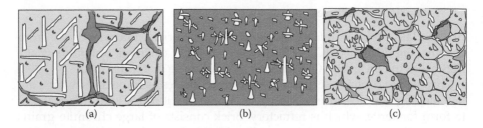

FIGURE 11.12
Precipitation of $MgFe_2O_4$ from MgO in basic refractory brick as (a) platelets parallel to (100) planes in MgO (500x); (b) dendritic precipitate (1000x); (c) spheroidal morphology (250x).

Periclase bricks are formed mainly from magnesite ore or seawater magnesia, and contain large MgO grains in a bond phase of fine MgO, together with some Mg_2SiO_4 and $MgAl_2O_4$ (Figure 11.12).

A minor class of basic brick is manufactured from olivine $(MgFe)_2SiO_4$. The major phase present is forsterite.

Dolomite $(MgCO_3 \cdot CaCO_3)$ is a raw material of high refractoriness, low cost, and wide availability. Calcined dolomite consists of a mixture of CaO and MgO solid solutions plus a minor amount of silicate phase.

11.5.4 Structural Clay Products

Structural clay products include materials such as building bricks, sewer pipes, drainpipes, and various kinds of tiles. The manufacture of these is characterized by inexpensive raw materials, efficient material handling methods, and a low cost for the resultant product. The main raw materials are the locally available clays having a variety of compositions and structures.

The clays used are glacial clays, shale, or alluvial clays. The clay minerals present are normally mixed with quartz, feldspar, mica, and other impurities. During firing, larger-grain quartz and other minerals are not normally affected. The impurities form a glassy phase. The structure consists of large grains of secondary constituents embedded in a matrix of fine-grained mullite and glass (Figure 11.13).

11.5.5 Glazes and Enamels

Glazes for ceramic ware and porcelain enamels for iron, aluminum, and jewelry metals are usually silicate glasses that may or may not include crystals or bubbles.

As a glaze is fired, its structure changes continuously. Materials initially present decompose and fuse. Bubbles are formed and rise to the surface, and reaction takes place with the body under the glaze. The interface between the glaze and body is rough because of the differences in solubility in the body constituents. Mullite crystals develop at the interface of porcelain

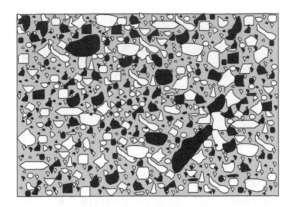

FIGURE 11.13
Microstructure of stoneware, ion-bombardment-etched (100x).

FIGURE 11.14
Crystals dispersed in a mat glaze (polished section, H_3PO_4 etch, 100x).

bodies fired at high temperatures. In a clear glossy glaze, the glass phase has a number of dispersed particles or porosity, and the surface is ideally perfectly smooth. Gloss is most commonly lost because bubbles rise and burst at the surface, forming small craters. Although the bubble content is initially high, the larger bubbles are rapidly eliminated during firing. The smaller bubbles are only slowly removed. When the fluid glazes, which tend to run, are used, even smaller bubbles are rapidly removed.

In mat glazes, the surface texture and low gloss result from the extensive development of fine crystals (Figure 11.14). For different glazes, the composition of the crystals is variable, but anorthite ($CaO \cdot Al_2O_3 \cdot 2SiO_2$) is the most common ingredient.

Porcelain enamels on metal are similar to glazes, in that they are basically a silicate glass coating, but they are fired for shorter periods and at lower temperatures. During firing, a ground coat containing nickel or cobalt oxide, or a nickel dip (of the metal in a nickel sulfate solution), is frequently used to

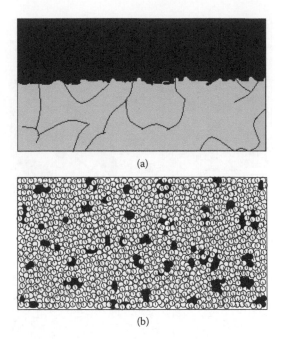

(a)

(b)

FIGURE 11.15

Microstructure of porcelain enamel: (a) cross-section of enamel–metal interface showing rough boundary (500x); (b) top view of bubble structure at metal–enamel interface provides space for gas evolution from metal on cooling (50x).

improve adherence of the enamel. For good results with porcelain enamels, it is found that the *bubble structure* is particularly important (Figure 11.15). Clays containing some organic impurities form fine bubbles adjacent to the metal surface. These bubbles provide reservoirs for hydrogen evolved from metal on cooling and prevent *fish scaling*, the breaking out of pieces of the coating on cooling.

11.5.6 Glasses

Liquid–liquid immiscibility is widespread in glass-forming systems. At magnifications above the range of optical microscopes, glasses may appear as optically homogeneous. However, on a scale of 30–50 AU up to a few hundred AU, phase separation could be observed in the glass structure. On the scale of microns to millimeters and larger, a number of types of defects are found in glasses.

11.5.6.1 Seed

Seeds are small gaseous bubbles that appear if the gas bubbles formed are not removed during melting.

11.5.6.2 Stone

Stones are small crystalline imperfections. They result from melting. When enough time is not given for complete dissolution of all the solid ingredients used to form the glass, the undissolved crystalline solid particles seen in glass are the stones. Figure 11.16 shows a stone.

11.5.6.3 Cord

Cords are amorphous inclusions. These have a different index of refraction than the matrix glass. Such an inclusion is shown in Figure 11.17. When the glass is in the molten state, and the different ingredients in the molten state are not mixed to give a homogeneous mixture, the unmixed liquid give rise to cord.

11.5.7 Glass-Ceramics

From the molten state during cooling, the composition is such that phase separation takes place. The separated phases are amorphous. In these amorphous phases, the one that is richer with nucleants starts crystallizing. Once the cooling is completed, at room temperature, the structure contains a crystalline phase along with an amorphous phase. The resulting material is called a glass-ceramic. Glass represents the amorphous phase, and ceramic represents the crystalline one. The microstructure of such a material is shown in Figure 11.18.

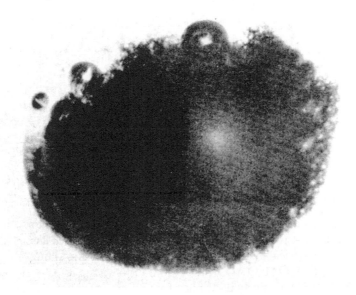

FIGURE 11.16
Refractory stone. (From W. D. Kingery et al., *Introduction to Ceramics*, 2nd Edn., John Wiley, New York, 1976, pp. 783–812.)

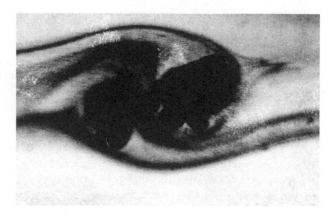

FIGURE 11.17
Cord in soda-lime-silica glass. (From W. D. Kingery et al., *Introduction to Ceramics*, 2nd Edn., John Wiley, New York, 1976, pp. 783–812.)

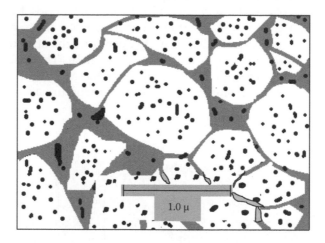

FIGURE 11.18
Submicrostructure of $Li_2O–Al_2O_3–SiO_2$ glass—ceramic nucleated with TiO_2—approximately 80% crystalline. Small rutile crystals shown in larger crystals of β-spodumene.

11.5.8 Electrical and Magnetic Ceramics

Electrical ceramics are used as insulators. As low-tension insulators, the composition that is most widely used are the triaxial porcelains. Glasses are also widely used for electrical insulation purposes. For low-loss and high-frequency applications, steatite ($MgO·Al_2O_3·SiO_2$), forsterite (Mg_2SiO_4), and alumina ceramics are used. Extensively used high-frequency insulators should possess good strength, high dielectric constant, and low dielectric losses. There are two main phases in them. Figure 11.19 shows the microstructure of a steatite porcelain. In this crystalline phase, enstatite ($MgO·SiO_2$) appears in a glassy matrix.

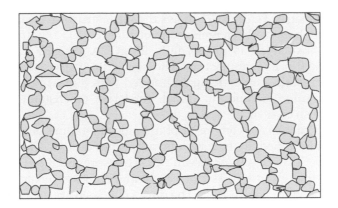

FIGURE 11.19
Microstructure of steatite porcelain (500x).

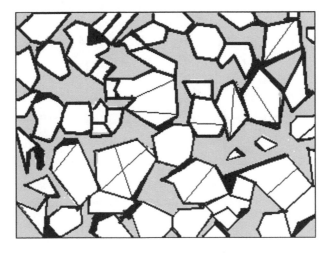

FIGURE 11.20
Crystal–liquid structure of a forsterite composition (150x).

Triaxial porcelains contain considerable amounts of alkalis derived from feldspar. Alkalis in feldspar help in the formation of flux during the production of porcelain. The presence of alkalis leads to high electrical conductivity and high dielectric loss. Electrical conductivity and loss of dielectricity are not desirable in an insulator. The low-loss compositions are made alkali-free by using alkaline earth oxides as fluxing constituents.

Forsterite ceramics contain Mg_2SiO_4 as the crystalline phase. This phase is bonded with a glassy matrix. In Figure 11.20, the microstructure of forsterite is shown.

The next type of insulating ceramics used in electrical applications is alumina. In this ceramic, Al_2O_3 forms the crystalline phase, which is bonded with a glassy matrix. Figure 11.21 shows the microstructure of such a ceramic silicate glassy phase has been removed in etching.

Cordierite insulator has the formula $2MgO \cdot 2Al_2O_3 \cdot 5SiO_2$. A variety of fluxes are used in its manufacture. The cordierite phase develops as prismatic habit crystals. These are again bonded with a glassy phase. In addition, some mullite, corundum, spinel, forsterite, or enstatite ($MgO \cdot SiO_2$) phases may also be seen.

Magnetic ceramics are composed of a single crystalline phase. The composition is determined by the magnetic properties desired. The compounds can be $FeNiFeO_4$, $BaFe_{12}O_{19}$, or $FeMnFeO_4$. These are produced with as high density and fine grain size as can be obtained. Nickel ferrite, a magnetic ceramic, is shown in Figure 11.22. The dark spots seen are etch pits.

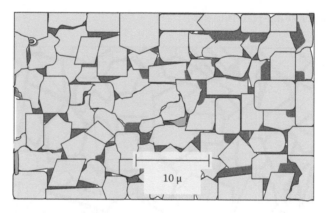

FIGURE 11.21
High-alumina porcelain and heavily etched to remove silicate-bonding phase (2300x).

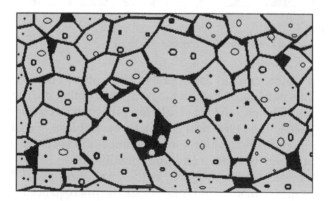

FIGURE 11.22
Microstructure of nickel ferrite; etch pits visible in grains result from sulfuric acid–oxalic acid etchant (600x).

11.5.9 Abrasives

Abrasives have three microconstituents. The essential constituent is a hard phase. These are incorporated as individual particles. They provide sharp cutting edges. A bonding phase holds these particles in a more or less tight grip. And a certain amount of porosity provides channels for air or liquid flow through the structure. This fluid flow brings down the temperature during the cutting or machining operation. For the hard abrasive grain, either aluminum oxide or silicon carbide is used. In either, abrasive individual grains are bonded to a wheel or paper or cloth with a strength depending on their proposed use. Bond materials include fired ceramic bonds and a variety of organic resins and rubbers. The microstructures of both the abrasives are shown in Figure 11.23.

11.5.10 Cement and Concrete

There are a wide variety of cementitious materials. Among these, the one that is of most economic importance and is most widely used is Portland cement. Portland cement is manufactured in rotary kilns. Various raw materials are used in its making. The overall composition is made up of two phases: tricalcium silicate ($3CaO \cdot SiO_2$) and dicalcium silicate ($2CaO \cdot SiO_2$). Figure 11.24 shows the microstructure of a

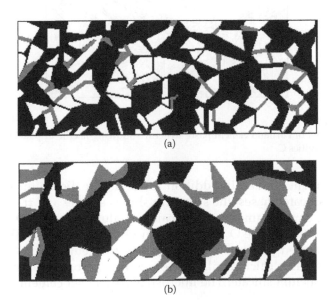

(a)

(b)

FIGURE 11.23
Abrasive products: (a) section of silicon carbide wheel, unetched (50x); (b) aluminum oxide wheel, unetched (100x). In both, light area is grain, gray area bond, and dark area porosity.

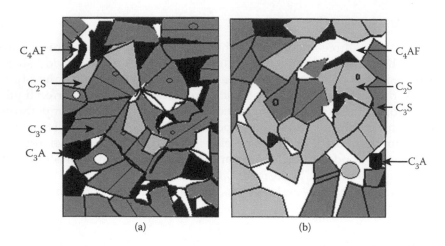

FIGURE 11.24
Micrographs of Portland cement clinker (835x): (a) type 1, high in 3CaO.SiO$_2$ (major gray phase is C3S, dark gray phase C3A, light gray phase C2S, white phase mainly C4AF); (b) type 11, containing nearly equal parts of 3CaO·SiO$_2$ and 2CaO·SiO$_2$ (gray C3S, light gray C2S, black C3A, white C4AF).

FIGURE 11.25
Simplified model of Portland cement paste structure showing needle or platelet gel particles and capillary cavities C.

Portland cement clinker. In addition to the major dicalcium silicate and tricalcium silicate phases, the structure shows silicate and tricalcium silicate phases, smaller amounts of tricalcium aluminate (3CaO·Al$_2$O$_3$), and millerite (4CaO·Al$_2$O$_3$·Fe$_2$O$_3$). On reaction with water, the clinker forms a complex hydrated product. This product is the cementitious material. It is a noncrystalline calcium silicate gel resulting from the tricalcium silicate and dicalcium silicate. The microstructure of set cement is shown in Figure 11.25.

In the gel phase itself, there are pore spaces between the individual gel particles. The excess water, when it leaves by evaporation, gives rise to large residual pores.

FIGURE 11.26
Interlocking crystalline network in pottery plaster, $CaSO_4 \cdot 2H_2O$ (courtesy W. Gourdin). (From W. D. Kingery et al., *Introduction to Ceramics*, 2nd Edn., John Wiley, New York, 1976, pp. 783–812.)

In addition to the cement gel, there is present in concrete an aggregate of crushed stone that acts as a filler material. The Portland cement paste serves to bond together the aggregate particles in much the same way as a bond material present in a refractory brick or an abrasive wheel.

Plaster of Paris is widely used as a cementing and mold material. The structure of set plaster is a highly crystalline reaction product. It is shown in Figure 11.26. The raw material for plaster of Paris is gypsum ($CaSO_4 \cdot 2H_2O$). It is formed by calcining gypsum and thereafter by rehydration. The reactions involved are:

Calcination: $CaSO_4 \cdot 2H_2O = CaSO_4 \cdot \frac{1}{2} H_2O + (3/2) H_2O$

Rehydration: $CaSO_4 \cdot \frac{1}{2} H_2O + (3/2) H_2O = CaSO_4 \cdot 2H_2O$

The individual crystals present are in the form of fine needles, so that the resulting structure corresponds to a felt-like arrangement in which there are very fine pores, and interlocking of the crystal needle provides sufficient strength. Residual porosity is also seen in the structure. The amount of this residual porosity depends on the amount of water present in the original mix. The larger the amount, the greater will be the porosity.

11.5.11 Microstructures of Some Special Compositions

These compositions are used for critical needs.

11.5.11.1 Cermets

Cermets are combinations of metals and ceramic materials. The compositions that are of most significance are carbides. They possess high-temperature strength and great hardness. Cemented TiC containing Ni as the metal is an example. Other examples are the oxide-based cermets. The valuable properties they have are high-temperature strength and reasonable stability in air at these temperatures. An example for this is a cermet made up of aluminum oxide and chromium. In the case of cemented carbide compacts, they consist of either spheroidal (Figure 11.27b) or prismatic carbide grains (Figure 11.27a), completely enclosed by the metal phase. The bond phase is liquid at the firing temperature, and completely wets and flows between carbide particles, forming thin films of the metal. In the aluminum oxide–chromium system, there is a continuous phase of both the oxide and chromium. The structure is shown in Figure 11.27c.

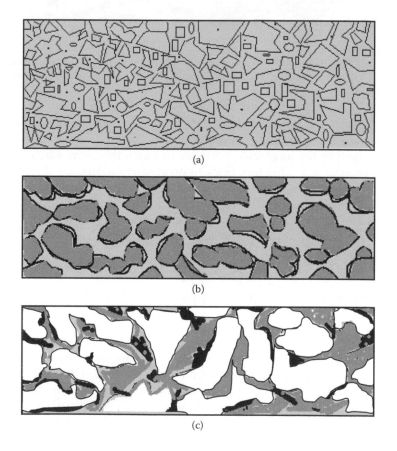

(a)

(b)

(c)

FIGURE 11.27
Metal–ceramic compositions: (a) 96WC-6Co (1500x); (b) 54 (1580x); (c) $30Al_2O_3$–70Cr (500x).

We have seen earlier the coatings of ceramics in the form of glazes and enamels. Additionally, coating of nonmetals can be applied by reactions from the vapor phase. The vapor deposits a coating on the surface. This can be achieved by flame-spraying oxide material through a high-density heat source. Fine-particle dispersion is formed on hitting the surface to develop a suitable coating. Flame-sprayed coatings have a porosity of 7%–10% and frequently show evidences of some layered structure during the buildup. Layers formed by evaporation of a solvent have very small crystals in the resultant coating.

Coatings developed from the vapor phase by reaction show large crystals. This is because the nucleation of new crystals in the surface takes place, and then they grow subsequently in the coating phase. Very often, the structure of coating is parallel to the underlying structure, in that new crystals are nucleated and grow on sites of crystals of the underlying material. Graphite coatings, for example, can be formed by passing hot CH_4, a gas, over a hot surface. The process is called *pyrolysis,* and the resulting graphite is called a *pyrolytic graphite.* In this process, the new pyrolytic graphite crystals form a deposit with their C axis normal to the underlying surface, in parallel bundles consisting of individual crystallites of nearly the same orientation. This structure is shown in Figure 11.28.

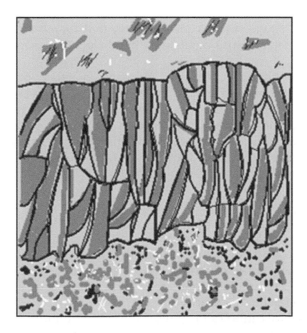

FIGURE 11.28
Pyrolytic graphite coating deposited on graphic rod at bottom from a methane–hydrogen atmosphere (150x).

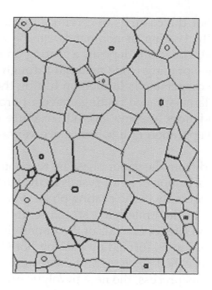

FIGURE 11.29
Microstructure of sintered Al_2O_3 is nearly pore-free, with only a few pores located within grains (500x).

11.5.11.2 Sintered Oxides

Pure, sintered, single-phase oxides are prepared for uses requiring high-strength, high-temperature capabilities, good electrical properties, or great hardness.

For example, pure Al_2O_3 is used as a tool material, possessing great hardness, low friction with metal, high-strength, and high-temperature capability. The resulting structure after its preparation is shown in Figure 11.29.

Another application of sintered oxides is in nuclear reactors—uranium oxide is used as a nuclear reactor fuel material. A large fraction of uranium atoms can be fissioned without degradation of the structure. Other oxides that are used in pure form after sintering are BeO, MgO, ThO_2, ZrO_2, and $MgAl_2O_4$.

11.5.11.3 Single Crystals

Single crystals of Al_2O_3 have been used as windows for heat resistance and good infrared transmission. They have also been used as high-density light-bulb enclosures, and as electronic device substrates. Another application of these have been in the form of rods and other special shapes as high-temperature refractory materials.

Single crystals of rutile (TiO_2), spinel ($MgAl_2O_4$), strontium titanate ($SrTiO_4$), and ruby (Al_2O_3 with some Cr_2O_3 in solid solution) have been used as synthetic jewel materials. Lithium niobate ($LiNbO_3$) is used as a laser host and as a substrate. The use of alkali and alkaline earth halide crystals for prisms and windows in optical equipment has been widespread for many years. Single crystals of calcium fluoride, lithium fluoride, and sodium chloride are commercially available.

11.5.11.4 Whiskers

Under certain conditions of growth, crystals form in a direction in which growth is rapid, developing filamentary crystals that are free from gross imperfection. These filamentary single crystals are called *whiskers*. An area of interest from a research point of view has been the structure and properties of whiskers of ceramic materials in which extremely high-strength values have been obtained.

Whiskers of aluminum oxide, several alkali halides, several sulfides, and graphite have been grown in the laboratory.

11.5.11.5 Graphite

Graphite is made from mixtures of coke and pitch. These mixtures are shaped and heat-treated to develop the graphite crystal structure. Graphite crystals are highly anisometric. They form platelets. The microstructure of graphite consists of grains of highly graphitized material in a matrix of very-fine-grained material that is more or less graphitized and more or less strongly crystallized. Figure 11.30 shows the microstructure of graphite.

11.5.11.6 High-Porosity Structure

High-porosity structures are useful for various insulating purposes. This kind of structure includes fibrous products such as glass wool, powdered insulated grain, and strongly insulating firebrick.

FIGURE 11.30
Microstructure of graphite (100x).

Problems

11.1 A typical porcelain body has the composition 50 clay–25 feldspar–25 quartz. Sketch an expected microstructure of such a body, indicating scale when (a) fired to achieve phase equilibrium (1450°C for 6 h) and (b) fired to 1300°C for 1 h.

11.2 Describe how you would experimentally determine the fractional porosity, fractional glass content, and fractional crystal content in steatite porcelain containing three phases (pores, glass, and $MgSiO_3$ crystals).

11.3 From a lineal analysis, estimate the fraction of porosity present in the porcelain in the following figure. What is the average pore radius?

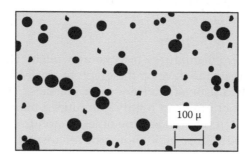

Bibliography

P. E. Doherty, R. M. Fulrath, and J. A. Pask (Eds.), *Ceramic Microstructures*, John Wiley, New York, 1968, pp. 161–185.

W. D. Kingery, H. K. Bowen, and D. R. Uhlmann, *Introduction to Ceramics*, John Wiley, New York, 1976, pp. 516–580.

T. C. Powers, Structure and physical properties of hardened Portland cement paste, *J. Am. Ceram. Soc.*, 41: 1, 1958.

F. Trojer and K. Konopicky, Die Kristallisationsformen von Magnesiumferrit bei Ausscheidung aus dem festen Zustand [The crystallization of magnesium ferrite forms during precipitation of the solid state], *Radex-Rdsch.* 7: 149–153, 1948.

D. Whatanasiriwech, K. Srijan, and S. Whatanasiriwech, Vitrification of illitic clay from Malaysia, *Appl. Clay Sci.*, 43: 57–62, 2009.

12

Production of Ceramic Powders

12.1 Introduction

Production of ceramic parts starts from powders. Hence, in this chapter, we shall consider the production of ceramic powders. These powders are prepared from the corresponding raw materials. The basic sources for the raw materials are:

a. Nature

b. Synthesis

In nature, the raw materials for ceramics occur in the form of minerals. The minerals are mined and beneficiated. These minerals usually contain oxides as their major constituents. If other types of ceramics, such as carbides, nitrides, or borides, are needed, then they are to be synthesized.

The final properties of the ceramic product will depend upon the powder characteristics. The important characteristics are shape, composition, and size of the powder particles. The characterization techniques also will be described in this chapter.

12.2 Raw Materials

Minerals for the production of ceramics are obtained from the earth's crust, which is its outermost layer. The crust is composed of silicates of Mg, Fe, Al, and Ca, and alkali metals plus Al, and free SiO_2. The minerals are mined from the earth and beneficiated. Physical beneficiation includes crushing and grinding of coarse rocks.

Chemical beneficiation includes processes of separating the desired mineral from waste products. For this, the mineral is dissolved in a solvent and filtered. A good example for chemical beneficiation is the purification of bauxite.

12.2.1 Silica

Silica is an important raw material for ceramics. The major use of silica is in glass manufacture. The silica content of optical glasses can be as high as 99.8%. A major source for silica is sand. In nature, sand is available by the natural disintegration of rock. It is also produced artificially by crushing rock. Quartz, the principal silica mineral, is a constituent of igneous rocks such as granite. Igneous rocks form when magma cools and solidifies. Magma is a molten material that originates deep within the earth. Quartz is also found in metamorphic rocks, comprising a major portion of sandstones, as well as in the pure form in veins running through other rocks. Metamorphic rocks are the result of structural and/or chemical transitions from their original form as a result of high temperatures and pressures deep beneath the earth's surface.

12.2.2 Silicates

The silicates, which are important as ceramic materials, are feldspar, kaolin, mica, and mullite. About 70% of feldspar produced goes into the making of glass. Kaolin is used in fine china, paper, and rubber. The use of mica is in terms of hundreds of thousands of tons per year. The use of mullite is again in hundreds of thousands of tons for refractory blocks.

12.2.2.1 Feldspar

Feldspar constitutes about 60% of the earth's crust. They are present in many sedimentary deposits and are found in igneous and metamorphic rocks. Sedimentary deposits form when small particles become cemented together. Feldspar is a source for alumina. It is also used in whiteware bodies as flux.

The processing of feldspar involves five steps. At the quarry, drilling and blasting are carried out first. The obtained mineral is transported for physical beneficiation by crushing and grinding. Froth flotation is then carried out to separate the mineral. In froth flotation, air is bubbled through the suspension of the mineral particles in water. The hydrophilic particles remain in the water, and the hydrophobic ones float, remaining at the air bubble–water interface. The floating ones are skimmed off from the surface, and the ones remaining in the suspension are filtered out. Various agents such as amino acids are used to enhance the relative wettability of the solids in the mixture. In the case of feldspar mineral, the different solids are mica, iron-bearing minerals, feldspar, and quartz.

12.2.2.2 Kaolin

Kaolin is a clay mineral. Clays are the primary ingredients in the making of traditional ceramic products. They are the layer silicates with a grain size of less than 2 μ. Mechanical and chemical weathering of feldspars in igneous and metamorphic rocks forms kaolin. Kaolin is a key ingredient in China clay. Primary kaolin deposits are located at the site of the original rock. Secondary kaolins were washed from the original weathering site, naturally beneficiated, and redeposited in large areas of pure kaolin. The main use of kaolin is in paper. It is white firing clay. It is composed of kaolinite, which is a hydrous aluminum silicate.

12.2.2.3 Mica

The mica group consists of 37 minerals. They are known as phyllosilicates. These have a platy texture. The name of the minerals has come from their texture. *Phyllon* in Greek means leaf.

There are two classes of micas—true and brittle. True micas contain univalent cations between each pair of layers. Cations are alkali metal ions, such as Na^+ or K^+. The layers separate in cleavage manner on application of a shearing stress. The layers represent basal planes in the crystal structure, and are flexible and elastic.

In brittle micas, the interlayer ions are divalent. An example for such an ion is Ca^{2+}. The cleavage is still cleavage. However, the interlayer bond here is stronger, and it breaks in a more brittle manner. These micas are rare.

Muscovite is the main silica used for applications. The reasons are that it is more abundant than others are, and it has better electrical properties. Its chemical formula is $KAl_2(Si_3Al)O_{10}(OH)_2$. Micas occur in igneous, sedimentary, and metamorphic rocks. The reason for their wide range of occurrence is because of their thermal stability.

The principal use for ground mica is as a filler and extender in gypsum wall board joint compounds, where it produces a smooth consistency, improves workability, and prevents cracking. It is also used in paints, molded rubber products including tires, and toothpaste.

12.2.2.4 Mullite

Mullite has the formula $3Al_2O_3 \cdot 2SiO_2$. It does not exist in nature. It is suitable for high-temperature applications. It is used in furnace lining and in other refractory applications in the iron, steel, and glass industries.

It is produced in two ways—by sintering and by fusing. Sintered mullite is obtained from a mixture of kyanite (Al_2OSiO_4), bauxite, and kaolin. Kyanite is a naturally occurring mineral found in metamorphic rocks. This mixture is prepared in the correct proportion. Then, it is sintered at temperature up to about 1600°C. The sintered product contains 85%–90% mullite, the balance being glass and cristobalite.

In the second method, the required amounts of alumina and kaolin are fused in an electric arc furnace at about 1750°C. This method gives rise to a higher-purity mullite. The fused product contains >95% mullite, the remaining being a mixture of alumina and glass.

12.2.3 Oxides

High-purity oxides suitable for parts fabrication are produced mostly by chemical processes. The important oxides are discussed in the following text.

12.2.3.1 Aluminum Oxide

Aluminum oxide or alumina (Al_2O_3) is also known as *corundum*. It is the most widely used inorganic chemical for ceramics and is produced from the bauxite mineral. The production method is called the *Bayer process*. Bauxite is hydrated aluminum oxide with iron oxide (Fe_2O_3), silica (SiO_2), and titania (TiO_2) as impurities. It is the result of the weathering of igneous aluminous rocks under tropical conditions.

A purity of 99.5% is obtained from the Bayer process, with the major impurity being Na_2O. The obtained crystal size range is 0.1–25 μ. The Bayer process is a seven-step one. The first step is *physical beneficiation*. In this step, the bauxite is ground to get a particle size of <1 mm. Such small-sized particles reduce the time for the chemical reaction in the next step.

- *Digestion*: In this step, the ground bauxite is treated with a sodium hydroxide solution. A temperature in the range of 150°C–160°C is maintained in this treatment. It is carried out at a pressure of 0.5 MPa. In the reaction that takes place, hydrated alumina is converted into sodium aluminate. The reaction is represented by Equation 12.1.

$$Al(OH)_3 \text{ (s)} + NaOH \text{ (aq.)} = Na^+ \text{ (aq.)} + Al(OH)_4^- \text{ (aq.)} \qquad (12.1)$$

- *Filtration*: The impurities in bauxite do not dissolve in the previous step. Therefore, they are filtered out from the solution.
- *Precipitation*: To the filtered sodium aluminate solution, fine gibbsite crystals are added. Gibbsite is a naturally occurring hydrated alumina [α-$Al(OH)_3$], and its crystals act as seeds for the precipitation of $Al(OH)_3$ from the solution. The pH of the solution is increased by bubbling carbon dioxide gas through it. This increases the precipitation.
- *Washing*: The precipitate is separated from the solution by filtration. This is then washed to reduce its sodium content.
- *Calcination*: The aluminum hydroxide obtained is then calcined. The temperature of calcination is 1100°C–1200°C, and this results in the formation of alumina. The reaction is represented by Equation (12.2). This product is obtained as agglomerates.

$$2Al(OH)_3 \text{ (s)} = Al_2O_3 \text{ (s)} + H_2O \text{ (g)} \qquad (12.2)$$

The size of these agglomerates is in the range of 5–10 μ.
- *Milling*: The last step in the Bayer process is milling. In this step, the alumina is obtained with the required particle size and distribution. The purity of the powder will be ≥ 99.5%. The main impurity is Na_2O.

12.2.3.2 Magnesium Oxide

The occurrence of magnesium oxide (MgO) or magnesia is in the mineral periclase. Periclase is a metamorphic mineral. It is formed by the breakdown of dolomite and other magnesium-bearing minerals. Periclase is not abundant in the earth's crust. Hence, it is not used for the manufacture of magnesia. It is manufactured from magnesite and magnesium hydroxide [$Mg(OH)_2$]. Magnesite is also called magnesium carbonate [$(Mg(CO)_3$].

The impurities in magnesite mineral are quartz, talc, mica, and magnetite. Since there are many impurities present, a number of beneficiation methods are used. They include crushing, screening, washing, magnetic separation, and froth flotation.

After beneficiation, the next step in the extraction is calcination. The calcination temperature is in the range of 800°C–900°C. The calcination reaction is given in Equation 12.3.

$$Mg(CO)_3 = MgO + CO_2 \qquad (12.3)$$

The result of calcination is very reactive fine-grained magnesia. Because of its reactivity, it is called *caustic magnesia*. Calcination at temperature above 1700°C causes sintering of the calcined powder. The resulting magnesium oxide becomes dead burned with particles grown in size.

Magnesium oxide is also produced from seawater and brines. About 1.28 g of magnesium ions is contained in 1 kg of seawater. A strong hydroxide is used to precipitate magnesium hydroxide from seawater, and the reaction for this precipitation is shown in Equation 12.4.

$$Mg^{2+} (aq.) + 2(OH)^- (aq.) = Mg(OH)_2 (s) \qquad (12.4)$$

The precipitated magnesium hydroxide is filtered out from the seawater solution, washed off any impurities, and calcined. The calcination reaction is given in Equation 12.5.

$$Mg(OH)_2 (s) = MgO (s) + H_2O (g) \qquad (12.5)$$

Magnesium-rich brine is also used for the production of magnesium oxide. The process involves the decomposition of magnesium chloride ($MgCl_2$) present in brine in the temperature range of 600°C–800°C. The decomposition takes place by reaction with steam, and the reaction is represented in Equation 12.6. The $Mg(OH)_2$ produced after the decomposition reaction is calcined as earlier to get MgO.

$$MgCl_2 + 2H_2O = Mg(OH)_2 + 2HCl \qquad (12.6)$$

Most of the MgO is used as a refractory lining in furnaces, to make a milky solution for ingestion, and for the manufacture of nonchrome spinel.

12.2.3.3 Zirconium Dioxide

The source for zirconium dioxide (ZrO_2, zirconia) is zircon ($ZrSiO_4$), which occurs in igneous rocks. A secondary source of zircon is from beach sands. In these sands, it is mixed with ilmenite, rutile, and monazite.

One way of producing zirconia from zircon is by dissociating it above 1750°C. The dissociation reaction is given in Equation 12.7.

$$ZrSiO_4 = ZrO_2 + SiO_2 \qquad (12.7)$$

When the dissociation is carried out with the help of plasma, it results in the melting of the products. On cooling these, ZrO_2 solidifies first. Later, when SiO_2 solidifies, it covers the solid ZrO_2. The cooling rate being high, because of the large difference between the temperature of the molten liquid and the surroundings, the liquid SiO_2 solidifies into a glass. This is separated from the ZrO_2 by dissolving SiO_2 in boiling sodium hydroxide. The residue is washed, and ZrO_2, which is retained in it, is separated by centrifuging.

Electric arc melting of $ZrSiO_4$ is the main production method for ZrO_2. The arc melting temperature is in the range of 2100°C–2300°C. At this temperature range, after the decomposition of zircon, the temperature being above the melting point of SiO_2, it will be in a liquid state. The separated solid ZrO_2 will be 99% pure.

12.2.3.4 Zincite

The mineral for zinc oxide (ZnO) is zincite. Zincite's color is red because it contains impurities. Commercial production of ZnO is not from zincite because it is not viable. ZnO can be formed by two ways:

1. By the oxidation of Zn vapor in air
2. By the reduction of sphalerite using C and CO

The formula for sphalerite is ZnS. It occurs in nature. The major use of zinc oxide is in the rubber and adhesive industries. The other uses are in the manufacture of latex paints, tiles, glazes, and porcelain enamels. It is also widely used in varistors.

12.2.3.5 Titania

The mineral for titania (TiO_2) is rutile. Rutile is a constituent of igneous rocks. It is an important constituent of beach sands.

Titanium dioxide is also produced from ilmenite. Ilmenite is also termed ferrous titanate ($FeTiO_3$). It is reacted with sulfuric acid. The temperature of the reaction is maintained in the range of 150°C–180°C. The result of this reaction is the formation of titanyl sulfate. The formula for this compound is $TiOSO_4$. The reaction is shown in Equation 12.8.

$$FeTiO_3 \text{ (s)} + 2H_2SO_4 \text{ (aq.)} + 5H_2O \text{ (l)} = FeSO_4.7H_2O \text{ (s)} + TiOSO_4 \text{ (aq.)} \quad (12.8)$$

The solution of $TiOSO_4$ is separated from the precipitated hydrated ferrous sulfate by filtration.

The solution is diluted with water at 90°C. Then, titanyl hydroxide gets precipitated, as shown in Equation 12.9. The precipitate is filtered out and washed. It is then calcined at about 1000°C.

$$TiOSO_4 \text{ (aq.)} + 2H_2O \text{ (l)} = TiO(OH)_2 \text{ (s)} + H_2SO_4 \text{ (aq.)} \quad (12.9)$$

This produces TiO_2.

12.2.4 Nonoxide Ceramics

Most of the nonoxide ceramics are synthesized, as they do not occur in nature. They can be synthesized in two ways:

1. By combining the corresponding metal with the required nonmetal. This direct combination is carried out at high temperatures.
2. By reducing the corresponding oxide with carbon, which is called *carbothermal reduction*, and then by combining the reduced metal with the required nonmetal.

Important nonoxide ceramics are carbides, nitrides, and borides. Some representative ceramics from these are discussed in the following text.

12.2.4.1 SiC

Among the nonoxide ceramics, SiC (silicon carbide) is the most commonly used. It is mostly used as an abrasive, the reason being its high hardness. Only diamond, boron carbide, and nitride have higher hardness than silicon carbide. It is synthesized as it is not available in nature. Its two crystalline forms can be synthesized by varying the temperature. In the temperature range of 1400°C–1800°C, we get β-SiC. α-phase is formed at temperature greater than 2000°C. β-phase is cubic whereas α is hexagonal.

The commercial production of SiC is called the Acheson process. In this process, sand containing 99.5% SiO_2 is mixed with carbon in the form of coke. The mixing is carried out in a large elongated furnace, which is known as a mound. Carbon electrodes are kept at the opposite ends in the mound, which can accommodate up to 3000 t of the mixture. Current is then passed through the mixture. The coke in the mixture acts as a resistor. Its resistance raises the temperature to about 2200°C. During one run of the mound, about 7 TJ of energy is consumed. The wattage is 9000–10000 kW.

The reaction in the mound is given in Equation 12.10.

$$SiO_2 \text{ (s)} + C \text{ (s)} = SiC \text{ (s)} + 2CO \text{ (g)} \tag{12.10}$$

Depending on the size of the mound and the capacity of the transformer, the heating is continued for 2–20 days. As size of the mound increases, the number of days will increase. As the capacity of the transformer increases, the number of days will decrease. After cooling, the mound is broken up. The high-purity green hexagonal silicon carbide crystals are separated out from the core. As the distance from the core increases, the purity of the silicon carbide decreases. The purity can be determined from the color of the crystals. If the color is light green, the crystals are 99.8% pure; dark green color shows that they are 99% pure; and 98.5% purity is indicated by black color. Around the core, the outer layer consists of a mixture of silicon carbide (with a purity of ≥97.5%), unreacted silicon carbide, and coke. This mixture is used in the next batch.

12.2.4.2 TiC

TiC (titanium carbide) is also not available in nature. It is produced by one of these two methods:

1. Carbothermal reduction of titanium dioxide
2. Direct reaction between titanium and carbon

The reaction temperature is in the range of 2100°C–2300°C. The reactions are represented in Equations 12.11 and 12.12:

$$TiO_2 + C = TiC + CO \tag{12.11}$$

$$Ti + C = TiC \tag{12.12}$$

12.2.4.3 AlN

There are two ways by which AlN (aluminum nitride) is manufactured industrially. In one method, Al is directly combined with N_2. The reaction is represented by Equation 12.13.

$$Al \text{ (l)} + \tfrac{1}{2} N_2 \text{ (g)} = AlN \text{ (s)} \tag{12.13}$$

In this method, aluminum powder is melted, and, through the melt, nitrogen gas is passed.

In the second method, Al_2O_3 is reduced in the presence of N_2 and C. The reaction is represented by Equation 12.14. In this method, a mixture of Al_2O_3 and C is prepared.

$$2Al_2O_3 \text{ (s)} + 3C \text{ (s)} + 2N_2 = 4AlN \text{ (s)} + 3CO_2 \tag{12.14}$$

This mixture is heated to a temperature above 1400°C in N_2 atmosphere. To ensure that complete conversion takes place, the mixture is prepared by taking fine powders of the two starting powders, and the mixture is made uniform.

In the final product of both the processes, oxygen to the extent of ~1 wt% and carbon less than 0.07 wt% are present as impurities. AlN crucibles are used to contain molten metals and salts.

12.2.4.4 Si₃N₄

Si₃N₄ (silicon nitride) is also produced synthetically. It exhibits allotropy. The lower-temperature α form and the higher-temperature β form are the two polymorphic forms of Si₃N₄. When the α form transforms to the β form, the crystals are elongated. The following routes are used for the production of silicon nitride:

a. Direct nitridation of silicon

b. Application of carbothermic reduction to SiO_2 in the presence of nitrogen

c. Reaction of silicon tetrachloride or silane with ammonia in the vapor phase

All the processes produce silicon nitride powder. Most of the commercially available powder is produced by direct nitridation. This reaction is carried out in the temperature range of 1250°C–1400°C. The reaction is shown in Equation 12.15.

$$3Si\ (s) + 2N_2\ (g) = Si_3N_4\ (s) \tag{12.15}$$

The powder produced by this reaction contains both α and β forms in the ratio of 90:10, respectively. Seeds of α form are mixed with the Si powder if we want to reduce the β form. β form is an undesired phase. The impurities present in the nitrided powder are iron, calcium, and aluminum. The purity can be increased by producing silicon powder using carbothermic reduction. The reaction for this reaction is given by Equation 12.16.

$$3SiO_2\ (s) + 6C\ (s) + 2N_2\ (g) = Si_3N_4\ (s) + 6CO\ (g) \tag{12.16}$$

It is carried out from 1200°C to 1550°C. This powder will contain impurities in the form of C and O. The surface area of the powder is high. To accelerate the reaction, seeds in the form of α-Si₃N₄ are used.

Vapor-phase reactions also produce high-purity powder. The reactions are represented by Equations 12.17 through 12.19.

$$SiCl_4\ (g) + 6NH_3\ (g) = Si(NH)_2\ (s) + 4NH_4Cl\ (g) \tag{12.17}$$

$$3Si(NH)_2\ (s) = Si_3N_4\ (s) + 2NH_3\ (g) \tag{12.18}$$

$$3SiH_4\ (g) + 4NH_3\ (g) = Si_3N_4\ (s) + 12H_2\ (g) \tag{12.19}$$

The as-produced powders from these reactions are amorphous. They can be converted to crystalline α-form by heating to 1400°C.

Si_3N_4 is used in gas turbine and diesel engine applications. Here, its high hot strength and excellent thermal shock, creep, and oxidation resistance are exploited.

12.2.4.5 ZrB₂

ZrB_2 (zirconium diboride) possesses excellent corrosion resistance. This property makes it useful for making crucibles to contain molten metals. In aluminum production, it is used as the cathode material. Aluminum is produced by molten salt electrolysis.

As earlier for the production of carbides and nitrides, the processes can be either a direct combination of the elements or a carbothermic reduction. The direct combination is given by Equation 12.20.

$$Zr\ (s) + 2B\ (s) = ZrB_2\ (s) \tag{12.20}$$

For a carbothermic reduction, either boron carbide (B_4C) or boron trioxide can be used with zirconium dioxide. Equations 12.21 and 12.22 show the reactions involved.

$$2ZrO_2 + C + B_4C = 2ZrB_2 + 2CO_2 \tag{12.21}$$

$$2ZrO_2 + 5C + 2B_2O_3 = 2ZrB_2 + 5CO_2 \tag{12.22}$$

High temperatures are involved in both the reactions. The atmospheres can be inert or vacuum.

12.2.4.6 WC

WC (tungsten carbide) is used in the metalworking, mining, and construction industries. In these, its wear resistance property is utilized. Machine parts and dies contain this material when they are required to be exposed to severe service conditions. The production method involves direct reaction between W powder and carbon. The reaction is given by Equation 12.23:

$$W + C = WC \tag{12.23}$$

12.2.4.7 Graphite

Graphite is a crystalline form of carbon. In contrast to the earlier nonoxide ceramics, this material is found in nature. It exists in marble, which is a metamorphic rock. It is mined by both opencast and underground minings.

Out of several methods of producing graphite, two are worth mentioning. The first and most common method is to heat nongraphitic carbons to temperature above 2500°C. For example, petroleum coke and coal tar pitch are mixed, and this mixture is calcined and heated to 3000°C. This will result in high-purity graphite. The high temperature involved causes the carbon atoms to regroup to form the graphite crystal structure.

The second method is based on chemical vapor deposition. Here, hydrocarbons are subjected to this process. It is carried out at temperature in the vicinity of 1800°C. At this temperature, the hydrocarbons decompose following Equation 12.24.

$$C_nH_m = nC + (m/2)H_2 \qquad (12.24)$$

Carbon deposits in the form of graphite since, at the temperature involved, graphite is the stable form of carbon. The gaseous hydrogen escapes from the reaction chamber.

The major applications of graphite are in the manufacture of electrodes, as lubricants, and for increasing carbon content of steel to make them medium- or high-carbon varieties. The graphite electrodes are used in arc furnaces. These furnaces produce steels by melting and refining. Battery electrodes are also made from graphite. Synthetic graphite is used in replacement heart valves.

Natural graphite is largely used as refractories. The use amounts to 45% of the total produced. The next major use is in brake linings, where the usage share is 20%.

12.3 Powder Production

All the methods available for the production of ceramic powders can be grouped into three. They are:

1. Mechanical
2. Chemical
3. Vapor phase

Generally, coarse-grained, naturally occurring minerals are the raw materials for mechanical methods. The particle size is brought down by different equipments to get the final size. When economy is needed and purity is not the primary concern, these methods are used. Traditional ceramic products are generally made from the powders obtained from these methods.

Chemical methods give rise to pure ceramic powders. Also, the various powder characteristics can be accurately controlled. Hence, these methods are used for making advanced ceramic products.

Vapor-phase methods are useful for the production of nonoxide and nano-size ceramic powders.

12.3.1 Milling

Milling is a mechanical method for the production of ceramic powders. Reduction of particle size is beneficial for sintering, which depends on the diffusion of atoms and molecules. This diffusion is inversely proportional to the square of the particle size.

There are many milling methods available. Out of these, the most common method is *ball milling*. In this, the balls constitute the milling media. These balls partially fill a barrel. The material of construction for the barrel and balls is the same. The hardness of the material should be equal or greater than the hardness of the mineral that will be milled in it. The barrel is rotated horizontally during the operation. During the rotation of the barrel, the balls undergo a cascading action. This is shown in Figure 12.1. The cascading action gives rise to shearing and crushing forces on the mineral grains.

The size of the balls depends on the final particle size required. The size ranges from 8 cm to 0.6 cm. As the size decreases, finer particles are obtained. Ball milling is a simple and inexpensive process. The media and containers can add impurities to the powder if they are of different materials. Only 2% of the total energy input goes into the reduction of particle size.

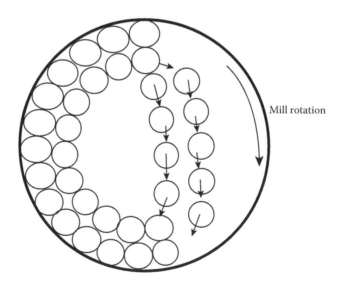

Mill rotation

FIGURE 12.1
Cascading action in ball milling.

Apart from the ball mill, there are other types of mills used for mechanical comminution—*fluid energy mill, vibratory mill,* and *attrition mill.*

The other name for the fluid energy mill is *jet mill*. In this, high-velocity fluid causes particle–particle impact. Usually, compressed air or superheated steam is used as the fluid. The particles are carried by the fluid into the milling chamber. The particle–particle impact is enhanced by the proper design of the chamber. The velocity with which the particles impact is near the velocity of sound. The contamination of the powder can be reduced by coating the milling chamber with a polymer. The particle size can be controlled by adjusting the fluid velocity. The size distribution can be made narrow by adjusting the velocity to become constant. Table 12.1 gives the use of fluid energy mills in the production of some powders.

From the milling chamber, the powders are to be separated from the fluid stream.

A drum containing the milling media is used in a vibratory mill. After loading the drum with the powder, the drum is vigorously vibrated. The collisions between media are more violent than those in the ball mill. This reduces the milling time. To reduce the contamination, polymer balls can be used. Any contamination can be burnt off during the heating step in the further processing of the powder.

There is a difference between *attrition milling*, sometimes called *agitated ball milling*, and ball milling. The difference is that, here, the milling chamber is not rotated. The input charge into the chamber contains a mixture of the slurry of the powder and the milling media. This mixture is continuously agitated at frequencies in the range of 1–10 Hz. Figure 12.2 shows a sketch of the mill. The chamber can also be placed horizontally. The agitator is located at the center of the chamber. Small spheres of the size of 0.2–5 mm constitute the milling media. It fills 60–90% of the inside volume of the milling chamber. The mill can be continuously operated. In this case, the feed is given from one end of the chamber. The product is collected from the other end. The efficiency of this mill is greater than the previous two mills. The solid content in the slurry can also be higher. The rapidity of

TABLE 12.1

Application of Fluid Energy Mill

Material	Mill Diameter, mm	Fluid	Material Feed Rate, kg/h	Average Particle Size, μm
Al_2O_3	203	Air	6.8	3
TiO_2	762	Steam	1020	<1
TiO_2	1067	Steam	1820	<1
MgO	203	Air	6.8	5
Dolomite	914	Steam	1090	<44
Fe_2O_3	762	Steam	450	2–3

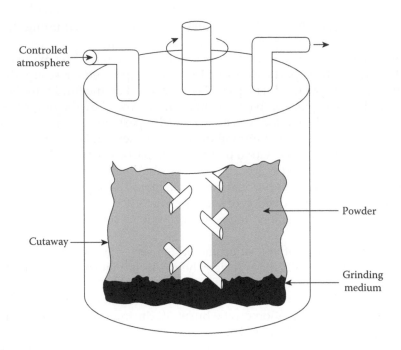

FIGURE 12.2
Sketch of an attrition mill.

the milling reduces the contamination. Further reduction of contamination can be brought about by lining the chamber with polymers or ceramics, and by the use of a ceramic agitator.

12.3.2 Spray Drying

Spray drying is another mechanical way of producing ceramic powders. Here, the raw material is dissolved in a suitable solvent. Ferrites, titanates, and other electrical ceramics are produced by this method. The prepared solution is atomized into a drying chamber. Figure 12.3 shows a centrifugal atomizer with co-current airflow. It is a rotary atomizer, and Figure 12.4 shows a nozzle atomizer. In ultrasonic atomization, the solution is passed over a rapidly vibrating piezo-electric membrane. The membrane has the frequency of ultrasonic waves. The droplet sizes range from 10 μm to 100 μm. The flow pattern of hot air in the drying chamber determines the moisture removal rate because the temperature of the particles is determined by this flow pattern. Once the drying is completed, the particles are collected by a stream of air passed through the chamber. This stream deposits the particles in a collector. When the particle size is below 0.1 μm, they get agglomerated.

Powder particle characteristics can be controlled by controlling the various parameters of the spray-drying process—the droplet size; solution

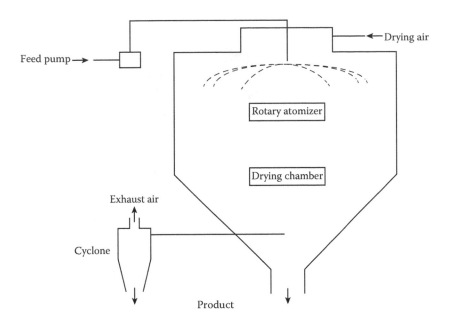

FIGURE 12.3
Centrifugal atomizer with co-current airflow.

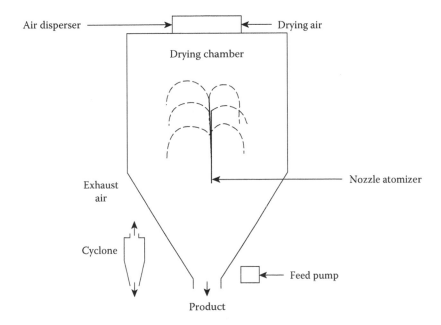

FIGURE 12.4
Nozzle atomizer using mixed-flow conditions.

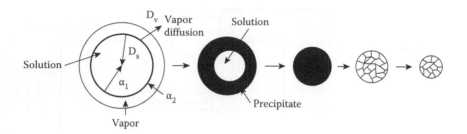

FIGURE 12.5
Stages in spray pyrolysis process.

concentration and composition; temperature and flow pattern of the air in the drying chamber; and the design of the chamber. Nitrates and acetates possess low decomposition temperatures. Decomposition will be needed when the dried product is not the final composition required. For example, if the final product required is oxide, it can be produced from its nitrate by decomposing it with the help of heat in the absence of air. This kind of decomposition is called *calcination*. Since nitrates and acetates can decompose at low temperatures, they are ideal for small-scale operations. Chlorides and oxychlorides easily dissolve in water. They are the ones, therefore, used in industries. The production rates here goes up to many hundreds of kilograms per hour. When the ceramic cannot be dissolved in water, other solvents can be used. For alkoxides, for example, alcohol can be used. By calcination, extremely fine particles can be produced.

Spray pyrolysis is a variation of the spray-drying process. In this process, the temperature is higher. Also, the environment in the chamber will be oxidizing. The admitted spray will be dried and the deposited compound will be directly decomposed in the chamber. The various stages of the process are shown in Figure 12.5. As the drying proceeds from the surface of the droplets, once a layer of solid formed, the remaining liquid inside the layer has to diffuse through it. Once the drying is over, the decomposition stage starts. After the completion of the decomposition, the size of the resulting particle becomes very small.

12.3.3 Sol-Gel Processing

Sol-gel processing is a chemical method. It is a more expensive process compared to the earlier two processes. The advantage of this process is that it produces powder with more surface area. This makes the powder sintered at a lower temperature, thereby compensating for some of the costs involved in the production of the powder.

The reactants in most sol-gel processes are the corresponding metal alkoxy compounds. Alkoxides are produced by reacting the metal in the required ceramic with alcohol. The general reaction can be represented as given in Equation 12.25.

$$mROH + Me = (RO)_nMe + (m/2)H_2 \qquad (12.25)$$

Here, R is an organic group. For example, if the alcohol is propanol, R is C_3H_7. Reaction rates are increased by the use of catalysts. In order to prepare aluminum isopropoxide, mercuric chloride will be needed. This is because there will always be an oxide layer on aluminum. This aluminum oxide layer prevents any reaction with the environment. The chloride ions in mercuric chloride attack and remove the oxide layer on the aluminum. The reaction with alcohol now becomes possible. The reaction is further enhanced by increasing the temperature to 80°C.

After forming the alkoxide, it is dissolved in an alcohol. This solution is called the sol. To this sol, water is added. Then the following reactions take place:

$$(RO)_2–Me–OR + H(OH) = (RO)_2–Me–OH + (OH)R \qquad (12.26)$$

$$(RO)_2–Me–OH + (RO)_2–Me–OR = (RO)_2–Me–O–Me–(RO)_2 + ROH \qquad (12.27)$$

The reaction shown in Equation 12.26 is a hydrolysis reaction, whereas the next one is a condensation reaction. The condensation reaction gives rise to a three-dimensional network structure by the continuous formation of the metal–oxygen bond in three directions. The suspension of the resulting alkoxide in alcohol is the gel. There is uniform mixing of alkoxide in alcohol at the molecular level. The resulting powder, therefore, is chemically homogeneous. The rate of the hydrolysis reaction can be increased by adding an acid, and that of the condensation reaction by adding a base. The gel is dried and calcined. The powders produced are amorphous. In order to convert the amorphous solid prepared from it, a crystallization step is carried out after sintering of the green body.

12.3.4 Precipitation

Precipitation depends on the principle that the dissolution of a solute in a solvent is greater at higher temperatures than that at room temperature. Figure 12.6 is a solubility-versus-temperature plot for many solutes. In this plot, all salts, except cerium sulfate, show increased solubility as temperature increases. The reason for the decreased solubility for cerium sulfate is that this salt has a negative heat of solution. Higher temperatures also cause faster dissolution. When the saturated solution is cooled to room temperature, it becomes supersaturated, and the excess solute gets precipitated out of the solution. When this precipitate is separated from the solution and dried, we get it in the form of a powder. To produce a saturated solution, we can

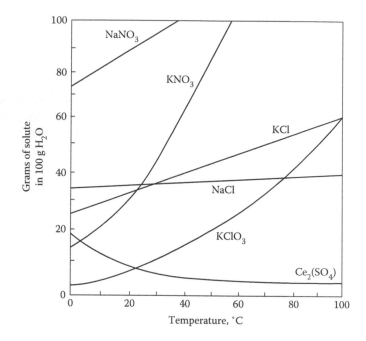

FIGURE 12.6
Solubility versus temperature for several ionic compounds.

also change the pH of the solvent. If decreasing pH causes more solute to dissolve, then increasing it causes precipitation.

When the concentration of the solution exceeds the threshold concentration for homogeneous nucleation, there will be a burst of tiny nuclei formation. When these numerous nuclei start growing, the concentration of the solution goes on decreasing after falling below the threshold concentration. This variation of concentration with time is shown in Figure 12.7. This diagram is the result of the work of LaMer and Dinegar, and is referred to as the *LaMer diagram*. Homogeneous nucleation takes place uniformly throughout the solution. This gives rise to uniform powder particles. If there is some heterogeneity in the solution, such as the presence of a dirt, then heterogeneous nucleation takes place, and the particles formed will be of different sizes.

The precipitation method can also be used to produce mixed powders. For example, for the production of ferrites, we require a divalent and a trivalent oxide. Nickel ferrite is an example for a magnetic ferrite. This is used to make memory chips in computers. In this ferrite, nickel oxide is the divalent oxide, and iron oxide, the trivalent one. To produce a mixture of these two oxides in the required proportion, we can take the sulfate salts of iron and nickel. These two compounds are mixed in the same proportion as that is needed in the ferrite—that is, a 1:1 mole ratio. They are then

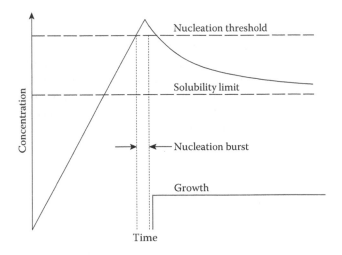

FIGURE 12.7
Concentration versus time for a solution in which the concentration is first increased to the point of nucleation and then declines as the precipitate grows.

dissolved in water by heating the water to 80°C. The salts are precipitated by adding ammonium hydroxide to take the pH to 11. This causes the precipitation of the salts in the form of hydroxides. The mixed precipitate is filtered and washed to remove any sulfate content. Drying this precipitate results in a mixed powder having a particle size in the range of 50 nm–1 μ. The powder is then analyzed for its composition. If any of the salts is less than the required percentage, it is made up by adding the corresponding pure powder.

12.3.5 Production of Nonoxide Powders

Nonoxides such as carbides and nitrides do not occur in nature. They are to be produced by chemical methods. SiC is produced, for example, by reacting SiO_2 with carbon in the form of coke. The reaction is represented by Equation 12.28.

$$SiO_2 \text{ (s)} + 3C \text{ (s)} = SiC \text{ (s)} + 2CO \text{ (g)} \qquad (12.28)$$

By taking both SiO_2 and coke in a powder form, the resulting SiC will also be in the form of a powder. The chemical reaction can also be a direct combination of the reactants. For example, Al and N_2 can be directly combined to form AlN. The reaction is represented by Equation 12.29.

$$Al \text{ (l)} + \tfrac{1}{2} N_2 \text{ (g)} = AlN \text{ (s)} \qquad (12.29)$$

The resulting solid will have to be mechanically comminuted to produce the powder.

The next type of chemical method to produce nonoxide powders is the liquid-phase reaction. An example for this method is the production of silicon nitride powder. For this, silicon tetrachloride and liquid ammonia are reacted. Silicon tetrachloride is liquid at room temperature. The reaction gives rise to silicon diimide. After this reaction, the obtained silicon diimide is decomposed to get the silicon nitride powder. Equation 12.30 shows the decomposition reaction.

$$3Si(NH)_2 = Si_3N_4 + 2NH_3 \qquad (12.30)$$

This decomposition reaction, which is a calcination process, is controlled to get the powder with the desired properties. By carrying out the process at low temperatures, we can produce fine-grained equiaxed powder particles. If the calcination temperature is above 1500°C, we get needle-like and coarse-grained hexagonal particles.

12.3.6 Production of Nanopowders

Production of nanopowders can be carried out by vapor-phase reactions. These reactions are more versatile, in the sense that we can produce both oxide and nonoxide powders. The powder produced is of high purity, it contains discrete particles, and its size distribution is narrow. Since the process is expensive, it is not used for the commercial production of powders.

One method of producing nanopowders by the vapor-phase reactions is by the use of a gas condensation chamber. The process is schematically shown in Figure 12.8. The material to be produced in the powder form is evaporated from two sources. The vapors produced are conducted to a tube surface through two funnels. Through this tube, liquid nitrogen is passed. It is called the liquid nitrogen cold finger. The vapors get condensed on the surface of this cold finger as powder particles. These particles are scrapped off later on, and are collected from the bottom of the gas chamber into a compactor. The compactor shapes the powder into a solid compact. Using this method, titanium dioxide powder has been produced with particle size in the range of 10–15 nm.

Another method of producing ceramic nanopowders is by the use of a plasma reactor. It is shown schematically in Figure 12.9. It consists of two parallel electrodes. The upper one is connected to an RF power source, which has a frequency of about 14 MHz. The other electrode is grounded. The reactor is filled with argon gas with a negative pressure. Therefore, between the electrodes, argon plasma is generated. The reactants in the form of gases are admitted into the reactor after the production of plasma. Under the action of the argon plasma, they get decomposed into atoms, ions, and electrons. Later, during their travel down the reactor, they recombine in the form of minute ceramic particles. The particle size of these particles will be less than 20 nm.

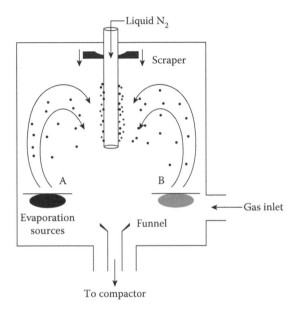

FIGURE 12.8
Schematic of a gas condensation chamber for nanoparticle synthesis.

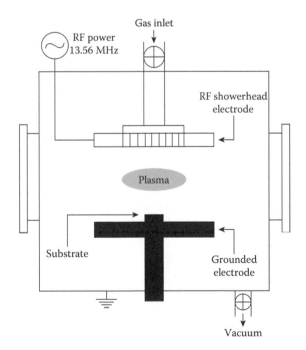

FIGURE 12.9
Schematic of a plasma reactor.

12.4 Powder Characterization

The important powder characteristics are particle size and its distribution. There are many methods to determine these characteristics. Each of these methods can analyze powders of a particular size range. The methods are listed in Table 12.2. The selection of a particular method depends primarily on the size range of the powder sample. When the size ranges overlap, the next consideration is the amount of sample. If the sample size is also the same, then comes the availability of the equipment. In case the required instrument is not available, then comes the cost of the instrument. The least costly instrument will be the choice. Finally, the choice of the instrument will also depend on the ease of operation of the instrument and its maintenance.

12.4.1 Microscopy

The use of microscope was discussed in connection with ceramic microstructures. Microscopy for particle characterization is a direct method. We can see the image of the particles directly and measure their size. When the size of the particle is greater than a micron, optical microscope can be used. The technique is called *visible light microscopy* (VLM).

In the first method by VLM, the total length of intercepts made by the particles on a line is found. This total length divided by the number of particles gives the average particle size. The measurements are to be carried out at least 15 times by focusing different areas of the slide on which the particles have been spread. A statistical analysis can be carried out to get the mean size of the particles and their size distribution. The tedious manual operation can be avoided by the use of a computer, by which image analysis can be carried out. The particle size distribution can be determined by plotting a histogram of measured sizes of the particles versus the number of particles having the same size.

TABLE 12.2

Methods for the Determination of Particle Size and Size Distribution

Method	Medium	Size Range (μ)	Weight (g)
Light microscopy	Air/oil	400–0.2	<1
Electron microscopy	Vacuum	20–0.002	<1
Sieving	Air	8,000–37	50
Gravity sedimentation	Liquid	100–0.2	<5
Centrifuging	Liquid	100–0.2	<1
Coulter counter	Liquid	400–0.3	<1
Fraunhofer scattering	Liquid/gas	1,800–1	<5

When the particles are of submicron size, electron microscopes are to be used. The larger sizes are measured by scanning electron microscopes, whereas the finer sizes are measured by transmission electron microscopes. The amount of sample required is very small. Hence, care should be taken to see that this small amount should represent the bulk powder mass.

12.4.2 Sieving

Sieving is used to separate the powder into fractions based on particle size. The sieves are arranged from top to bottom in the order of decreasing size of their apertures. At the bottom of this arrangement is a pan to collect the powders passing through all the sieves. This arrangement constitutes the sieve set. Table 12.3 shows the sieve numbers with their corresponding aperture sizes. A sample size of 50 g is put into the topmost sieve of the sieve set. The set is shaken for 30 min after keeping it on a sieve shaker. After that, each sieve is taken and the content in it is weighed. The mass fractions are then calculated, and they are plotted against the aperture size of the sieves. The resulting curve is called the fractional weight curve, and it gives an idea about the particle size distribution.

TABLE 12.3

Aperture Sizes of US Standard Sieves

Sieve Number	Aperture (μ)	Sieve Number	Aperture (μ)
3.5	5,660	60	250
4	4,760	70	210
5	4,000	80	177
6	3,360	100	149
7	2,830	120	125
8	2,380	140	105
10	2,000	170	88
12	1,680	200	74
14	1,410	230	63
16	1,190	270	53
18	1,000	325	44
20	841	400	37
25	707	600	30
30	595	1,200	15
35	500	1,800	9
40	420	3,000	6
45	354	8,000	3
50	297	14,000	1

12.4.3 Sedimentation

A particle of diameter d, when it is dropped into a liquid, attains a constant velocity under the action of two forces. This velocity is called the terminal velocity of the particle. They are the gravitational force, which is determined by the weight of the particle, and the upward thrust of the liquid, as a result of the equivalent volume of the liquid displaced by the particle. These two forces are represented by Figure 12.10. The terminal velocity is given by Stokes' equation, as given in Equation 12.31:

$$v = d(\rho_s - \rho_l)g/18\eta \tag{12.31}$$

Here, ρ_s is the density of the particle; ρ_l, that of the liquid; g, the acceleration due to gravity; and η, the viscosity of the liquid. The diameter d is calculated after measuring the terminal velocity.

Gravity sedimentation is useful for particle sizes in the range 0.2–100 μ. When size becomes less than this, that is, of the order of 10^{-2}, centrifuge

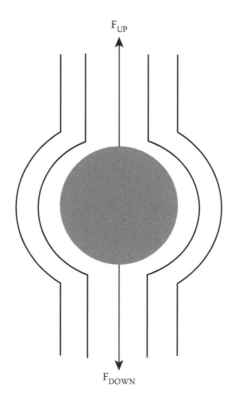

FIGURE 12.10
Illustration of the force balance during the settling of a particle in a Newtonian fluid with laminar flow.

sedimentation is employed. The terminal velocity should correspond to streamlined flow. Streamlined or laminar flow is decided by the Reynold's number. This number is given by Equation 12.32.

$$Re = v\rho_l d/\eta \qquad (12.32)$$

If Re is less than 0.2, then the flow is laminar. If the particles are of almost the same size, then the measurement of the terminal velocity is carried out in the following manner. A suspension of the particles is prepared in a long flask, called the *sedimentation flask*. Once the suspension becomes stagnant, a clear plane separating the suspension and a clear liquid will become apparant. The velocity of this plane will correspond to the terminal velocity of the particles. The time taken by this plane for moving from one graduation on the flask to another is found by using a stopwatch. The distance between the two graduations divided by the time gives the terminal velocity.

When the particles have a range of sizes, then the turbidity of the suspension is measured at some height of the sedimentation flask. This is done by measuring the intensity of an incident beam of light and its transmitted intensity. The ratio between these two intensities is given by Beer–Lambert law, which can be stated as given in Equation 12.33.

$$I/I_0 = \exp(-KAcx) \qquad (12.33)$$

In this equation, I is the intensity of the transmitted light, I_0 is the incident light intensity, K is called the extinction coefficient, A is the projected area per unit mass of the particles, c is the mass of the particles representing the concentration of the particles, and x is the distance of travel by light in the suspension.

If the mass of particles contains two sizes, then two transmitted lights will be detected, when light passes through the mass. Let us say that L is the distance from the top surface of the suspension at which the light source is located. The plot of turbidity versus time is shown in Figure 12.11. The turbidity is expressed in terms of nephelometric turbidity units (NTu). The instrument nephelometer measures the turbidity in terms of the reflected light from the passing particles. The time of travel is obtained from the plot and is used to calculate the terminal velocity.

12.4.4 Coulter Counter

Coulter counter measures the number and size of particles as they are caused to flow through an orifice. For this to happen, again a suspension of the particles is prepared. A tube containing a narrow orifice on its side near its bottom is kept standing in the suspension. One electrode is kept inside the tube,

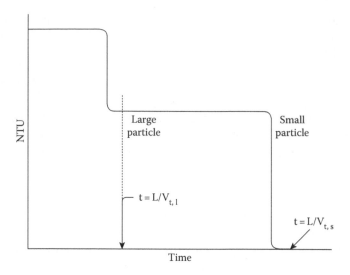

FIGURE 12.11
Result of sedimentation measurements using turbidity for two particle sizes in a solution:
V_t is terminal velocity.

and another is kept outside it. These are connected to a constant voltage source. When a particle passes through the orifice, it causes a change in the resistance of the electric circuit. This change is a measure of the size of the particle. Figure 12.12 shows the setup and a plot showing the change in the resistance of the circuit when two particles A and B passed through the orifice.

12.4.5 Fraunhofer Scattering

Fraunhofer scattering takes place in the case of particles that are larger than the wavelength of incident light. The intensity and angle of diffraction (scattering) depend on the size of the particle. This dependence is shown in Figure 12.13. In this figure, the diffractions caused by two particles differing in size are compared. A larger particle diffracts the incident light with a smaller diffraction angle, but the intensity of the diffracted beam is greater. Thus, by measuring diffraction angle or intensity, the particle size can be determined. The intensity is proportional to d^2, where d is the diameter of the particle. The relation between the particle size and diffraction angle is given in Equation 12.34.

$$\sin \theta = 1.22 \, \lambda/d \tag{12.34}$$

In the case of He–Ne laser, the wavelength value is 0.63 μ. Using this, particle sizes in the range of 2–100 μ can be measured. The measurement by this technique gives rise to good accuracy and speed. The sample size needed is <5g. The method can be automated.

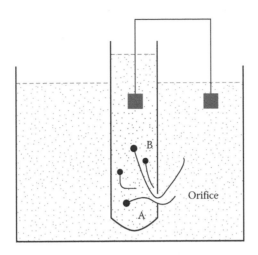

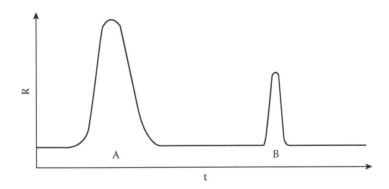

FIGURE 12.12
Results of Coulter counter measurements for two particle sizes A and B; R is the resistance between the electrodes, shown as shaded squares.

12.4.6 x-Ray Diffraction

The width of the peaks obtained when x-rays are used to diffract them by particles depends on the particle size. The relation between the particle size and width is given by Equation 12.35.

$$d = 0.9\lambda/\beta \cos \theta \tag{12.35}$$

In this equation, d is the particle size, λ is the wavelength of x-rays, β is the width of the diffraction peak, and θ is the Bragg angle. This equation is called the Scherrer equation. The dependence of particle size on the peak width can be gauged from Figure 12.14. Here, $ZrO_2/3mol\%Y_2O_3$

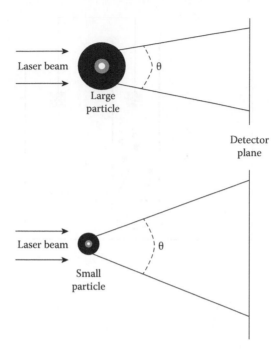

FIGURE 12.13
Scattering of light by large and small particles.

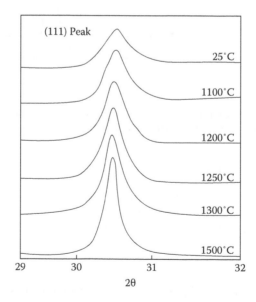

FIGURE 12.14
Illustration of x-ray line broadening for a ZrO_2/3 mol% Y_2O_3 powder prepared by hydrothermal synthesis.

powder diffraction peaks from {111} planes are shown. The temperatures mentioned are the calcination temperatures. As this temperature increases, the particle sizes become more coarsened, with the result that the peaks become narrower. The broadening of the peaks is a function of two factors. One is the broadening due to the instrument, and the other is that due to the particle size. When the size becomes greater than about 0.1 μ, the peaks are so narrow that the instrumental broadening and particle size broadening are indistinguishable. Hence, x-ray diffraction is useful for sizes less than 0.1 μ.

12.4.7 Surface Area

Surface area is another important powder characteristic. It is important in sintering. Sintering depends on the diffusion of atoms across the interparticle boundary. As the particle surface area increases, interparticle boundary increases, and thereby sintering is enhanced. From surface area, the particle size can also be calculated. The surface area can be measured by adsorbing a monomolecular layer of a gas. This kind of adsorption takes place at subzero temperatures. The pressure of gas before and after adsorption is measured. The difference in pressures is related to the mass of the gas adsorbed. From this mass, the surface area is calculated.

The preceding principle is utilized in the Brunauer, Emmet, and Teller (BET) method. This method is used to find the particle size by finding the surface area. In this method, the experiment is conducted at liquid nitrogen temperature. The gas used for adsorption is also nitrogen. The BET equation is used to calculate the volume of the nitrogen adsorbed V_a. This is given by Equation 12.36.

$$P/[V_a(P_0 - P)] = (V_m C)^{-1} + (C - 1)P/[P_0 V_m C] \qquad (12.36)$$

In this equation, P is the pressure of the gas at which the adsorption takes place, P_0 is the saturation vapor pressure for the adsorbate at the adsorption temperature, V_m is the adsorbate volume per unit mass of solid for monolayer coverage at relative pressure P/P_0, and C is the BET constant.

During the experiment, the P/P_0 is maintained in the range of 0.05–0.2. One nitrogen molecule requires 0.162 nm^2 area for covering. Therefore, from the volume of adsorbed nitrogen molecules, the total surface area of the sample can be calculated. The value is expressed in m^2/g.

A plot of $P/V_a(P_0 - P)$ versus P/P_0 gives a straight line. From this straight line, V_m and C are determined. The value of V_m is used to calculate the specific surface area, S, from Equation 12.37.

$$S = N_A \sigma V_m / V' \qquad (12.37)$$

Here, N_A is the Avogadro's number, V' is the molar volume, and $\sigma =$ 0.162 nm^2 for nitrogen molecule.

In the case of a spherical particle, its radius can be calculated from Equation 12.38.

$$R = 3/\rho S \tag{12.38}$$

In this equation, r is the radius of the particle, and ρ is its density.

12.4.8 Particle Shape

The particle shape influences factors such as the packing of powder, compacting and sintering characteristics, and the mechanical strength of the sintered product. Irregularly shaped particles have reduced apparent density and flow rate. They have good compacting and sintering characteristics. Spherical particles have the maximum flow rate, but reduced compacting characteristics.

Particle shape is observed by optical or electron microscopes, depending on the size of the particles. The shape is expressed quantitatively by the shape factor. Shape factor is calculated from the knowledge of the length and breadth of the particles, or surface area. The ratio of length to breadth or surface area to particle size gives the shape factor.

12.4.9 Particle Composition

Another important powder characteristic to be determined is its composition. The importance of composition analysis can be gauged by considering the production of red lead. Red lead has the formula Pb_3O_4. On its production, it may contain free lead and litharge. Litharge has the formula PbO. These impurities are to be kept within a limit in red lead. Otherwise the throughput from the furnace will be affected. The impurities are maintained within the limit by controlling the furnace temperature. To fix the temperature, the output from the furnace should be analyzed for impurities. This analysis is carried out on an hourly basis in the factory. Industrially, the analysis is done by either the gravimetric or volumetric method. These methods are simple to perform.

Apart from the common wet chemistry, based on precipitation and titration, which are the ways of doing gravimetric and volumetric analyses, respectively, there are many other sophisticated techniques for analyzing the composition. These are enumerated in Table 12.4. The selection of a method depends on many factors. First, it depends on the type of the material. If the sample requires to be dissolved in a solvent, then we have to find a solvent for it. Some techniques can also use agglomerated particles. The size of the sample will decide the type of the instrument that can be used. Some impurities will not be detected by some instruments. The selection

TABLE 12.4

Sophisticated Techniques for Composition Analysis

Technique	Sample Description	Application
Atomic emission spectroscopy	5 mg powder	Elemental analysis up to ppm level
Flame emission spectroscopy	Solution sample	Alkali elements and Ba to ppm level
Atomic absorption spectroscopy	Solution sample	Elemental analysis up to ppm level
X-ray fluorescence	Solid/liquid	Elements, Z>11, up to 10 ppm
Gas chromatography/mass spectrometry	Vapors and gases	Identification of compounds
Infrared spectroscopy	mg powder dispersed in transparent liquid	Identification and structure of organic and inorganic compounds
X-ray diffraction	Powder	Identification and structure of crystalline phases, up to 1%
Nuclear magnetic resonance	5 mg for H, 50 for C	Identification and structure of organic and inorganic compounds

of an instrument will also be dependent on the level of impurity in the sample. Also, the selected instrument should be available. If it is not available, either another facility will have to be approached, or it will have to be purchased. For purchase, the cost will have to be considered. From Table 12.4, we can infer the following:

1. X-ray fluorescence cannot be used for analyzing elements having atomic numbers less than 12.
2. Flame emission spectroscopy is useful if the alkali elements are present at ppm levels.
3. Nuclear magnetic resonance can be used for hydrogen analysis, though it is expensive.
4. For phase analysis in terms of identification and quantity, x-ray diffraction is useful.

12.5 Batch Determination

For forming, batches of the ceramic powder and processing additives are used. Processing additives are added to produce the particle dispersion and flow behavior required for forming. The various additives belong to the following classes:

a. Liquid solvent
b. Surfactant

c. Deflocculant

d. Coagulant

e. Flocculant

f. Plasticizer

g. Foaming agent

h. Antifoaming agent

i. Lubricant

j. Bactericide/fungicide

Except the solvent, all other additives get eliminated during the processing, especially during the firing stage. Liquids are used to wet ceramic particles, to provide a viscous medium, and to dissolve salts, compounds, and polymeric substances. Water is the principal liquid used. For materials that react with water, nonaqueous liquids are used. Examples for such liquids are trichloroethylene, alcohols, ketones, and refined petroleum oil or liquid wax.

A surfactant is added to reduce the surface tension of the liquid solvent. This improves the wetting of the ceramic particles, and their dispersion. Its molecule has got a polar and a nonpolar end. It can be an ionic or an anionic end. Examples for these types of surfactants are shown in Figure 12.15.

Deflocculants are additives that prevent the close approach of particles. They adsorb on the particles and increase the repulsive forces. Coagulants do the opposite function. They reduce the repulsive forces or steric hindrance and promote particle agglomeration. When the agglomeration is produced by bridging action, the substances used are called *flocculants*, known

$$CH_3 - (CH_2)_{11} - O - SO_3^- Na^+$$

Hydrophobic   Hydrophilic

Sodium dodecyl sulphate ionic surfactant

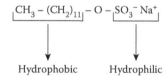

Octyl phenoxy polyethoxy ethanol nonionic surfactant

FIGURE 12.15
Examples of ionic and anionic surfactants.

more commonly called as *binders*. They improve the green strength of the ceramic product.

Plasticizers are added during ceramic processing to modify the visco-elastic properties of a condensed binder-phase film on the particles. When processing is done after making slurry, there will be a tendency for the formation of bubbles during the formation of the slurry. This tendency is reduced by adding an antifoaming agent. A foaming agent does the opposite. It increases the stability of the bubbles. A lubricant helps in reducing the friction between the ceramic particles and the dies in which they are formed. Bactericides and fungicides are added as preservatives to prevent the enzymatic degradation of the binders.

Different types of particle size distributions are used for forming ceramics. Batch composition gives the proportions of the constituents. Many parameters determine the consistency of a batch. They are:

a. The amount, distribution, and properties of the liquid phase
b. The amount, sizes, and packing of the particles
c. The amount, distribution, and types of additives adsorbed on the surface of the particles
d. The interparticle forces between the particles

There are five consistency states. They are:

1. Bulky powder
2. Agglomerates
3. Plastic body
4. Paste
5. Slurry

Batch compositions are determined by experiments and are finally given in the form of a table. Example for such a batch composition is given in Table 12.5.

TABLE 12.5

Batch Composition of Slurry for Spray Drying

Component	Weight (%)	Weight (g)
Alumina powder	73.22	2994
Water	23.23	950
Polyacrylate deflocculant	2.45	100
Polyvinyl alcohol binder	0.98	40
Ethylene glycol plasticizer	0.12	5

Problems

12.1 Assuming that Figure 12.14 was recorded using Cu–Kα radiation, plot the change in particle size as a function of annealing temperature.

12.2 The calcination of $CaCO_3$ is described by the following equation:

$$CaCO_3(s) = CaO(s) + CO_2(g)$$

The standard free energy of this reaction is given by:

$$\Delta G^0 = 182.50 - 56.16T$$

Here, T is the absolute temperature. The partial pressure of CO_2 in air is 3×10^{-3} atm.

If the calcination is carried out in air, at what temperature will $CaCO_3$ decompose?

12.3 The critical speed of rotation for a ball mill, defined as the speed required to take the balls just to the apex of revolution, is equal to $(g/a)^{1/2}/2\pi$ revolutions per second, where a is the radius of the mill, and g is the acceleration due to gravity. Determine the rotation speed for a ball mill with a radius of 5 cm that is operating at 75% of the critical rotation speed.

12.4 Al_2O_3 spheres of 1 micron are surrounded by excess ZnO powder to form the zinc aluminate. It is found that 25% of the Al_2O_3 is isothermally reacted to form $ZnAl_2O_4$ during the first 30 min. Determine how long it will take for all the Al_2O_3 to be reacted on the basis of the Jander equation.

12.5 Calculate the volume of HCl gas when producing 1 kg of Si_3N_4 by a vapor-phase reaction

The vapor-phase reaction is given by:

$$3SiCl_4\ (g) + 4NH_3\ (g) = Si_3N_4\ (g) + 12HCl\ (g)$$

Bibliography

C. B. Carter and M. G. Norton, *Ceramic Materials—Science and Engineering*, Springer, New York, 2007, pp. 345–356, 359–370, 378.

M. N. Rahaman, *Ceramic Processing and Sintering*, 2nd Edn., Taylor & Francis Group, LLC, New York, 2003, pp. 118–120.

J. S. Reed, *Principles of Ceramics Processing*, 2nd Edn., John Wiley, New York, 1995, pp. 65, 135, 137, 140, 142, 150, 172–173, 201, 213, 231–232, 238.

A. K. Sinha, *Powder Metallurgy*, Dhanpat Rai & Sons, New Delhi, 1982, pp. 29–30.

13

Forming Processes

13.1 Introduction

Forming processes give shape to the product. The powders are the starting materials for shaping a ceramic product. The powders produced by different methods illustrated in Chapter 12 are first given preconsolidation treatment. They are subjected to batching. Each batch is then taken for shaping. Again, there are different shaping methods. The selection of a particular method depends on many factors. They are:

a. The size and dimensional tolerances of the product
b. The requisite microstructural characteristics
c. The levels of reproducibility required
d. Economic considerations
e. The required shape

Shaping methods in ceramics can be grouped into three—pressing, casting, and plastic forming. *Pressing* is also termed as *compaction*. Compaction involves the application of an external pressure to a powder contained in a die. The pressure can be applied uniaxially or isostatically. Uniaxial pressing is used for the production of simple shapes. For complex shapes, isostatic pressing is employed.

Casting in ceramics is different from solid casting in metals. Here, a suspension or a slurry of ceramic powder is prepared and poured into a mold. This mold absorbs the liquid and leaves the solid with its internal shape imparted to it. The same method is used in powder metallurgy to produce porous parts. The slurry can also be used to produce tapes, where the method is called *tape casting*.

In *plastic forming*, the ceramic powder is mixed with a binder or a plasticizer. This incorporation of a binder makes the mixture plastic and makes it possible to be formed under the use of pressure. This method again is used in powder metallurgy.

13.2 Pressing

Uniaxial pressing, when carried out at room temperature, is called *dry pressing*. The high-temperature pressing is hot pressing. The alternatives to these two variations in isostatic pressing are cold and hot isostatic pressing. These four methods will be discussed in the following text.

13.2.1 Dry Pressing

Dry pressing is a uniaxial pressing. This means that the pressure is applied in one direction to compress the powder in a die. The powder assumes the inside shape of the die. There are three steps involved in this type of compaction—filling the die with the powder, pressing, and ejection of the compact from the die. The different stages are shown schematically shown in Figure 13.1. In automatic pressing, the die is filled with the help of a feed shoe. This is connected to a feed hopper through a flexible pipe. The hopper stores the powder before the powder falls down under gravity to the feed shoe. The feed shoe is vibrated to get a uniform filling of the die. The bottom of the die is closed with the help of the bottom punch before the powder is filled. Once the die is filled, the feed shoe is moved away from the die mouth. Then the upper punch is brought down for pressing. Before the ejection step starts, the upper punch is withdrawn. Then the lower punch is pushed up to get the compact ejected.

The pressure can be applied in single or two directions. The compaction in single direction is called *single-ended compaction*, and the second

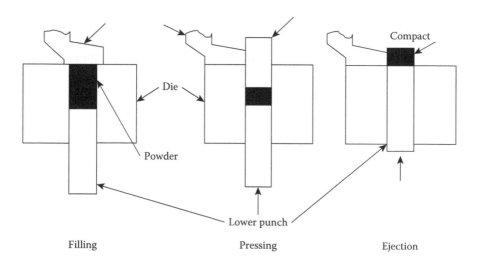

Filling   Pressing   Ejection

FIGURE 13.1
Different stages in single-ended dry pressing.

method is called *double-ended compaction*. In single-ended compaction, only the top punch moves during pressing, whereas in double-ended compaction, both the top and bottom punches move. In dry pressing, sometimes up to 5% binder is mixed to hold together the powder particles in the compact. A particle size distribution is desirable for dry pressing. This allows the small particles to go into the interstices among the larger particles and gives greater green density for the compact. Depending upon the material to be pressed and the press capacity, the pressures can go as high as 300 MPa to get maximum density.

The reasons for the widespread use of dry pressing are:

a. Simple process
b. Low investment cost

The size and shape of the part and the type of the press determine the production rate. For large parts, the rate goes down to a mere 15 parts per minute. Small parts may be produced at a rate of several thousands per minute. The presses can be mechanical or hydraulic. Mechanical presses operate at greater speeds than hydraulic ones.

13.2.2 Hot Pressing

When the dry pressing is carried out at high temperatures, the process is called *hot pressing*. For this to be done, the tooling, consisting of the die and the punches, are to be enclosed inside a furnace, as shown in Figure 13.2. In this method, the tooling will have to withstand the high temperatures involved.

The powders used for hot pressing need not be of any specific size distribution, and they need not be uniformly mixed. The densification can be

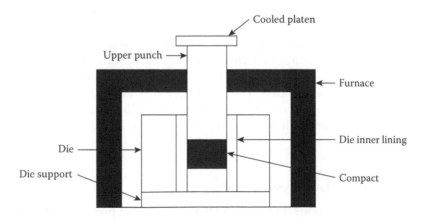

FIGURE 13.2
Schematic of a hot press.

achieved at a temperature lower than that used for pressureless sintering. Grain growth and secondary recrystallization can be prevented by this lower-temperature operation. Even very hard materials can be compacted by this method. This is possible because the strength of the materials decreases at high temperatures.

The disadvantage of hot pressing is that the life of the tooling is short, and they are expensive. The selection of die material for hot pressing depends on the temperature and pressure involved. Molybdenum-based alloys can withstand 1000°C at pressure of about 80 MPa. At the same pressure, ceramic dies made of Al_2O_3, SiC, and Si_3N_4 are useful up to 1400°C. For lower pressures, of the order of 10–30 MPa, graphite dies are used up to 2200°C. However, graphite dies require reducing atmospheres. The advantages of using graphite dies are:

a. They are easily machined.

b. They are inexpensive.

c. They possess hot strength that increases with temperature.

d. They possess good creep strength.

e. Thermal conductivity is excellent. Therefore, heat is transferred efficiently to the charge from the furnace.

f. The coefficient of thermal expansion is low. Hence, compact dimensions can be controlled accurately.

Only simple parts, in the shape of flat plates, blocks, and cylinders, are made by hot pressing. We go for this method only when the parts require small grain size, low porosity, or low impurity level, owing to the high cost and low productivity.

13.2.3 Isostatic Pressing

When we require complex shapes, we go for isostatic pressing. In this operation, pressure is applied from all sides uniformly with the help of a fluid. Thus, the problem of density variation, which could be there in dry pressing, can be avoided. The mold here should be flexible. When the pressing is carried out at room temperature, it is called *cold isostatic pressing* (CIP), and when carried out at high temperatures, it is called *hot isostatic pressing* (HIP).

13.2.3.1 Cold Isostatic Pressing

There are two variations for CIP—wet bag and dry bag processes. Figure 13.3 shows the schematic of a wet bag process. The powder is contained in a rubber mold. If a hollow part is required, then a metal mandrel is inserted

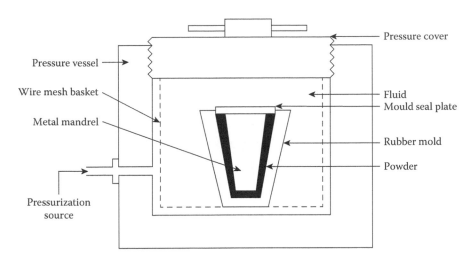

FIGURE 13.3
Schematic of cold isostatic pressing.

inside the mold. The mold is then sealed and kept inside a wire mesh basket. The whole assembly is kept inside a pressure chamber, which is then sealed. The fluid is then admitted through an opening in the chamber under high pressure. The pressure can be in the range of 20 MPa–1 GPa, depending on the capacity of the press and the material to be pressed. The fluid reaches the mold through the wire mesh basket and exerts the same pressure on all its sides. After allowing for sufficient time for compaction, the pressure of the fluid is brought down to the atmospheric pressure. The whole assembly is then removed from the pressure chamber. The mold is then opened, and the part is taken out.

The wet bag process gives rise to a wide range of shapes and sizes. Tooling costs are low. The part size can be in the range of 150 mm to 2 m in diameter, and 460 mm to 3.7 m in height. However, the shape and dimensional control are poor. Because of this, green machining is needed. The cycle times are between 5 and 60 minutes. Hence, production rates are low.

Figure 13.4 shows the schematic of the dry bag process. This process differs from the earlier one, in that the mold is an integral part of the press. There is a channel to distribute the fluid in the mold. The process can be fully automated. Hence, the production rate is as high 1 part per second.

This process is commercially used for the production of spark plugs. The various steps in the production are shown in Figure 13.5. The powder is filled into the mold at atmospheric pressure. An inner pin is inserted into the mold to give shape to the inner cavity of the plug. The mold is then closed, and oil is admitted under high pressure. After compaction, it is drained out. The part is taken out and machined.

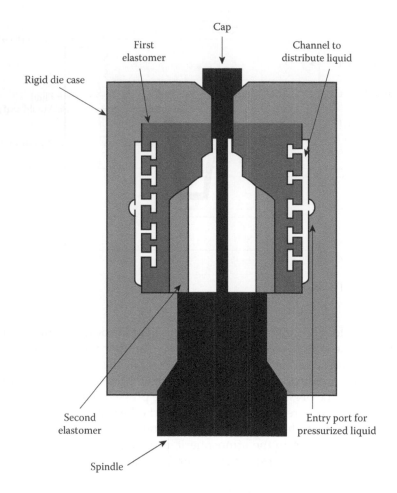

FIGURE 13.4
Schematic of a dry bag process.

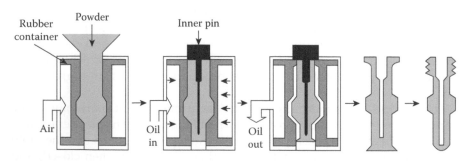

FIGURE 13.5
Making spark plugs.

13.2.3.2 Hot Isostatic Pressing (HIP)

HIP (hot isostatic pressing) uses pressurized hot fluid. Temperatures can go up to 2000°C, and the pressure can vary between 30 and 1000 MPa. Normally, a gas is used as the fluid. Figure 13.6 shows the sketch of the process. A furnace is fixed inside the pressure chamber and is maintained at the required temperature. Pressurized gas, when it enters the chamber, is raised to the furnace temperature. The pressure can further increase when the gas is heated. The excess pressure is released with the help of a pressure-release valve.

HIPing can be done using a flexible mold in which the powder is contained, or to a sintered part to increase its density. Because of the high temperatures involved, the flexible mold is made from a metal whose melting temperature is greater than the HIPing temperature. This method has been modified for producing small parts. In this method, green compact is made by the dry pressing or injection molding process. This compact is then encapsulated in glass. This encapsulated compact is then HIPed. Figure 13.7 shows the different steps of this method.

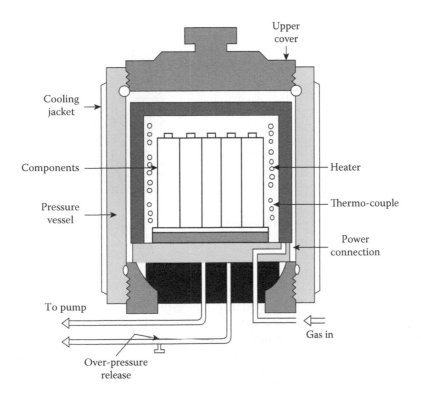

FIGURE 13.6
Sketch of hot isostatic pressing.

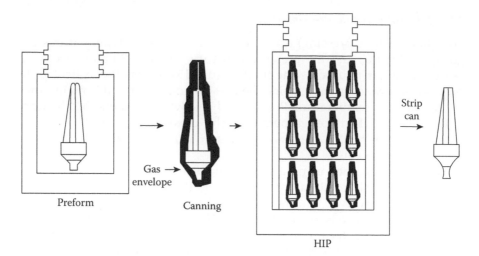

FIGURE 13.7
Modified HIPing.

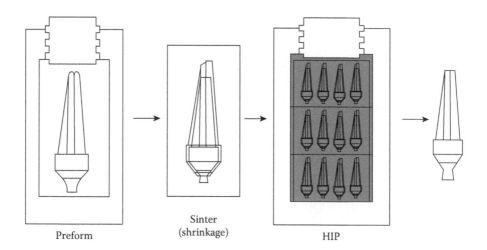

FIGURE 13.8
Steps in the sinter-plus-HIP method.

The second method is called the sinter-plus-HIP method. In this, the powder is first shaped using dry pressing or injection molding. The green compact is sintered, so that all the open pores become closed. After sintering, the compact is HIPed without any encapsulation. These steps are shown in Figure 13.8. Alumina-based tool bits and silicon nitride nozzles are made by HIPing. The advantage of HIPing is that even 100% dense parts can be made. The high cost involved causes it to be used only for the manufacture of special materials such as piezoelectric ceramics.

13.3 Casting

The casting involving a slip is called *slip casting*. A slip is a suspension of very fine particles in a liquid. The particles are made very fine so that they do not settle. Their high surface area ensures that electrostatic forces are more predominant than the gravitational force. They are prevented from agglomeration with the help of deflocculating agents. Examples for these are sodium silicate and soda ash. Slip casting can be used for the production of solid and hollow parts. For the production of solid parts, the liquid in the slip is allowed to be absorbed fully by the plaster of Paris mold. When this happens, a solid with shape of the mold cavity is obtained. Afterwards, the mold is dried along with the solid. The solid is then obtained by cutting the mold open, if the design does allow the lifting of the part from the mold. The solid will have its dimensions lesser than the dimensions of the mold cavity. Also, it will be highly porous. It is densified by the next step, which is sintering. Sintering further shrinks the casting. Hence, the mold cavity is given enough allowances to accommodate all the three shrinkages—shrinkage due to the liquid absorption by the mold, the drying shrinkage, and the sintering shrinkage.

In order to produce hollow parts, after sufficient thickness is obtained for the solid, the remaining slip inside the solid is drained off. Because of this step, this method is also called *drain casting*. The process is shown schematically in Figure 13.9.

Slip casting is low-cost production process. No large initial investment is needed. Complex shapes such as gas turbine rotors can be obtained. The slip can penetrate any intricate regions of a complex mold. Large articles such as wash basins are made using this method.

13.4 Plastic Forming

Extrusion and injection molding are two examples for plastic forming.

13.4.1 Extrusion

In extrusion, a plastic mass is forced through a die. In the case of clays, this plastic mass is prepared by mixing them with water. In the case of materials containing no clay, the mix is prepared with certain additives called *plasticizers*. For example, if we want to extrude alumina, alumina powder is mixed with methylcellulose, and then water is added to get the required consistency for this extrusion. Figure 13.10 shows the extrusion of a rod and a tube. Thus, extrusion yields a product with a uniform cross-section and a large length-to-diameter ratio. After the extrusion process, the green extrudate is fired.

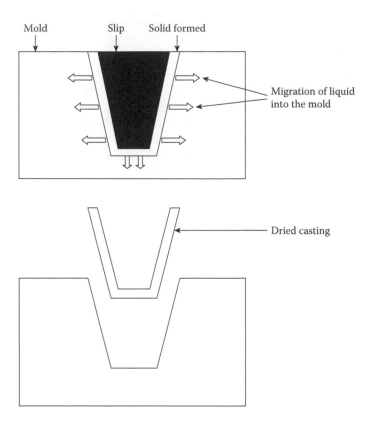

FIGURE 13.9
Schematic of drain casting.

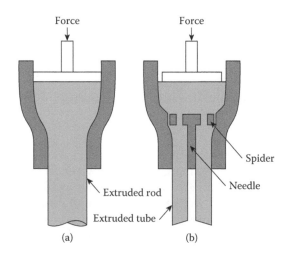

FIGURE 13.10
The extrusion of a rod and a tube: (a) rod and (b) tube.

13.4.2 Injection Molding

This process is more popular in polymers. For the injection molding of ceramics, a thermoplastic polymer is required to the extent of 40 vol.%. This is called the *binder*. For example, to injection mold silicon carbide, ethyl cellulose is one such binder that is used. During the firing process, a binder removal soak, called *binder burnout*, is given to eliminate the polymer from the final product.

Figure 13.11 illustrates the process. The mixture of the ceramic powder and the binder is taken in the hopper of the injection-molding machine. As this mixture falls down into the barrel of the machine, it is conveyed to the nozzle by rotating a screw. As it is conveyed through the barrel, it is heated by the barrel heaters provided on the barrel the form of band heaters. When the mixture reaches the nozzle, the binder would have melted. This molten plastic containing the ceramic powder is injected into a mold. The mold here can have intricate shape and is in the form of a split mold. After the mold is filled, it is allowed to cool. The solid formed is removed by opening the split mold. The whole machine is made up of a metal. The molded part is then fired. During firing, the binder is removed. This causes large shrinkage for the product, to the extent of 15–20%. Hence, allowance should be given to the mold cavity to compensate for this shrinkage, so that the final product has the required dimensions. Final machining will still be needed to get the final dimensions. Because of the thorough uniform mixing, the final product acquires uniform density.

The initial investment on the machine and the mold is rather high. However, the production is rapid, and the molds are reusable and permanent. The machine operates at low temperatures, just above the melting temperatures of the binders.

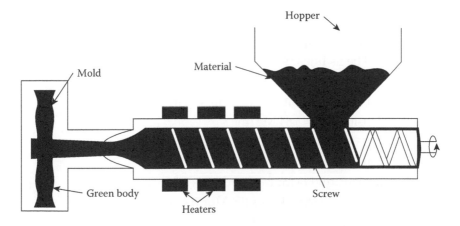

FIGURE 13.11
Cross-sectional side view of a screw-type injection-molding machine.

13.5 Green Machining

Green machining is the machining process done before firing. Since ceramics are generally very hard, machining may become difficult after firing. The tool for machining should be harder. Such tools can be very expensive and will require frequent replacement because of the wear and tear associated with the process. The difference in hardness between the product and the tool will not be very high, as both are ceramics. Hence, the wear of the tool becomes a problem. The cutting speed cannot be very high because, again, the tool is ceramic, which is a brittle material. For example, if we want to green machine silicon compact, before its nitridation to form silicon nitride, the cemented carbide tool can only be worked with a cutting speed of 30 m per minute. The speed decides the material removal from the work piece. As the speed is less, the material removal rate is less. Hence, productivity reduces.

Green machining is applied to spark plug insulators. This step was shown in Figure 13.5.

Problems

13.1 For a dry-pressed ceramic part, the total volume shrinkage is 50% after firing. If the shrinkage during sintering is 45%, what is the shrinkage during the binder removal?

13.2 For an extruded ceramic part, the total volume shrinkage is 50% after firing. If the shrinkage during sintering is 45%, and that during drying is 3%, what is the shrinkage during the binder removal?

13.3 If the total volume shrinkage for a ceramic part, having an as-formed volume of 50 g/cc, is 50%, what will be its final volume?

13.4 If the total isotropic volume shrinkage for a ceramic part is 50%, what will be its linear shrinkage?

13.5 Two different processes are used to make the alumina product.

Process 1: Green density $(D_g) = 2.43$ Mg/m^3

Fired density $(G_f) = 3.84$ Mg/m^3

Process 2: Green density $(D_g) = 2.74$ Mg/m^3

Fired density $(D_f) = 3.90$ Mg/m^3

The true density (D_t) for the product is 3.96 Mg/m^3. What are the—

13.5.1 Volume shrinkages (Sv_s) for the two processes?

13.5.2 Linear shrinkages (Sl_s) for the two processes?

13.5.3 Fired density ratio (D_r) of the two products?

Bibliography

C. B. Carter and M. G. Norton, *Ceramic Materials—Science and Engineering*, Springer, New York, 2007, pp. 412–421.

J. S. Reed, *Principles of Ceramics Processing*, 2nd Edn., A Wiley-Interscience Publication, John Wiley, New York, 1995, pp. 411–412, 416–417.

14

Thermal Treatment

14.1 Introduction

In the manufacturing of ceramic materials, three thermal treatments are important: calcination, sintering, and glazing.

14.2 Calcination

Calcination was discussed in Section 12.2, "Raw Materials." It was required for some ceramics to be obtained in the desired compound. For example, in the preparation of alumina, it was mentioned that aluminum hydroxide was calcined to get the alumina. Calcination is the process of heating in the absence of any atmosphere. This heating decomposes the compound into a solid and a gas or vapor. In the case of aluminum hydroxide, it decomposes into aluminum oxide and water vapor.

14.3 Sintering

Sintering is another thermal treatment. It improves the strength of a green compact along with other properties. It is a necessary step in the manufacture of ceramic products. It has got parallels in the metal field. In the processing of metals from their powders, which is called the *powder metallurgy* technique, sintering is an essential step. In ceramics, the separate step in which sintering takes place is called *firing*. It can also take place during hot pressing or hot isostatic pressing, because both these processes take place at high temperatures. What is really taking place is interparticle bonding. This happens because of the migration of atoms to the interparticle boundary area. Once they reach these areas, they cross over the boundaries. Once this happens, the boundaries disappear, and the adjacent particles get bonded.

14.3.1 The Process

In sintering, interparticle bonding is allowed to take place at a temperature below the melting point of the material. In order to facilitate this bonding, certain additives are added. These are called *sintering aids*. These sintering aids may get melted in the process. The particles may be crystalline or amorphous.

Figure 14.1 shows how the particle size distribution matters in sintering. When particles are of different sizes, there will be more areas of contact. This will facilitate greater sinterability. When the particles are of uniform size, the maximum density achievable is 74%. If there are smaller-sized particles, the interstices can be filled with them, and the density can be increased.

14.3.2 Mechanisms

Ceramics are mostly incompressible, in contrast to metals. During compaction, metallic particles plastically deform, and the curved surfaces on them are flattened. In ceramics, the compaction is achieved by the breaking down of larger particles. Because there is no flattening taking place, the particle surfaces will mostly be curved. During sintering, the tendency will be for the curved surfaces to become flattened. This is facilitated by the pressure difference existing between the inside and outside of a curved surface. This pressure difference is given by:

$$\Delta P = 2\gamma/r \tag{14.1}$$

Here, γ is the surface energy, and r is the radius of the spherical shape. This radius of curvature is positive or negative depending on whether the

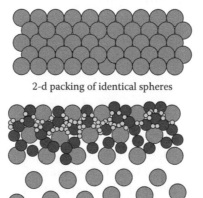

2-d packing of identical spheres

2-d packing of three different sized spheres

FIGURE 14.1
Effect of particle size on density.

center of the radius is inside or outside the material, respectively. As a result, the pressure on a concave surface is negative. For a small spherical particle, its surface-to-volume ratio is very large.

Sintering occurs in ceramics during the firing step, giving rise to consolidation, interparticle porosity removal, shrinkage, and growth of some grains at the expense of others. To understand the process of sintering, let us consider two spherical particles in contact (Figure 14.2a). After sometime, there are two radii to be considered, as shown in Figure 14.3a. The first one is that

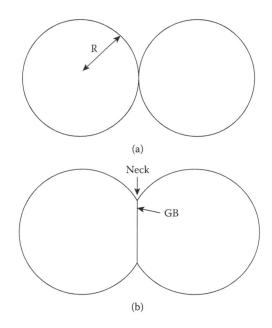

(a)

(b)

FIGURE 14.2
Coalescence of two particles in contact during sintering: (a) two spherical particles in contact; (b) formation of grain boundary.

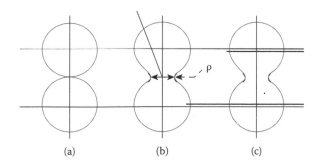

(a) (b) (c)

FIGURE 14.3
(a) and (b) formation of two curvatures during sintering; (c) decrease of interparticle distance.

of the neck, which is formed at the area of contact between the particles. To this area, atoms from the inside of the particles diffuse. The diffusion takes place because of the concentration differences existing between the bulk of the particle volume and the surface. This diffusion results in the filling up of vacancies at the particles' interface. The radius of this neck is shown as x in the figure. The other radius is that of the curvature resulting because of the neck formation. This radius is represented by ρ in the figure.

As the neck grows, the particle surface at the area of contact disappears. In the case of a crystalline material, the disappearance of the particles' boundaries results in a grain boundary formation. This is shown in Figure 14.2b. In the case of an amorphous solid, the boundaries totally disappear. As evident from Figure 14.3c, the center-to-center distance decreases as the neck grows. Thus, sintering causes shrinkages, which give rise to densification.

Another effect of sintering is the reduction of porosity. This is again due to the migration of atoms toward the pores. Thus, there is material transport in sintering, due to which the neck grows and the pores are filled. This migration is the result of one or more of the following mechanisms:

    a. Surface diffusion
    b. Volume diffusion
    c. Evaporation and condensation
    d. Grain boundary diffusion
    e. Plastic flow

Surface diffusion takes place along the surface of the particles. Volume diffusion is the movement of the atoms through the bulk volume of the particles. The source for this diffusion can be either the surface or the grain boundary. That means atoms come from these two sources and diffuse through the bulk volume of the particles. In the case of volatile materials, at the temperature of sintering, the material evaporates from the convex surface away from the neck region and is deposited at the neck region. This happens because, at the convex surface, the vapor pressure is lesser than that at the neck. Grain boundary diffusion is the migration of the atoms along the grain boundaries toward the neck region. Plastic flow is the result of the movement of dislocations as a result of the sintering temperature.

When we consider three spherical particles in contact, we can see that an open pore gets closed during sintering when the boundaries are formed. This is shown in Figure 14.4. The directions of migration of atoms by different mechanisms, except the plastic flow, are also shown.

Figure 14.5 shows the different-sized grains. The larger-sized grains have more number of sides than the smaller-sized ones. When the number of sides is less than six, the sides have a convex surface; when it is greater than six, they have a concave surface.

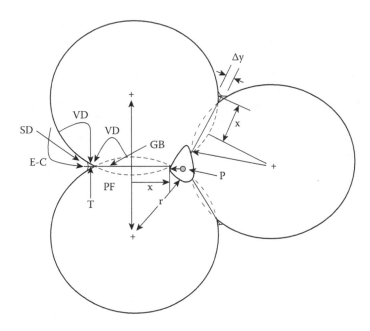

FIGURE 14.4
Sintering of three spheres.

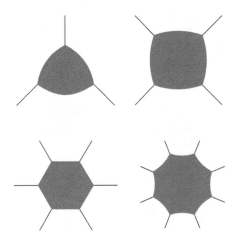

FIGURE 14.5
Grains with different number of sides.

The pressure difference existing between the outside and inside of a curved grain boundary gives rise to a free energy change. The extent of this free energy change is given by Equation 14.2.

$$\Delta G = 2\gamma\, V/r \tag{14.2}$$

Here, V is the molar volume. Because of this free energy difference, the boundary tends to move toward the center of curvature. For the grains having their surfaces convex, their centers of curvature are inside those grains. Those grains tend to shrink, meaning that their size gets decreased as the sintering proceeds. Finally, all such small-sized grains may get eliminated toward the final stages of sintering. The reverse happens in the case of grains having greater than six sides—they tend to grow. Those grains having six flat sides, neither grow nor shrink. In the case of smaller grains, their surface to volume is greater. As the system tends to decrease its total energy, it acquires a lower-energy state by the elimination of smaller-sized grains and by the growth of larger sized ones. The rate of movement of the boundary depends on the radius of curvature and on the rate of diffusion of atoms across the boundary. As the radius of curvature decreases, the movement rate increases.

14.3.2.1 Diffusion Mechanism

From the list of sintering mechanisms given earlier, the evaporation and condensation mechanisms predominate only when the material is volatile at the sintering temperature. The plastic flow will predominate when the compact is highly stressed. This will happen more in metals than in ceramics. In metals, their powders are more ductile, and the compact will hence consist of plastically deformed particles. In all other cases, diffusion is the main mechanism. It was also stated earlier that diffusion in powder compacts can take place through three different pathways. When diffusion of atoms takes place along these three ways to the necks and pores, there is an equivalent flow of vacancies from these sources in the opposite directions.

Out of the three ways of diffusion, it is volume diffusion that predominates. When volume diffusion results in the growth of the neck, this can be represented by Equation 14.3.

$$(x/r) = (40\gamma a^3 D^*/kT)^{1/5} r^{3/5} t^{1/5} \tag{14.3}$$

In this equation, x is the neck radius at any time t, a^3 is the volume of the diffusing vacancy, D^* is the self diffusion coefficient of the material, k is Boltzmann constant, and T is the sintering temperature.

We have also seen that the interparticle distance decreases due to sintering. Due to the particle centers coming closer, there is shrinkage after sintering. This shrinkage can be represented by Equation 14.4.

$$\Delta V/V_0 = 3\Delta L/L_0 = 3(20\gamma a^3 D^*/\sqrt{2kT})^{2/5} r^{6/5} t^{2/5} \tag{14.4}$$

Here, ΔV is the volume shrinkage during a time interval of t, V_0 is the initial volume, and ΔL is the change of interparticle distance during the same time interval. Volume change is three times the linear change, as evident from the equation. The change of interparticle distance is shown in Figure 14.3b.

The change of relative density as a function of time is given in Figure 14.6. Relative density is the ratio of the actual density of the compact to the theoretical density of the material. Two things are evident from these curves:

1. The densification rate decreases to a very low value after initial large increase. This indicates that there is no need to sinter for a long time.
2. As the particle size decreases, the time required to attain the maximum density decreases.

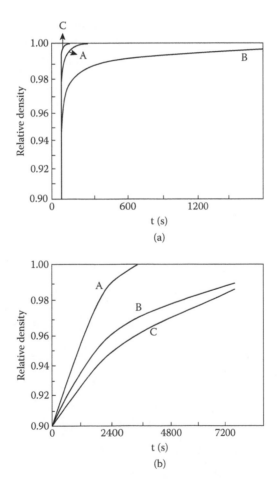

FIGURE 14.6
Relative density as a function of time: (a) for 0.8 μ; (b) for 4 μ; the three curves are drawn according to three different equations.

Now let us consider the movement of a curved grain boundary. The boundary is shown in Figure 14.7. Here, the boundary is perpendicular to the paper. The atoms move across the grain boundary during sintering. Since the grain on the left side has a boundary with a convex surface, the boundary tends to move to the left as a result of the movement of atoms to the right. In the case of an ionic solid, the ions will be moving. The movement of ions need not be in the same direction. The frequency of atomic jumps in the forward direction is given by Equation 14.5.

$$f_{AB} = \nu \exp\left(-\Delta G^*/RT\right) \tag{14.5}$$

Here, f_{AB} represents the frequency of jumps from a position A to a position B, and ν is the frequency factor. It has a value equal to about 10^{13} Hz for solids. ΔG^* is the activation barrier for the jump. It represents the required energy that the atom should acquire for the jump to be made possible. This energy is acquired from the heat supplied during the sintering operation. The model for the jump is shown in Figure 14.8. The ion in position A is shown to possess a free energy value greater (by ΔG) than the free energy value of an ion in position B. When the ion in position B acquires this energy plus the activation energy required for the ion in position A,

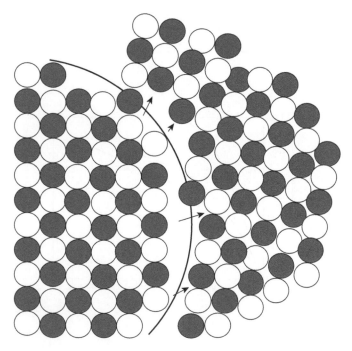

FIGURE 14.7
Atomistic model for grain boundary migration.

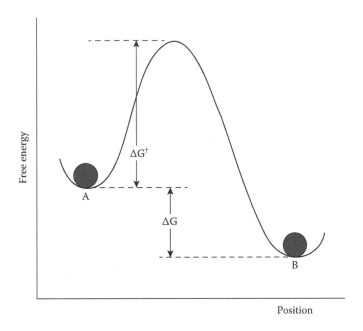

FIGURE 14.8
Activation energy model for grain boundary migration.

then that ion can jump back to position A. The frequency of this reverse jump is given in Equation 14.6.

$$f_{BA} = v \exp\left[-(\Delta G^* + \Delta G)/RT\right] \tag{14.6}$$

The growth rate, which is same as the net movement of the grain boundary, is given by the product of the distance of jump and the net frequency of jump. If λ is the distance of jump, then the growth rate is given by:

$$U = \lambda \left(f_{BA} - f_{AB}\right) \tag{14.7}$$

Writing ΔG in terms of ΔH and ΔS, and simplifying Equation 14.7, we get:

$$U = v \lambda \left[\gamma V/RT(1/r_1 + 1/r_2)\right]\exp(\Delta S^*/R)\exp(-\Delta H^*/RT) \tag{14.8}$$

It can be seen from the equation that the grain growth rate exponentially increases with temperature.

14.3.3 Liquid-Phase Sintering

Liquid-phase sintering (LPS) is a process in which there is a liquid presence at the sintering temperature. This is shown in Figure 14.9. The liquid phase enters all the vacant spaces, such as triple junctions, quadruple junctions,

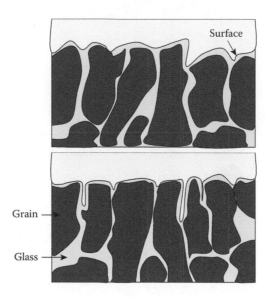

FIGURE 14.9
Liquid-phase sintering.

and grain boundaries. If there is enough liquid, even the surface of the compact will be covered. This is seen in the figure. During cooling, when the freezing point of the liquid is crossed, the liquid solidifies. Since there is a large shrinkage associated with solidification, the surface liquid will be drawn inside the compact. This is also evident from Figure 14.9.

The density obtained could reach 100%, as all the pores can get filled in LPS. The densification takes place in three stages. In the first stage, the rearrangement of the solid particles takes place. This is made possible because the mobility of these particles is facilitated by the liquid. The capillary forces play a major role in this rearrangement. This is because the spaces between the solid particles are of the dimensions of capillaries. In the second stage, reprecipitation of the solid takes place. If the liquid formed is that of a glass composition, this liquid can dissolve some solid in it. The reprecipitation may take place elsewhere on the grain. This is because diffusion is facilitated in the glass phase with a rate greater than that of grain boundary diffusion. Also, the crystal–glass interfacial energy varies with the orientation of the interface.

The second stage further depends upon many other factors, such as the composition of the glass. This composition may vary within the compact, and this is why dissolution and reprecipitation take place. Another factor is the viscosity of the glass. Viscosity changes with temperature. As temperature increases, viscosity decreases. Viscosity also depends on the composition.

In the third stage, the grains coarsen. This is a time-dependent phenomenon, and the rate is low. It is similar to the grain growth that takes place in solid-state sintering. LPS is utilized in the applications of capacitors, cutting tools, friction materials, grinding materials, metalworking tools, porcelain, refractories, and varistors. Cemented carbide is a common example of a cutting tool that is made by LPS. In this, cobalt metal forms the liquid phase during sintering. It cements the solid tungsten particles. The formation of liquid phase in porcelain was discussed when we considered the microstructure of triaxial whiteware compositions.

14.4 Glazing

Glazing is a temperature-assisted coating process done on ceramics. When ceramics are used to coat metals, the process is called *enameling*. The purposes of glazing are twofold: aesthetics and protection. Glazing gives rise to a smooth and shining surface that improves the aesthetics of the product. This smoothness and shining is because the glaze assumes a glass structure. The glaze is impermeable. Hence, fluids cannot enter the body of the part, and therefore the corrosive fluids cannot attack the part. Impermeability is also the property of glass. Glazing is done to most of the potteries and whitewares.

Glazing was introduced by the Egyptians by around 3000 BC or earlier. This was done to seal the pores of earthenware hermetically, which is done by firing. For this, an aqueous suspension of finely ground quartz sand, mixed with sodium salts, is prepared. The salts are sodium carbonate, sodium bicarbonate, sodium sulfate, and sodium chloride. This suspension is coated on the fired product. Afterward, the product is again fired. The temperature of refiring is usually lower than that of the original firing. During the refiring process, the particles in the suspension will fuse together, after the removal of the liquid by evaporation. On fusing, they form a glass in the liquid state and spread on the surface of the product. On cooling, we get the coating in the form of a glass layer.

Two types of glazes have been in use on earthenware since hundreds of years. They are lead and tin glazes. Lead glaze is transparent, and tin glaze is white and opaque.

14.4.1 Lead Glaze

The lead in the lead glaze reduces the fusion point of the glaze mixture. This allows the refiring to be done at much lower temperatures than the initial firing. During the period 475–221 BC, the lead oxide (PbO) content was about 20%. During 206 BC to 200 AD, the PbO content was increased to 50%–60%. Lead (Pb) can leach out of the glaze when in contact with certain solvents. Being a poisonous substance, there is a limit fixed for the amount of lead that may be released. The US Food and Drug Administration limit for lead from small hollowware is 2 ppm. Standard tests for determining the amount of leached lead use 4% acetic acid. The glazed item is immersed in this solution at 20°C for 24 h. Afterward, the acid is analyzed for lead content by flame atomic absorption spectrometry.

Easy lead leaching from glazes can be prevented by the use of fritted glazes. Here, the glaze composition is first melted and powdered. This powder is then used for glazing.

14.4.2 Tin Glaze

Tin glazing was used in the second millennium BC to decorate bricks. This did not sustain for long. In the ninth century AD, it came into existence again. When its use spread to a Spanish island called Majorca, tin glazing came to be known as *Majolica*. When the same process was developed in Faenza in Italy, it was known as *Faience*. In 1584, it got developed in Delft in the Netherlands. The tin-glazed product was then known as *Delftware*. When the growth in the use of sanitary ware took place in the nineteenth century, tin glazing became industrially important.

14.4.3 Glass as Glaze

Glazing uses the viscous properties of glass to form a smooth continuous layer on a ceramic substrate. Pots were the first to be glazed.

Let us be familiar with some terms in pottery that are associated with glazing.

14.4.3.1 Underglaze

This is the first coating on a pot. It is prepared by mixing an opacifying agent such as oxide, carbonate, sulfate, etc., with a flux. This mixture is calcined and then mixed with a liquid. This liquid mixture is applied to get a smooth layer. Then it is decorated. On this, a clear glaze is coated. Afterward, the whole coating is fired by keeping the coated pot in a furnace.

14.4.3.2 Craze Crawling

This is the separation of glaze from the pot surface. It happens sometimes during firing. The reason is the dewetting that takes place when the glaze is in the liquid phase.

14.4.3.3 Crackle Glazes

These are the glazes that contain a pattern of hairline cracks. These cracks are sometimes deliberately produced for aesthetics. These are incorporated by adding higher concentrations of alkali metals. From the firing temperature, the coated ceramic is fast cooled to get a fine pattern.

Celadon, tenmoku, raku, and copper glazes are examples of crackle glazes.

Celadon glazes were produced 3,500 years ago. The color of the glaze varies from light blue to yellow and green. The different colors are obtained by adding Fe_2O_3 in concentrations of 0.5–3.0 wt%.

Tenmoku glaze is dark brown or black in color. These colors are produced again by varying the Fe_2O_3 content from 5 to 8 wt%. There is a variety called oil-spot tenmoku in this type of glaze. This is produced by producing bubbles in the glaze during firing. The firing time is controlled such that the bubbles leave spots on the glaze surface. If the firing is carried out under reducing atmosphere, the Fe_2O_3 is reduced to FeO. The original Fe_2O_3 is an intermediate in the glaze. When it becomes FeO, it functions as a modifier. The color will change accordingly.

Raku glazes have a metallic appearance. They are produced by first firing in the usual way. Afterward, a reducing atmosphere is given by keeping it inside sawdust. From this reducing atmosphere, they are quenched to room temperature. During application, they change color as oxidation slowly takes place in them.

Copper glazes may contain up to 0.5 wt% $CuCO_3$ initially. On firing, $CuCO_3$ decomposes to CuO. The furnace atmosphere is filled with CO to reduce CuO to Cu. Thus, the glaze contains copper particles in it. These copper particles give a red color to the glaze.

14.4.3.4 Crystalline Glazes

These are glass-ceramic glazes. They contain an amorphous and a crystalline phase. The crystals are allowed to nucleate and grow while the glaze is slowly cooled from the firing temperature. The shape of the crystals will be similar to that of platelets, as it is growing in glaze having approximately 0.5 mm thickness. To facilitate the nucleation of the crystals, TiO_2 is added as a seed. The nucleation is called *rutile break-up*. The crystal formed is actually $PbTiO_3$ in this case. In order to get $PbTiO_3$, PbO is added to the extent of 8–10 wt%. The remaining composition is constituted by SiO_2 and Al_2O_3.

Another type of crystals is the willemite crystals. Willemite is a rare zinc mineral (Zn_2SiO_4). In order to get these crystals, zinc is added. The zinc addition is controlled, as it reduces the viscosity of the molten glaze. Zinc acts as a modifier in the glaze.

14.4.3.5 Opaque Glazes

Glazes are made opaque by adding crystals to the molten glaze. Examples of such crystals are those of SnO_2 and zircon ($ZrSiO_4$). To get white color for the zircon glaze, ZnO is used. TiO_2 gives yellow color to the glaze. There are three ways of making opaque glazes. They are:

a. By forming crystals
b. By trapping a gas
c. By causing liquid–liquid separation

Wollastonite ($CaSiO_3$) crystals can be formed by a thermal treatment of the glaze. Gases such as fluorine or air can be trapped inside a glaze during the cooling of the glaze from the firing temperature. The glaze composition is adjusted to get two immiscible liquids during firing. These two liquids separate and solidify to get an opaque glaze. When the crystals are very fine, we get the matt glaze. Lime matt contains fine wollastonite glaze. Willemite crystals form the zinc matt. Calcite ($CaCO_3$) is added to SiO_2-based glaze to form wollastonite. Calcite is also known as *whiting*. Matt glazes can also be formed by adding much of very fine powders of crystalline oxides. These oxides remain unchanged during firing. An example for such glaze is the satin or vellum glaze. This contains small crystals of 18% SnO_2 or ZnO and 4% TiO_2 in a high-lead glaze fired at 1000°C.

Color in glazes can be imparted by nanoparticles. Gold nanoparticles are used to give golden color, and copper nanoparticles give red color. In the case of copper particles, as given earlier, they can be produced during the making of glaze inside it.

14.4.3.6 Colored Glazes

In order to get consistent colors, ceramic pigments are used. These are needed in the production of sanitary ware and tiles. Where change of color is required, then metal oxides are used. Such variability is needed in pottery crafts. CoO gives deep blue color if it is less than 1%. Cr_2O_3 dissolves to the extent of 1.5% in the glaze. It gives green color. It can also give other colors with the help of other oxides. Red color can be obtained by combination with Pb. In combination with Zn, Cr^{6+} gives brown color. If both Pb and Zn are present, Cr^{6+} changes its color to yellow. Similarly, MnO_2 gives a brown glaze. However, the color can be changed to red, purple, or black, depending on the concentration of Na. In the case of CuO, turquoise can be produced if 1%–2% of it is added to an Na-rich glaze. If the percentage is increased to 3, green/blue is obtained. If the percentage is more than 3%, a metallic luster results. Bright yellow glaze is produced by the addition of CdS/CdSe. If Pb is present in this glaze, then it forms PbS, and the glaze becomes black. Cd-based colors are stabilized by adding zircon. In combination with V, zircon produces vanadium zircon blue [$(V,Zr)SiO_4$]. If praseodymium is present, then it gives praseodymium zircon yellow [$(Pr,Zr)SiO_4$]. Uranium gives a dark brown glaze.

14.4.3.7 Salt Glaze

To produce this glaze, salt is allowed to react with the pot surface made of clay. It is done by throwing salt on to the pot surface during the firing of the pot. Salt corrodes the surface, and the corroded layer is the salt glaze. The process is actually high-temperature soda corrosion of the fired clay.

14.4.3.8 Enameling

When glazing is done on metallic surfaces, the process is called *enameling*. It also refers to glazing on top of a glazed surface. When the organic component in a paint is replaced by glass, the paint is known as *enamel paint*. This kind of paint is everlasting.

Problems

14.1 Estimate the mean firing shrinkage between particles, Δl, when the mean particle diameter is 1 μ and the measured firing shrinkage is 44%. Assume body centered cubic packing of particles.

14.2 The diametral shrinkage is 16.3%. Calculate the volume shrinkage, assuming the shrinkage is isotropic.

14.3 Given below is the number of grains intersecting on a line of 5 cm when found at a magnification 100×:

   3, 6, 5, 4, 3, 5, 4, 6, 3, 6

   Find the grain size.

14.4 A quartz substrate made by dry pressing has a green density after drying of D_g = 1.48 Mg/m^3, and a fired density of D_f = 2.47 Mg/m^3. Assuming that shrinkage is isotropic, what is the percent volume and percent linear shrinkage? What is the porosity of the fired material? The theoretical density (Dt) of quartz is given as 2.65 Mg/m.

14.5 Hot isostatic pressing (HIP) is carried out at 30 MPa on a ceramic material. The same material was prepared by conventional sintering to get the same pore diameter of 8 μ. If the surface energy (γ_s) of the material is 650 mN/m, what are the driving forces for both the processes?

Bibliography

C. B. Carter and M. G. Norton, *Ceramic Materials—Science and Engineering*, Springer, New York, 2007, pp. 20–21, 390–392, 427–433.

J. S. Reed, *Principles of Ceramics Processing*, 2nd Edn., John Wiley, New York, 1995, pp. 552, 558, 620–621, 623.

15

Mechanical Properties

15.1 Introduction

Ceramics work better in compression than in tension. A number of defects are incorporated into them during their manufacture from powders. These defects play an important role in decreasing their tensile properties and in causing brittleness. Plastic deformation is almost negligible because dislocations cannot move easily. Fracture takes place mostly along grain boundaries. A comparison of the tensile properties of three classes of materials is shown in Figure 15.1. For ceramic, its modulus of elasticity and fracture strength are the highest; it does not have plastic deformation; and it is the least tough. Metal possesses the highest toughness, an intermediate modulus of elasticity, and shows plastic deformation. Elastomers have the highest ductility, intermediate toughness, and the least strength.

The strength of ceramics is affected by many factors as shown in Figure 15.2. During which stage in the manufacture process these factors are incorporated into ceramics is also shown in Figure 15.2. Certain raw materials decide their chemical composition. If certain chemical reactions take place during firing, then the final composition is decided at that stage. How two of those factors, porosity and grain size, affect two of the mechanical properties is shown in Figure 15.3. Though density and porosity are inversely proportional, their effect on strength differs. Pores affect the strength to a greater extent than does density. This is because pore size also affects the strength. Cracks nucleate at the pores because of stress concentration. And the extent of stress concentration is decided by the size of the pores. The largest pore will lead to the fracture of the material. Similarly, the range of grain size is another factor that affects the strength. This effect is shown in Figure 15.4. As grain size decreases, strength increases. But the rate of increase is greater the larger the grain size range. Strength is lower with larger grain size because larger grains contain more defects. The reason for different rates of decreasing strength is that in addition to the preexisting flaws causing brittle facture, there is a competing fracture mechanism that links dislocations and crack nucleation to subsequent failure. Dislocations are generated more often in smaller-sized grains.

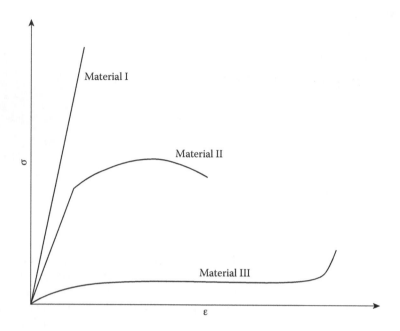

FIGURE 15.1
Comparison of a ceramic, a metal, and an elastomer with respect to their tensile properties.

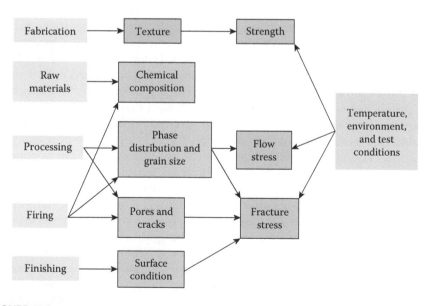

FIGURE 15.2
Factors affecting the various strength values of ceramics.

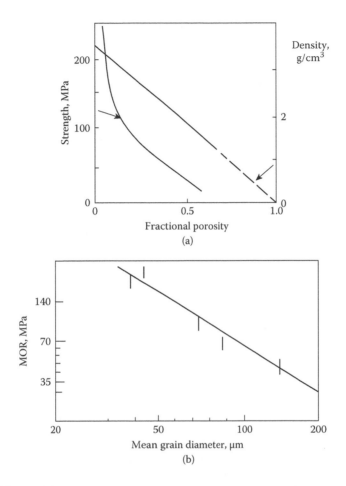

FIGURE 15.3
(a) Effect of porosity and density on the strength of a polycrystalline alumina; (b) effect of grain size on the modulus of rupture (MOR) of Beryllia. (From M. W. Barsoum, *Fundamentals of Ceramics*, Taylor & Francis, 2003, pp. 187–189, 245–246, 356–368, 380–386, 396–399, 401–415, 435–436.)

15.2 Tensile and Compressive Strengths

High performance structural ceramics possess high temperature capabilities with the ability to carry a significant tensile stress. Ideally, ceramics are more suited for load-bearing applications. In these applications, their compressive strength becomes important. Tensile and compressive strengths of ceramics will be discussed in this section. The American Society for Testing of Materials (ASTM) has devised many standards for testing ceramics for their various properties. The standards useful for determining tensile and compressive strengths are given in Table 15.1.

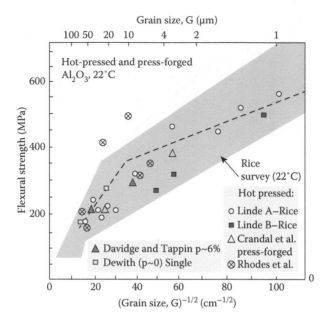

FIGURE 15.4
Variation of flexural strength as a function of grain size; data is taken from different authors as given in the figure.

TABLE 15.1

ASTM Standards for Tensile and Compressive Properties of Ceramics

Sl. No.	Number	Description
1.	C1273-95	Test method for tensile strength of monolithic advanced ceramics at ambient temperature
2.	C1366-97	Test method for tensile strength of monolithic advanced ceramics at elevated temperature
3.	C1424-99	Test method for compressive strength of monolithic advanced ceramics at ambient temperature

15.2.1 Testing in Tension and Compression

Tensile testing is not widely used for ceramics for the following reasons:

a. Ceramics are inherently brittle.

b. A dog bone shape is difficult to make in ceramics.

c. In instruments using threaded samples, machining to get such samples in ceramics is tricky. Samples may also break at the grips.

d. There is a need to align the samples perfectly; otherwise bending stresses may affect the test.

Machining needed to make the sample may introduce surface flaws. Fracture of the samples normally takes place by the cracks nucleated at the surface flaws. Perfect alignment is necessary because ceramics fail after about 0.1% strain. Because of these difficulties, ceramics are more commonly tested in bending rather than in tension.

Tensile strength of a ceramic is important in certain situations. For example, in the growth of single crystals by the Czochralski method, the maximum weight of the crystal that can be produced is determined by the tensile strength of the material.

Ceramics give much higher values in compression. The ratio of compressive strength (σ_c) to bending strength (σ_b) is given in Table 15.2 for some ceramics. Because the strength depends on the grain size, the grain size values are also quoted.

Compressive strength is important for concrete. Concrete is a structural material that is subjected to compressive load. It is a ceramic-based composite where the reinforcements are also ceramics. Here, cement is the matrix; stones and sand are the reinforcements. The mixture of stones and sand is known as the aggregate. When the pressure due to load is beyond the compressive strength of concrete, it is crushed. Crushing is due to propagation of stable crack growth. Unstable crack growth under tension and compression in a brittle material is shown in Figure 15.5. Type B cracks grow stably under compressive stress. As shown, they change their direction under the action of tensile and compressive stresses developed on their two sides. Under tension, those lie perpendicular to the tensile stress direction propagate and ultimately result in fracture.

In tension, the fracture results from the critical flaw size. In the case of failure under compressive stress, it is the average stress value that determines the failure.

TABLE 15.2

Ratio of Compressive Strength to Bending Strength

Sl. No.	Ceramic	Grain Size, μ	σ_c/σ_b
1.	TiB_2	20–50	4–6
2.	ZrB_2	20–50	4–6
3.	B_4C	1	7
4.	WC	1–6	4–6
5.	Al_2O_3	1–100	4–30
6.	$MgAl_2O_4$	1	7
7.	ThO_2	4–60	13–17
8.	UO_2	20–50	5–18

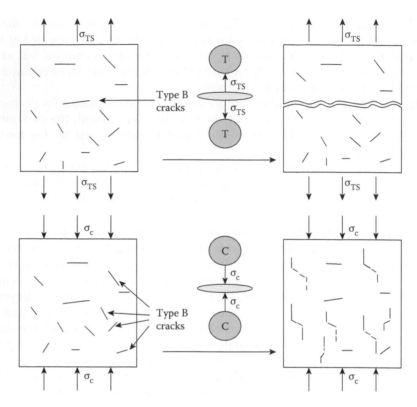

FIGURE 15.5
Unstable crack growth under tension and compression in a brittle material.

15.2.2 Testing in Bending

As mentioned earlier, a bend test is more commonly used for ceramics than a tensile test. The advantages of conducting bend tests are:

a. Lower cost

b. Sample geometry simple

The geometries used in bend tests are rectangular or cylindrical. There are two ways of testing, three- and four-point loading. These are shown in Figure 15.6. Out of the two, four-point loading is preferred because an extended region with constant bending exists between the inner rollers. The loads are applied through the rollers. A bend test determines the modulus of rupture (MOR). The formula for calculating the MOR is:

$$\sigma_r = 6M_r/Wh^2 \tag{15.1}$$

Here, σ_r is the modulus of rupture, M_r is the moment in the beam at rupture, h is the height of the beam, and W is its width. These dimensions are

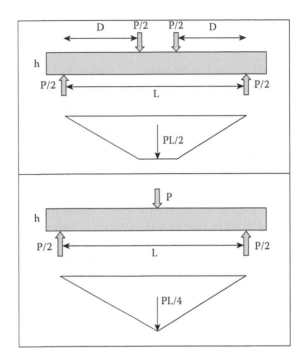

FIGURE 15.6
Set-ups for bend test.

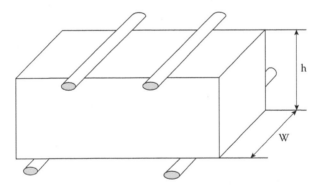

FIGURE 15.7
Dimensions of a specimen used in bend test.

marked in the specimen sketch shown in Figure 15.7. In the case of a three-point bend test, the value of M_r is given by Equation 15.2, and for four-point test, by Equation 15.3. These values are shown in Figure 15.6.

$$(M_r)_3 = PL/4 \tag{15.2}$$

$$(M_r)_4 = PL/2 \tag{15.3}$$

15.3 Creep

Creep failure was discussed in Section I of this volume. Creep is the slow and continuous deformation of a solid with time. It occurs at temperatures greater than $0.5T_m$. T_m is the melting temperature of the solid in degrees Kelvin. This phenomenon depends on many factors—stress, time, temperature, grain size and shape, microstructure, volume fraction, and viscosity of glassy phases at the grain boundaries as well as dislocation mobility.

15.3.1 Measurement

To measure creep, a load is attached to a specimen of the material. The load is selected such that it gives rise to a stress that is below the yield strength of the material. As soon as the load is attached, an instantaneous elongation takes place corresponding to the elastic strain produced. This elongation is noted and is also noted at every small interval of time. This process continues until the fracture of the sample takes place. The data obtained is used to plot the creep curve. From the elongation at every instant of time, the corresponding strain is calculated. For calculation, the following equation is used:

$$\varepsilon = \Delta L/L_O \tag{15.4}$$

Here, ΔL is the elongation, and L_o is the original length. A typical creep curve is shown in Figure 15.8. Three regions can be identified in the creep curve. In the first region, the rate of increase of strain decreases with time. This region is characterized by *primary creep*. In the second region, there is a constant rate of strain increase. This can be identified by a constant slope for the creep curve in this region, which is called the *secondary creep*

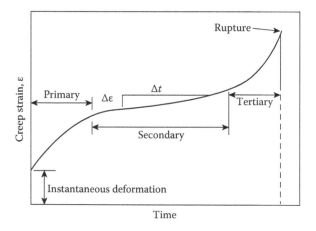

FIGURE 15.8
Typical creep curve.

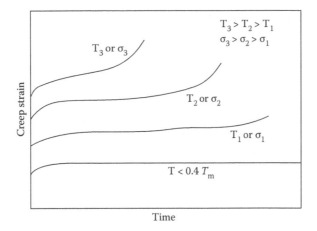

FIGURE 15.9
Effect of stress and temperature on creep.

or *steady-state creep* region. In the third region, the creep rate increases with time until fracture takes place. This region is called the *tertiary creep* region.

Figure 15.9 shows the effect of temperature and stress on creep. Both these factors increase creep as they increase. In addition, the instantaneous strain also increases with both stress and temperature. The time to failure decreases.

If the steady-state creep rate is plotted against stress, a straight line is obtained. Hence, the formula connecting the two can be written as:

$$\dot{\varepsilon} = d\varepsilon/dt = \Gamma\sigma^p \tag{15.5}$$

Here, Γ is a temperature-dependent constant, and p is the *creep law exponent* having values between 1 and 8. For p > 1, we have *power law creep*. There are three mechanisms—*diffusion*, *viscous*, and *dislocation* creeps.

15.3.2 Diffusion Creep

For creep to take place, there should be atom movement. Also, there should be some driving force.

15.3.2.1 Driving Force for Creep

Change of Helmholtz free energy is given by:

$$dA = -SdT - pdV \tag{15.6}$$

For a constant temperature, Equation 15.6 changes to Equation 15.7.

$$dA = -pdV \tag{15.7}$$

On rearranging this equation, we get:

$$p = -\partial A/\partial V \qquad (15.8)$$

Multiply both sides by atomic volume, Ω, and rearrange them:

$$p\Omega = -(\partial A/\partial V)\Omega = -\partial A/\partial(V/\Omega) = -\partial A/\partial(N) \qquad (15.9)$$

Here, N is the number of atoms. $\partial A/\partial N$ is the excess chemical potential per atom, and is given by Equation 15.10 where μ_o is the standard chemical potential of atoms in a stress-free solid.

$$\partial A/\partial(N) = \bar{\mu} - \bar{\mu}_o \qquad (15.10)$$

The applied stress p produces a stress σ, let us say. The excess chemical potential is due to this stress. Thus, substituting p by σ in Equation 15.9, then putting the value of $\partial A/\partial N$ in Equation 15.10 and writing this equation in terms of the chemical potential of the stressed solid, we get:

$$\bar{\mu} = \bar{\mu}_o - \sigma\Omega \qquad (15.11)$$

The physical meaning of Equation 15.11 can be understood by considering the model depicted in Figure 15.10. In this figure, it is shown that a solid is subjected to two unequal compressive pressures, P_A and P_B. P_A is greater than P_B. These pressures give rise to two equivalent stresses, σ_{11} and $-\sigma_{22}$, respectively. According to convention, compressive stresses are indicated by a negative sign. Let us assume that under the action of the pressure P_A, an

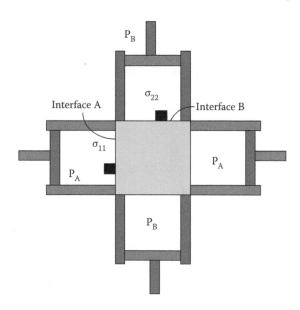

FIGURE 15.10
Model illustrating the meaning of Equation 15.11.

atom from the surface on which this pressure acts goes to a vacancy below that surface. The work done on the system for this diffusion is given by the force applied multiplied by the distance. This will be same as the pressure applied multiplied by the volume. Pressure applied is the same as the stress. Because P_A is greater than P_B, the surface on which P_B acts will experience a tensile force. Under this force, let us say that an atom from below the surface on which P_B acts diffuses to the surface. Both these diffusions are shown in Figure 15.10 schematically. In this case, the work is done by the system. The net work is given by the work done by the system minus that done on the system. Equation 15.12 gives the mathematical expression for this work.

$$\Delta W_{A \to B} = \Omega(P_B - P_A) = \Omega(\sigma_{11} - \sigma_{22}) \tag{15.12}$$

In the case $\sigma_{11} = -\sigma_{22}$, we get Equation 15.13

$$\Delta W = -2\Omega\sigma \tag{15.13}$$

This equation gives the extent of driving force available for an atom to diffuse from a region under a compressive stress to that under a tensile stress.

Example 1

Consider Figure 15.10. Let P_A be equal to 30 MPa, and P_B be equal to 15 MPa.
 a. Calculate the energy change for an atom that diffuses from surface A to surface B. Let the molar volume be 15 cm^3/mol.
 b. Assuming the applied stress to be 110 MPa, calculate the value of $\sigma\Omega$.
 c. Assuming the elastic modulus to be 160 GPa, compare this value with the elastic strain energy of an atom subjected to the same stress.

Solution

 a. Given that the molar volume = 15 cm^3/mol
   Therefore, $\Omega = 15 \times 10^{-6}/N_{Av} = 15 \times 10^{-6}/6.023 \times 10^{23} = 2.49 \times 10^{-29}$ m^3
   Given, $\sigma_{11} = -30$ MPa, and $\sigma_{22} = -15$ MPa
   Net energy recovered = $\Delta W_{A \to B} = \Omega(\sigma_{11} - \sigma_{22})$
   $= 2.49 \times 10^{-29} (-30 + 15) \times 10^6$
   $= -3.74 \times 10^{-22}$ J/atom
   $= -3.74 \times 10^{-22} \times 6.023 \times 10^{23}$
   $= -225$ J/mol
 b. $\sigma\Omega = 110 \times 10^6 \times 2.49 \times 10^{-29} = 2.74 \times 10^{-21}$ J/atom
   $= 2.74 \times 10^{-21} \times 6.023 \times 10^{23}$ J/mol
   $= 1650$ J/mol
 c. The elastic energy associated with a volume Ω is given by:
   $U_{ee} = (\Omega/2)(\sigma^2/Y) = (2.49 \times 10^{-29}/2)[(110 \times 10^6)^2]/160 \times 10^9$
   $= 9.42 \times 10^{-25}$ J/atom
   $= 9.42 \times 10^{-25} \times 6.023 \times 10^{23}$
   $= 0.57$ J/mol
   $\sigma\Omega$ is found to be three orders of magnitude greater than the elastic energy.

Now, at atomistic level, let us discuss the mechanism of creep. We have seen that under the action of a stress, atoms migrate to vacancies. Hence, vacancy concentration affects creep. Vacancy concentration, C_o, under a flat and stress-free surface is given by Equation 15.14

$$C_o = K' \exp(-Q/kT) \tag{15.14}$$

Here, K' is a constant and contains an entropy of formation of vacancies. Q is an enthalpy of formation. Considering the difference in the chemical potential between a flat surface and undersurface atoms, $\Delta\mu$, under the action of a normal stress, σ_{11}, we can write the vacancy concentration as Equation 15.15.

$$
\begin{aligned}
C_{11} &= K' \exp - (Q + \Delta\mu)/kT \\
&= K' \exp(-Q/kT) \times \exp(-\Delta\mu/kT) \\
&= C_o \exp(-\Delta\mu/kT) \text{ from Equation 15.14} \\
&= C_o \exp(\sigma_{11}\Omega/kT) \text{ using Equation 15.11}
\end{aligned} \tag{15.15}
$$

Similarly, for a normal stress of σ_{22}:

$$C_{22} = C_o \exp(\sigma_{22}\Omega/kT) \tag{15.16}$$

Equation 15.16 minus Equation 15.15 gives Equation 15.17 for $\sigma\Omega \ll kT$.

$$\Delta C = C_{22} - C_{11} = C_o(\sigma_{22}-\sigma_{11})\Omega/kT \tag{15.17}$$

For $\sigma_{11} = -\sigma_{22} = \sigma$, Equation 15.17 becomes Equation 15.18.

$$\Delta C_{t-c} = C_{tens} - C_{comp} = 2\sigma\Omega C_o/kT \tag{15.18}$$

And σ_{11} becomes tensile stress and σ_{22} compressive stress. Equations 15.17 and 15.18 show that vacancy concentration is greater where there is tensile stress compared to where there is compressive stress. This is shown in Figure 15.11a. The concentration gradient created causes atoms to migrate to the vacancies. This diffusion takes place from the region of fewer vacancies to that of greater vacancies, as shown in Figure 15.11b. Figure 15.11c shows the resultant creep deformation.

15.3.2.2 Diffusion in a Chemical Potential Gradient

Atoms diffuse from areas of higher free energy to areas of lower free energy. Higher free energy is associated with higher chemical potential. Consider the diffusion of an atom from its initial position to the next nearest vacancy. The initial, intermediate, and final positions are given in Figure 15.12a.

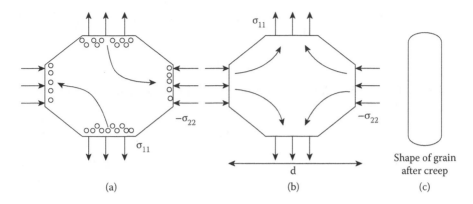

FIGURE 15.11
Diffusion of atoms under a pressure gradient in a grain: (a) vacancy concentration below the surfaces where stresses are applied; (b) diffusion paths of atoms indicated by curved arrows; and (c) grain that has undergone creep.

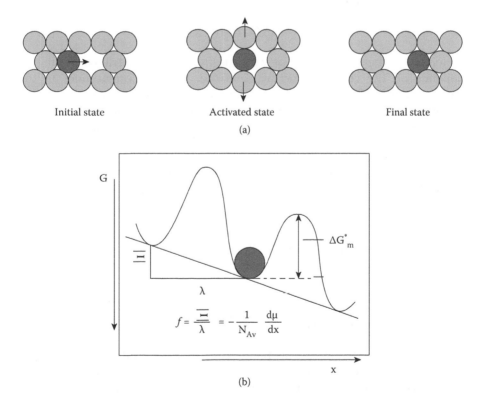

FIGURE 15.12
Diffusion of an atom to the nearest vacancy: (a) The initial, intermediate, and final positions; (b) variation of free energy versus distance.

Let us say that the atom is diffusing under a chemical potential gradient of $d\mu/dx$ per mole. This gradient per atom (f) is given by Equation 15.19.

$$f = -(1/N_{Av})(d\mu/dx) \tag{15.19}$$

Here, N_{AV} is the Avogadro's number, and f is also obtained from Figure 15.12b as:

$$f = \Xi/\lambda \tag{15.20}$$

In this equation, Ξ is the free energy difference between two sites; λ is the interatomic distance. Using Equations 15.19 and 15.20, we get an equation for the energy difference between two adjacent atom positions as given by Equation 15.21.

$$\Xi = f\,\lambda = -\lambda d(\mu/N_{Av})/dx \tag{15.21}$$

The rate of jump of an atom in the forward direction, that is, from left to right in Figure 15.12b, is given by:

$$v_{forward} = \Lambda\alpha\zeta v_0 \exp(-\Delta G_m^*/kT) \tag{15.22}$$

Here, Λ is the site fraction of vacancies, α is a geometric constant that depends on the crystal structure, ζ is the coordination number of the vacancy, v_0 is the frequency of vibration of the atoms, and ΔG_m^* is the free energy change required for the diffusion to take place. The geometric constant takes into account that only a fraction of the total jumps, ζv_0, are jumps in one direction. In the case of cubic lattices, ζ is 6, hence, only one-sixth of the total jumps are in one direction. Hence, the value of α is 1/6. It is also possible for the atoms to diffuse backward, provided they acquire sufficient energy. The frequency for the backward jump is given by:

$$v_{backward} = \Lambda\alpha\zeta v_0 \exp[-(\Delta G_m^* + \Xi)/kT] \tag{15.23}$$

The net rate of diffusion in the x-direction is given by Equation 15.24.

$$\begin{aligned} v_{net} &= v_{forward} - v_{backward} \\ &= \Lambda\alpha\zeta v_0\{\exp(-\Delta G_m^*/kT) - \exp[(-\Delta G_m^* + \Xi)/kT]\} \\ &= \Lambda\alpha\zeta v_0 \exp(-\Delta G_m^*/kT)\{1 - \exp[-(\Xi/kT)]\} \end{aligned} \tag{15.24}$$

In this equation, Ξ is chemical potential and kT is thermal energy. Chemical potential energy is in general very much less than thermal energy. Hence, Ξ/kT is very much less than 1. Therefore, $\exp(-\Xi/kT)$ reduces to $(1-\Xi/kT)$. Then Equation 15.24 becomes:

$$v_{net} = (\Lambda\alpha\zeta v_0\Xi/kT)\exp(-\Delta G_m^*/kT) \tag{15.25}$$

The average drift velocity v_{drift} is given by Equation 15.26.

$$v_{drift} = \lambda v_{net} = (\lambda \Lambda \alpha \zeta v_0 \Xi / kT) \exp(-\Delta G_m^* / kT) \tag{15.26}$$

Substituting the value of Ξ from Equation 15.21, we get:

$$v_{drift} = (\lambda^2 f \Lambda \alpha \zeta v_0 \Xi / kT) \exp(-\Delta G_m^* / kT) \tag{15.27}$$

The flux of atoms is given by Equation 15.28.

$$J_a = c_a v_{drift} = (c_a / kT)[(\lambda^2 f \Lambda \alpha \zeta v_0) \exp(-\Delta G_m^* / kT)]f \tag{15.28}$$

Here c_a is the concentration of the atoms diffusing. The term in square brackets is the expression for the diffusion coefficient. There the flux becomes:

$$J_a = (c_a D_a / kT)f \tag{15.29}$$

Consider now the diffusion in creep, and the associated Figure 15.11. Let us say that $\sigma_{11} = -\sigma_{22} = \sigma$. The chemical potential difference per atom between the top and side faces of the grain boundary is equal to the energy recovered by virtue of the diffusion of the atoms. That is:

$$\Delta\mu = -2\sigma\Omega \tag{15.30}$$

This chemical potential difference acts over an average distance of d/2, d being the grain diameter. Expressing this in the form of an equation, we get:

$$f = -d\bar{\mu}/dx = 4\sigma\Omega/d \tag{15.31}$$

The negative sign of $d\mu/dx$ indicates that the chemical potential due to the compressive stress applied on the sides of the grain decreases as the distance increases. Substituting the value of f from Equation 15.31 in Equation 15.29, it becomes:

$$J_a = (c_a D_a / kT)*(4\sigma\Omega/d) \tag{15.32}$$

The total number of atoms transported in a time t, crossing through an area A, is given by:

$$N = J_a A t \tag{15.33}$$

The volume associated with these atoms is ΩN. The strain from the diffusion of these many atoms from the two opposite sides of the grain is given by Equation 15.34.

$$\varepsilon = \Delta d/d = 2(\Omega N/A)/d \tag{15.34}$$

Substituting the value of N from Equation 15.33 in Equation 15.34, we get:

$$\varepsilon = 2(\Omega J_a t)/d \tag{15.35}$$

Combining Equations 15.33 and 15.35, we get an equation for the strain rate as:

$$\dot{\varepsilon} = 8\sigma\Omega D_a / kTd^2 \tag{15.36}$$

While combining, the product of the concentration and the atomic volume was noted to be 1. Equation 15.36 follows what is called the *Nabarro-Herring creep*. This mechanism of creep is predominant at higher temperatures where volume diffusion is faster than grain boundary diffusion. At lower temperatures, grain boundary diffusion predominates. Hence, Equation 15.36 is modified as:

$$\dot{\varepsilon} = \psi\sigma\Omega\delta_{gb}D_{gb} / kTd^3 \tag{15.37}$$

The number 8 in Equation 15.36 is replaced by a constant ψ whose value is approximately equal to 14π. Again D_a is replaced by $D_{gb}\delta_{gb}/d$. Subscript "gb" represents the grain boundary. The grain boundary width is given as δ_{gb}. Then, δ_{gb}/d^3 becomes the grain boundary cross-sectional area over which the diffusion takes place. Equation 15.37 represents the equation for *Coble creep*.

The total creep rate is the sum of the Nabarro-Herring creep and the Coble creep. In case solids contain more than one species, as in the case of ceramics, one should use a complex diffusion coefficient. This is given approximately by Equation 15.38 [1].

$$D_{complex} \approx \left(D_d^M + \pi\delta_{gb}^M D_{gb}^M\right)\left(D_d^X + \pi\delta_{gb}^X D_{gb}^X\right)$$

$$/\pi\left[a\left(D_d^M + \pi\delta_{gb}^M D_{gb}^M\right) + b\left(D_d^X + \pi\delta_{gb}^X D_{gb}^X\right)\right] \tag{15.38}$$

Here, superscripts M and X stand for the cation and anion of the ceramic MX, respectively. D is the bulk diffusion coefficient.

15.3.3 Viscous Creep

Viscous creep takes place when the ceramic contains a glassy phase. This phase softens and flows when temperature is raised to a high value. Out of many mechanisms proposed, three are the most notable.

15.3.3.1 Solution Precipitation

In the presence of an amorphous phase at the interface among crystalline grains, the crystalline phase tends to dissolve at high temperatures due to the action of a compressive force. Because of the differential stress existing in the material, the viscous glass phase formed moves to the region of tensile stress, and precipitation from this phase takes place there.

15.3.3.2 Viscous Flow

The viscosity of the glass phase decreases with an increase in temperature. When the viscosity is low, the glass phase can flow under the action of a stress. The viscosity is also related to the volume fraction of the glass phase. The relation is shown in Equation 15.39.

$$\eta_{eff} = (a\ const.)\,\eta_i/f_v^3 \tag{15.39}$$

Here, η_{eff} is the effective viscosity that is affected by the volume fraction of the phase, f_v. The intrinsic viscosity that is independent of the volume fraction is denoted as η_i. The strain rate, which is same as the creep rate, is dependent upon the shear stress and the effective viscosity.

$$\dot{\varepsilon} = \tau/\eta_{eff} \tag{15.40}$$

15.3.3.3 Cavitation

Cavitation is the formation of cavities. This phenomenon has been found to take place in ceramics containing a glass phase. The final creep fracture in this case results from the accumulation of cavities. The factors controlling this kind of creep are the microstructure, volume of glass phase, temperature, and applied stress. These factors give rise to bulk and localized damage.

Bulk damage refers to the formation of cavities throughout the material. This happens as a result of long exposure under low stresses. Figure 15.13 shows the bulk damage that has taken place in alumina. Localized damage is the result of high stresses that cause creep damage in short time periods. Here, cavities nucleate and grow at defects that act as stress concentrators. The concentrated stress at these defects nucleates the cavities. Two mechanisms lead to crack growth as a result of cavitation:

1. Direct extension of a crack
2. Damage ahead of a crack tip and its growth

The extension of a crack takes place along grain boundaries by diffusion or viscous flow. In the second mechanism, cavities nucleate and coalesce ahead of a crack tip. This extends the crack. Ahead of the extended crack, at its tip, new cavities nucleate and coalesce. These processes continue to take place until creep failure results. Cavity nucleation under tensile stress is shown in Figure 15.14 schematically.

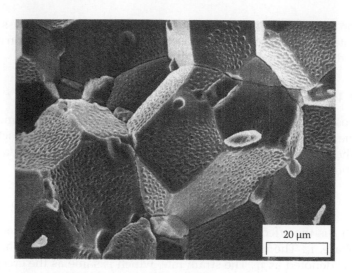

FIGURE 15.13
Bulk damage in alumina.

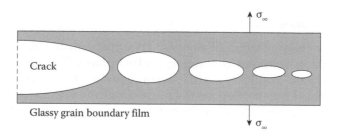

FIGURE 15.14
Cavity nucleation in viscous grain boundary film under a tensile stress.

15.3.4 Dislocation Creep

Dislocation creep takes place with the combined movement of rows of atoms or ions under the action of a stress. High temperature facilitates this movement. The extra rows of atoms or ions, called the dislocations, move along the slip planes by gliding or climbing. The steady-state creep rate by this mechanism can be represented by:

$$\dot{\varepsilon}_{ss} = \mathbf{b}[\rho v(\sigma) + d\rho(\sigma)\lambda/dt] \tag{15.41}$$

In this equation, $\mathbf{b}$ is the Burger's vector, ρ the dislocation density, $v(\sigma)$ the average velocity of a dislocation at an applied stress σ, and λ is the average distance they move before they are pinned.

15.3.5 Generalized Creep Expression

The steady-state creep can be expressed by a general formula [2] given by Equation 15.42.

$$\dot{\varepsilon} = [(const.)DGb/kT](b/d)^r(\sigma/G)^p \tag{15.42}$$

In this equation, G is the shear modulus, r is the grain size exponent, and p is the stress exponent as indicated in Equation 15.42. Based on Equation 15.42, the creep behavior of ceramics can be divided into two regimes:

1. A low-stress, small-grain-size regime where the creep rate is a function of grain size and the stress exponent is unity. Rearranging Equation 15.42, we get:

$$(\dot{\varepsilon}kT/DGb)(d/b)^r = (const.)\sigma/G \tag{15.43}$$

A plot of the LHS of this equation versus the RHS should yield a straight line with a slope of 1. Figure 15.15 shows such a plot using the experimental values. All the values lie in a band and the boundaries of this band are almost parallel straight lines. The dotted line is

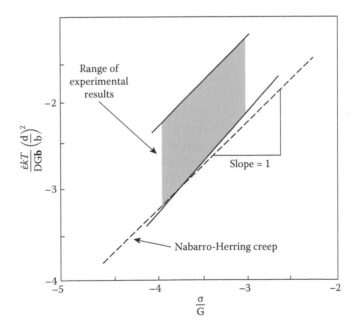

FIGURE 15.15

Normalized creep rate versus normalized stress for alumina. The dotted line is the predicted line according to Nabarro-Herring creep. (From W. R. Cannon and T. G. Langdon, *J. Mater. Sci.*, 18, 1–50, 1983.)

the predicted curve following Nabarro-Herring creep. This is found to be almost parallel to the band and also close to it. Thus, it can be concluded that at low stress and small grain regime, Nabarro-Herring creep is applicable for alumina.

2. When the stress is high, the creep rate becomes independent of grain size. That is, r becomes equal to zero. Now, Equation 15.43 becomes:

$$(\dot{\varepsilon}kT/DGb) = (const.)\sigma/G \tag{15.44}$$

Using a logarithm to plot large values, we get Equation 15.45.

$$\log(\dot{\varepsilon}kT/DGb) = A + \log(\sigma/G) \tag{15.45}$$

Here, A is a constant. This equation is plotted for a number of ceramics in Figure 15.16. It is seen that all the plots result in straight lines, with slopes lying in the range of 3–7. This proves that at stress levels, the creep rate is independent of grain size. This is not true for smaller grain sizes. This effect is shown in Figure 15.17. As the grain size decreases, the creep rate increases. The mechanism in this case happens to be diffusion because paths for diffusion decrease for smaller grain sizes.

15.4 Fracture

15.4.1 Introduction

Failure in ceramics and refractories was discussed in Section I. Here, we shall consider the mechanical fracture resulting from the application of a load. Because ceramics are brittle materials, the fracture is fast. Fast fracture is a characteristic of brittleness. This type of fracture is sudden and occurs without any warning such as plastic deformation.

In Section I, it was shown that the theoretical fracture strength of a material is about one-tenth of its Young's modulus. Young's modulus of ceramics is in the range of 100–500 GPa. Thus, the expected fracture strength is in the range of 10–50 Gpa. Only ceramics that do not contain any defect can achieve these high values of fracture strength. Practically all ceramics contain a certain amount of defects; therefore, their strength values are much lower. Also, strength values can differ by 25% from sample to sample. Fracture toughness decides a material's resistance to fracture.

15.4.2 Fracture Toughness

Fracture toughness is determined by the flaw sensitivity of the material.

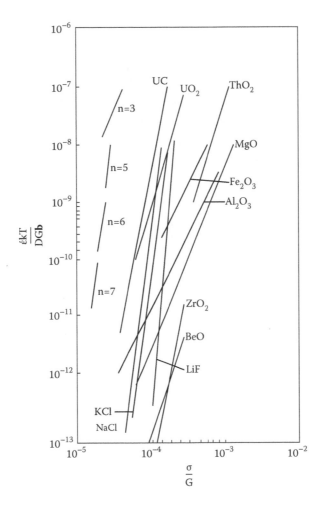

FIGURE 15.16
Creep data for a number of ceramics at high stress values. (From W. R. Cannon and T. G. Langdon, *J. Mater. Sci.*, 18, 1–50, 1983.)

15.4.2.1 Flaw Sensitivity

The presence of flaws causes a material to be vulnerable to fracture. How a flaw can lead to fracture can be seen by considering Figure 15.18. When there is no flaw, the applied force is uniformly distributed over the entire cross-sectional area. The stress experienced by the material is given by the force divided by the cross-sectional area. When there is a crack in the material, the cross-sectional area decreases correspondingly and increases the stress in the material. The force lines passing through the area above the crack can become continuous and pass to the area below the crack only

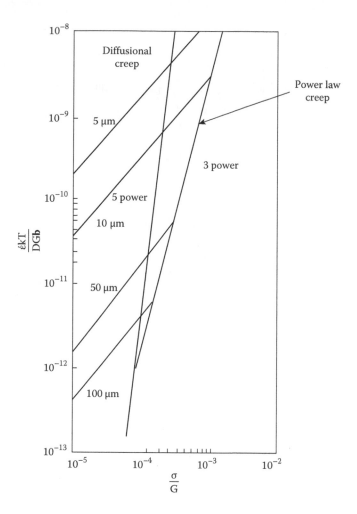

FIGURE 15.17
Effect of grain size on creep. (From W. R. Cannon and T. G. Langdon, *J. Mater. Sci.*, 18: 1–50, 1983.)

by passing through the crack tip. These force lines increase the stress at the crack tip. The stress reaches the fracture strength at the crack tip first. Hence, the crack grows by the breakage of the interatomic bond at the crack tip. The cross-sectional area further decreases, leading to more stress concentration reaching the fracture stress in less time and again to crack extension; this cycle continues in a fraction of a second, finally resulting in sudden fracture.

The stress at the crack tip can be obtained from the formula:

$$\sigma_{tip} = 2\sigma_{app}\sqrt{(c/\rho)} \tag{15.46}$$

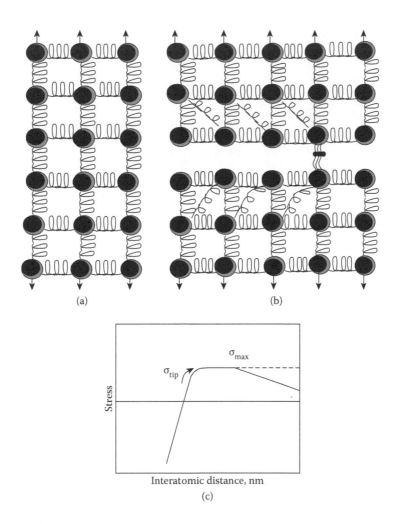

FIGURE 15.18

Illustration of flaw sensitivity: (a) Uniform stress when there is no flaw; (b) stress concentration when a crack is created; and (c) variation of stress with respect to interatomic distance.

The applied stress is given by σ_{app}. The letters "c" and "ρ" are specified in Figure 15.19. The letter "c" represents crack length, and "ρ" is the radius of curvature. This σ_{tip} is a function of the type of loading, sample, crack geometry, and so forth [3].

When the σ_{tip} reaches the fracture stress (σ_f), the failure takes place. It was mentioned that fracture stress is about one-tenth of the materials' Young's modulus. When the applied stress reaches this value, fracture takes place. Thus Equation 15.46 can be written in terms of σ_f as:

$$\sigma_f \approx (Y/20)\left(\sqrt{(\rho/c)}\right) \tag{15.47}$$

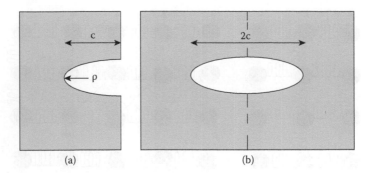

FIGURE 15.19
(a) Surface crack of length c and radius of curvature ρ; (b) interior crack of length 2c.

Here, Y is the Young's modulus. The following are the inferences from this equation:

a. Fracture stress decreases as the radius of curvature of the crack decreases.

b. It is inversely proportional to the square root of the crack size.

The first inference means that sharp cracks are more vulnerable than blunt cracks. The two inferences have been proved experimentally.

15.4.2.2 Griffith Criterion

Griffith [4] has evolved a criterion for fracture based on the energy release rate. When a crack propagates, two new surfaces are created. The energy supplied by means of applied stress is utilized in creating these surfaces. The rate at which this energy is released exceeds the rate at which the energy is consumed; that point, at which it exceeds, marks the critical condition for fracture.

When a parallelepiped (Figure 15.20) is stretched by means of an applied stress, the energy acquired by the material is utilized in elastically stretching the interatomic bonds. This increases the internal energy to an extent given by the plot of stress versus strain (Figure 15.21). This internal energy is in the form of elastic stored energy as given by Equation 15.48.

$$U_{elas} = \left(\frac{1}{2}\right)\varepsilon\sigma_{app} = \left(\frac{1}{2}\right)\sigma_{app}^2 / Y \tag{15.48}$$

Because this energy is per unit volume, the total internal energy in the parallelepiped of volume V_0 is given by:

$$U = U_0 + V_0 U_{elas} = U_0 + V_0\sigma_{app}^2 / 2Y \tag{15.49}$$

U_0 is the internal energy before stressing.

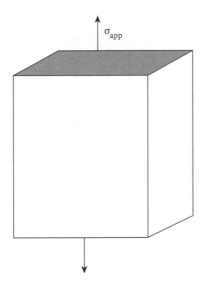

FIGURE 15.20
Stretching of a parallelepiped.

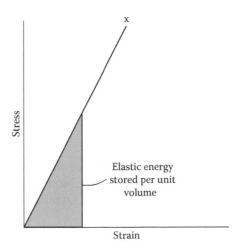

FIGURE 15.21
Elastic stored energy as a result of applied stress.

Suppose there is a surface crack of size "c" in the solid sample shown in Figure 15.20, a small volume around the crack will be relieved of the strain energy. Thus, the total internal energy now can be written in the form of Equation 15.50.

$$U = U_0 + V_0 \sigma_{app}^2 / 2Y - \left(\sigma_{app}^2 / 2Y \right) \pi c^2 t / 2 \qquad (15.50)$$

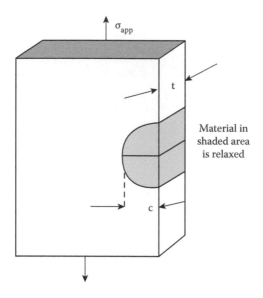

FIGURE 15.22
Strain energy released volume.

Here, "t" is the thickness of the slab. The third term is the *strain energy released*. The affected volume containing the crack is shown in Figure 15.22.

For the formation of a crack, *surface energy* is required. A crack formation gives rise to two surfaces. The energy expended in the creation of the crack shown in Figure 15.22 is given by:

$$U_{surf} = 2\gamma ct \tag{15.51}$$

In this equation, γ is the surface energy of the material. The "2" in the equation represents the two surfaces created. Now, the total internal energy of the parallelepiped has become the sum of Equations 15.50 and 15.51. This sum is given by Equation 15.52.

$$U_{total} = U_0 + V_0\sigma_{app}^2 / 2Y - \left(\sigma_{app}^2 / 2Y\right)\pi c^2 t / 2 + 2\gamma ct \tag{15.52}$$

Equations 15.50 through 15.52 are plotted in Figure 15.23, which shows a critical crack size. At this size, the total energy reaches a maximum. Cracks having sizes less than this size requires more energy to grow, whereas those having greater size release energy as they grow. Therefore, when the cracks exceed the critical size, they propagate spontaneously resulting in fracture. This critical crack size depends on the applied stress. The plot shown in Figure 15.24 is for a higher stress than that used in Figure 15.23. It shows that the critical crack size decreases with an increase in applied stress.

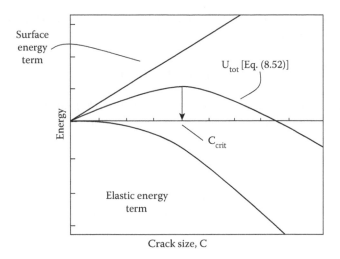

FIGURE 15.23
Plot showing the critical crack size.

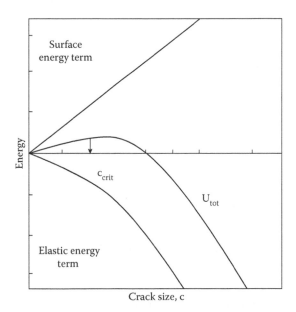

FIGURE 15.24
Plot showing the critical crack size for higher applied stress.

The maximum value of the internal energy can be obtained by differentiating Equation 15.52 equating the result to zero. Thus, we get:

$$\sigma_f \sqrt{(\pi c_{crit})} = 2\sqrt{(\gamma Y)} \qquad (15.53)$$

Here, σ_f represents σ_{app} of Equation 15.52. The criterion for fracture can be written as Equation 15.54.

$$\sigma_f \sqrt{(\pi c_{crit})} > \sqrt{(2\gamma Y)} \qquad (15.54)$$

The LHS is called the stress intensity factor (K_I), and the RHS is the *critical stress intensity factor* (K_{Ic}) or *fracture toughness*. Hence, the criterion for fracture is:

$$K_I > K_{Ic} \qquad (15.55)$$

In order to also accommodate the plastic deformation taking place in certain materials into fracture toughness, the fracture toughness is written in the form of Equation 15.56.

$$K_{Ic} = \sqrt{(YG_c)} \qquad (15.56)$$

G_c is the *toughness* of the material. Its unit is joules per square meter.

Example 2

A sharp edge crack of 80 μ is present in a ThO_2 plate. It is under a tensile load of 120 MPa.
   a. Find out whether the plate will fracture.
   b. If the crack is an internal one, what does that say about its survival under the same tensile load? Given: $(K_{Ic})_{ThO_2} = 1.6$ MPa.m$^{1/2}$

Solution
   a. $K_I = 120\sqrt{(3.14 \times 80 \times 10^{-6})} = 120 \times 15.85 \times 10^{-3} = 1.9$ MPa.m$^{1/2}$
      This value being greater than 1.6 MPa.m$^{1/2}$, the plate will fracture
   b. $K_I = 120\sqrt{[3.14 \times (80 \times 10^{-6}/2)]} = 120 \times \sqrt{(3.14 \times 40 \times 10^{-6})} = 120 \times 11.21 \times 10^{-3} = 1.3$ MPa.m$^{1/2}$

Because $K_I < K_{Ic}$, the plate will survive. Next, we consider the tests for determining fracture toughness.

15.4.2.3 Measurement of Fracture Toughness

There are two methods generally adopted for the measurement of fracture toughness (K_{Ic}). Both methods depend on using specimens of specific geometry and the contain known initial crack length.

The criterion for fracture given in Equation 15.54 can be written in a general form as:

$$\psi\sigma_f \sqrt{(\pi c)} \geq K_{Ic} \qquad (15.57)$$

Here, ψ is a dimensionless constant that depends on the sample shape, crack geometry, and crack size relative to sample dimensions. The crack should be atomically sharp. The problem lies in producing such a minutely sharp crack in a macroscopic specimen. The value of ψ for different specimen and crack geometries can be obtained from fracture mechanics handbooks.

The first of the two common methods for testing fracture toughness is called the *single-edge notched beam* (SENB) test. The set-up for this test is shown in Figure 15.25. The load is applied through the top two rollers until the specimen fractures. This fracture load is noted as F_{fract} and substituted in Equation 15.58. In this equation, ξ is a calibration factor. The drawback of this test is that we may not always obtain an atomically sharp crack.

$$K_{Ic} = 3\sqrt{c(S_1 - S_2)}\xi F_{fract} / 2BW^2 \qquad (15.58)$$

The second common test uses a chevron notch specimen. In this specimen, the initial crack is chevron-shaped. As the crack grows by the increasing load, the crack front constantly widens. This causes the crack growth to be stable prior to fracture. Because the crack grows with the increasing load, the growth is the result of the atomically sharp crack formation. Thus, there is no need to precrack the specimen. The fracture toughness [5] is obtained from:

$$K_{Ic} = (S_1 - S_2)\xi^* F_{fract} / BW^{3/2} \qquad (15.59)$$

Here, ξ^* is the minimum compliance function.

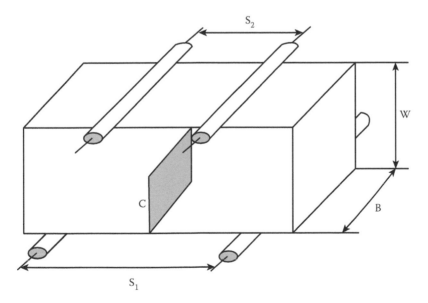

FIGURE 15.25
Schematic of single-edge notched beam specimen.

15.5 Toughening Mechanisms

There have been many attempts to increase the toughness of ceramics because they are inherently brittle. All these attempts are an effort to increase the G_c value given in Equation 15.56. The basic approaches are crack deflection, crack bridging, and transformation toughening. These approaches are discussed further here.

15.5.1 Crack Deflection

Polycrystalline ceramics are found to be tougher than single crystalline ones. For example, polycrystalline alumina possesses a K_{Ic} value of 4 MPa.m$^{1/2}$. Single crystalline alumina has only 2.2 Mpa.m$^{1/2}$. The major cause for this difference is found to be crack deflection. This mechanism is shown schematically in Figure 15.26. Cracks more easily propagate through grain boundaries. Grain boundaries have different orientations. Hence, cracks are deflected to different angles as they propagate. This increases the path length required for the fracture and, consequently, increases the energy to be expended. The stress intensity factor at the crack tip (K_{tip}) is related to the angle of deflection (θ) and the applied stress intensity (K_{app}) by Equation 15.60.

$$K_{tip} = [\cos^2(\theta/2)]K_{app} \qquad (15.60)$$

Taking an average θ value of 45°, the K_{Ic} value of polycrystalline material should be greater by 1.25 over that of a single crystal. If we apply this value to alumina, the polycrystalline alumina should have a K_{Ic} value of 3.45 Mpa.m$^{1/2}$. This says that crack deflection does not account for all of the increase. So, let us consider the next mechanism.

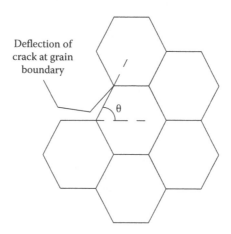

Deflection of crack at grain boundary

θ

FIGURE 15.26
Mechanism of crack deflection.

15.5.2 Crack Bridging

Crack bridging results from the presence of a strong reinforcing phase. This phase connects the two crack surfaces. The reinforcing phase can be in the form of whiskers, continuous fibers, or elongated particles. Figure 15.27 [6] shows how elongated particles can cause crack deflection and bridging. In this figure, the first, second, and fourth particles cause crack deflection, and the third causes crack bridging. Figure 15.28 [7] shows how continuous fibers can cause crack bridging. These fibers remain unbroken, and they carry part of the applied load. Thus, stress intensity at the crack tip is reduced.

The K_{Ic} for a fiber-reinforced composite is given by Becher [7]:

$$K_{IC} = \sqrt{\left[Y_c G_m + \sigma f^2 \left(r V_f Y_c \gamma_f / 12 Y_f \gamma_i \right) \right]} \qquad (15.61)$$

Here, the subscripts c, m, and f represent composite, matrix, and fiber, respectively. And Y, V, and σ are the Young's modulus, volume fraction, and fracture strength, respectively. Also, r is the radius of the fiber, G is the toughness, and γ_f/γ_i is the ratio of the fracture energy of the fiber to that of the interface. From this equation, we can determine that in order to increase fracture toughness, we must:

a. Increase the volume fraction of fiber

b. Increase the Y_c/Y_f ratio

c. Increase the γ_f/γ_i ratio

The third factor can be increased by decreasing the interfacial energy. An aftereffect of this decrease is the pullout of the fibers. Because this pullout consumes energy, it further increases the fracture toughness.

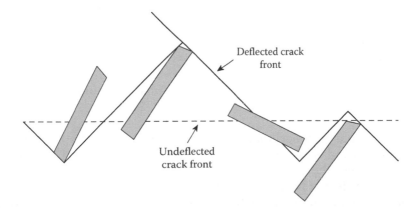

FIGURE 15.27
Mechanism of crack bridging and deflection by rod-shaped grains. (From A. G. Evans and R. M. Cannon, *Acta Metall.*, 34, 761–800, 1986.)

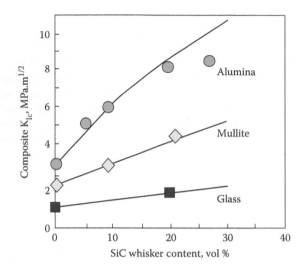

FIGURE 15.28
The effect of SiC whisker content on toughness enhancement in different matrices. (From P. Becher, Microstructural design of toughened ceramics, *J. Am. Ceram. Soc.*, 74, 255–269, 1991.)

Figure 15.28 shows the effect of the whisker volume fraction on fracture toughness. The solid lines are the ones obtained by using Equation 15.61. The data points represent the experimental values. The plots show good agreement between the two.

15.5.3 Transformation Toughening

If we can induce a phase transformation ahead of a crack tip with the help of applied stress such that the transformed phase is tough, then we can increase the toughness of the material. This mechanism was originally discovered in zirconia [8]. Zirconia at 1 atmospheric pressure undergoes the following transformations:

$$\text{Monoclinic} \leftrightarrow \text{Tetragonal} \leftrightarrow \text{Cubic} \leftrightarrow \text{Liquid}$$
$$1170°C \qquad 2370°C \quad 2680°C$$

The tetragonal to monoclinic phase is a diffusionless transformation. It is a shear process and is associated with a large volume increase. This transformation toughens the matrix [6,9].

The mechanism of transformation toughening is shown in Figure 15.29. Zirconia is heated to above 1170°C in order to partially transform to the tetragonal phase. The transformation temperature and time are controlled to obtain fine precipitate particles in the tetragonal phase. Being fine in size, these particles are restrained from transforming back to the

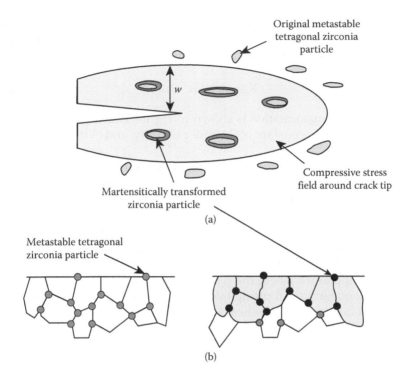

FIGURE 15.29
Mechanism of transformation toughening. (a) shows the compressive stress field around crack tip, and (b) shows the stabilized zirconia particles.

monoclinic phase by the surrounding matrix phase, when the material is cooled to room temperature. Later, when a crack is formed under a tensile stress, there will be stress concentration at the crack tip. Under the action of this stress, the tetragonal phase particles shear and transform to the monoclinic phase. This transformation expands the matrix in a small volume surrounding the crack. This volume is shown in Figure 15.29a. The tetragonal-to-monoclinic transformation is accompanied by about 4% volume expansion. The shear strain associated with this expansion is about 7%.

When the particles near the surface transform, they give rise to a compressive stress on the surface. This compressive stress tends to close any crack already formed or prevent any nucleation of a fresh crack. The compressive layer is shown schematically in Figure 15.29b.

The effect of dilation strain is to reduce the stress intensity at the crack tip. The extent of the decrease in stress intensity is called the shielding factor K_s. This is related to the crack tip stress intensity K_{tip} by Equation 15.62.

$$K_{tip} = K_a - K_s \qquad (15.62)$$

Here, K_a is the stress intensity due to applied stress. If the volume fraction of the transformable phase is V_f, and the width of the zone of transformation is w, then the shielding intensity factor is given by Equation 15.63 [10].

$$K_s = A'YV_f\epsilon^T\sqrt{w} \tag{15.63}$$

The width of transformation is shown in Figure 15.29a. In Equation 15.63, A' is a dimensionless constant of the order of unity, and ϵ^T is the transformation strain.

As the temperature increases, the driving force for transformation decreases. The compressive surface layer can be formed by abrading the surface. There are three kinds of toughened zirconia:

1. *Partially stabilized zirconia* (PSZ). In this kind of zirconia, the cubic phase is formed initially. This cubic phase is partially stabilized with the help of MgO, CaO, or Y_2O_3. Then, this material is heated to a temperature where the tetragonal phase is stable. By this heat treatment, tetragonal precipitates are formed from the partially stabilized cubic phase. The heat treatment is controlled in such a way that it results in fine precipitates. Only when a stress is applied, do these precipitates transform to develop a toughened zirconia matrix.

2. *Tetragonal zirconia polycrystals* (TZPs). In this material, Zirconia is 100% in the tetragonal phase. To stabilize this phase until transformation takes place, there will be small amounts of Y_2O_3 and other rare earth additives. TZPs are among the strongest ceramics, with a flexural strength exceeding 2000 MPa.

3. *Zirconia-toughened ceramics* (ZTCs). These are ceramic matrix composites. The matrix is alumina, mullite, or spinel. The reinforcements are tetragonal or monoclinic zirconia particles.

Problems

15.1 Estimate the size of the critical flaw for a glass that failed at 105 MPa if $\gamma = 1$ J·m^2 and Y = 70 GPa.

15.2 What is the maximum stress that a glass will withstand if it contains a largest crack of 100 μ and a smallest crack of 10 μ. Take $\gamma = 1$ J·m^2 and Y = 70 GPa.

15.3 Ten rectangular test specimens of MgO were tested in three-point bending. The bars were 1 cm wide and 0.5 cm high and were tested over a 5-cm span. The failure loads for each are given in ascending order: 140, 151, 154, 155, 158, 165, 167, 170, 173, and 180 kg. Calculate the MOR for each sample and the average MOR for this group of samples.

15.4 The tensile fracture strengths in MPa of hot-pressed silicon nitride (HPSN), reaction-bonded silicon nitride (RBSN), and chemical vapor deposited silicon carbide (CVDSC) are listed below:

HPSN: 521, 505, 500, 490, 478, 474, 471, 453, 452, 448, 444, 441, 439, 430, 428, 422, 409, 398, 394, 372, 360, 341, 279

RBSN: 132, 120, 108, 106, 103, 99, 97, 95, 93, 90, 89, 84, 83, 82, 80, 80, 78, 76

CVDSC: 386, 351, 332, 327, 308, 296, 290, 279, 269, 260, 248, 231, 219, 199, 178, 139

15.4.1 Rank the specimens in the order of increasing strength.

15.4.2 Calculate the mean strength and standard deviation of the strength distributions.

15.4.3 Determine the survival probability for each specimen.

15.4.4 Plot $-\ln \ln(1/S)$ versus $\ln \sigma$.

15.4.5 Determine the Weibull modulus for each material.

15.4.6 Estimate the design stress for each material.

15.5 When the ceramic shown in the figure here was loaded in tension, it fractured at 20 MPa.

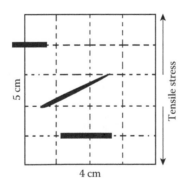

4 cm

The heavy lines denote the cracks. Estimate K_{Ic} for this ceramic. State all assumptions.

15.6 Evans and Charles [11] proposed the following equation for the determination of fracture toughness from indentation:

$$K_{Ic} \approx 0.15(H\sqrt{a})\,(c/a)^{-1.5}$$

Where H is Vickers hardness in Pa. a and c are defined in the figure here (side and top views of a crack developed from a Vickers indentation).

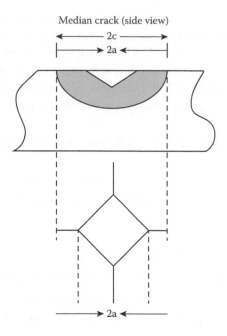

Median crack (side view)

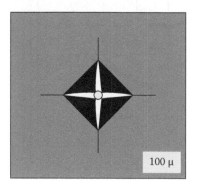

A photomicrograph of a Vickers indentation in a glass slide and the cracks that emanate from it is shown here:

100 μ

Estimate the fracture toughness of this glass if its hardness is ≈5.5 GPa.

15.7 The average MOR of a ceramic material was found to be 100 MPa with a Weibull modulus of 10. Estimate the stress required to obtain a survival probability of 95%. State all assumptions.

15.8 Kschinka et al. [12] tested 44 glass spheres 0.51 mm in diameter and found the average strength of these spheres was 435 MPa. What would be the design stress for these spheres to ensure a 99% survival probability and an m value of 6.82?

15.9 The grain boundary diffusivity is given by

$$D_{gb} = 100 \exp(-40kJ/RT)$$

and the bulk diffusion coefficient is

$$D_{latt} = 300 \exp(-50kJ/RT).$$

Determine which diffusion will dominate at 900 K and at 1300 K? At what temperature will both be equal?

15.10 Typical crack growth data for a glass placed in a humid environment are listed below:

Stress intensity, K_I, MPa·m$^{1/2}$	Crack velocity, v, m/s
0.4	1E-6
0.5	1E-4
0.55	1E-3
0.6	1E-2

In the following equations, calculate the values of A′, A, and n:

$$v = A (K_I/K_{Ic})^n$$
$$v = AK_I^n$$

Given: $K_{Ic} = 0.7$ MPa·m$^{1/2}$

References

1. A. G. Evans, and T. G. Langdon, Structural ceramics, *Prog. Mater. Sci.*, 21: 171–285, 1976.
2. W. R. Cannon and T. G. Langdon, Creep of ceramics, *J. Mater. Sci.*, 18: 1–50, 1983.
3. C. E. Inglis, Stresses in a plate due to the presence of cracks and sharp corners, *Trans. Inst. Nav. Architect.*, 55: 219, 1913.
4. A. A. Griffith, The phenomenon of rupture and flow in solids, *Phil. Trans. R. Acad.*, A221: 163, 1920.
5. J. Sung and P. Nicholson, Valid KIC determination via in-test subcritical pre-cracking of chevron-notched bend bars, *J. Am. Ceram. Soc.*, 72(6): 1033–1036, 1989.
6. A. G. Evans and R. M. Cannon, Toughening of brittle solids by martensitic transformations, *Acta Metall.*, 34: 761–800, 1986.
7. P. Becher, Microstructural design of toughened ceramics, *J. Am. Ceram. Soc.*, 74: 255–269, 1991.
8. W. Y. Yan, G. Reisner, F. D. Fischer, Micromechanical study on the morphology of martensite in constrained zirconia, *Acta mater.*, 45: 1969–1976, 1997.

9. G. Patridge, Strengthening glasses and ceramics: Composite technology, *Advanced Materials*, 5(6): 468–473, 1993.
10. R. M. McMeeking and A. G. Evans, The mechanics of transformation toughening in brittle materials, *J. Am. Ceram. Soc.*, 63: 242–246, 1982.
11. A. G. Evans and E. A. Charles, Fracture toughness determinations by indentation, *J. Am. Ceram. Soc.*, 59: 317, 1976.
12. B. A. Kschinka, S. Perrella, H. Nguyen, and R. C. Bradt, Strengths of glass spheres in compression, *J. Am. Ceram. Soc.*, 69: 467, 1986.

Bibliography

M. W. Barsoum, *Fundamentals of Ceramics*, Taylor & Francis, New York, 2003, pp. 187–189, 245–246, 356–368, 380–386, 396–399, 401–415, 435–436.
C. B. Carter and M. G. Norton, *Ceramic Materials—Science and Engineering*, Springer, New York, 2007, pp. 289–292, 296–298, 307.

16

Thermal and Thermo-Mechanical Properties

16.1 Introduction

Ceramics are widely used as insulators, primarily in various kinds of furnaces. The class of ceramics used for this purpose is called the refractories. Apart from heat resistance, these refractories should also possess a thermo-mechanical property called refractoriness under load (RUL). This chapter focuses on the various thermal and thermo-mechanical properties of ceramics.

16.2 Thermal Properties

16.2.1 Introduction

The thermal properties that are of importance in ceramic applications are heat capacity, thermal expansion, and thermal conductivity. Heat capacity is a measure of the heat required for changing the temperature. It is therefore important in the heat treatment and use of ceramics. The economy of running a furnace is decided by the amount of heat needed to reach a particular temperature. Thermal expansion becomes important in ceramic composites. The different constituents in a composite should have close thermal expansion coefficients in order to prevent accumulation of thermal stresses in the composite. Thermal conductivity is an undesirable property in the case of insulators. Generally, ceramics possess low thermal conductivity.

16.2.1.1 Heat Capacity

Heat capacity is defined as the heat required to raise the temperature of 1 g of the material by 1°C in centigrade-gram-second (CGS) system. When 1 mole of the substance is considered, it is called molar heat capacity. Depending on whether the measurement is taken at constant volume or pressure, we have heat capacity at constant volume (c_v) or at constant pressure (c_p). The expressions for the two kinds of heat capacities are given by Equations 16.1 and 16.2.

$$c_v = (\partial Q/\partial T)_v = (\partial E/\partial T)_v \qquad (16.1)$$

$$c_p = (\partial Q/\partial T)_p = (\partial H/\partial T)_p \qquad (16.2)$$

E is the internal energy, and H is the enthalpy of the material. The difference between the two heat capacities is given by:

$$c_p - c_p = \alpha^2 V_0 T/\beta \qquad (16.3)$$

Here, α is the volume thermal expansion coefficient, V_0 is the molar volume, and β is the compressibility. The formulae for α and β are given by Equations 16.4 and 16.5.

$$\alpha = (1/v)(dv/dT) \qquad (16.4)$$

$$\beta = -(1/v)(dv/dp) \qquad (16.5)$$

For condensed systems such as solids and liquids, the difference between c_p and c_v is negligibly small at ordinary temperatures.

16.2.1.2 Thermal Expansion

Thermal expansion is measured in two ways: volume and linear thermal expansion coefficients. The formula for the volume thermal expansion coefficient has already been given in Equation 16.4. The formula for the linear thermal expansion coefficient is given by Equation 16.6.

$$a = (1/l)(dl/dT) \qquad (16.6)$$

16.2.1.3 Thermal Conductivity

The thermal conductivity is obtained from Equation 16.7.

$$dQ/dt = -kA(dT/dx) \qquad (16.7)$$

Here, dQ is the amount of heat flowing normal to the area A in time dt; dQ/dt is called the heat flux. For a given area, the heat flow is proportional to the temperature gradient, $-dT/dx$. The proportionality constant k is called the thermal conductivity.

If q is the heat flux through a slab, it is given by:

$$q = -kA(T_2 - T_1)/(x_2 - x_1) \qquad (16.8)$$

Here, x_1 and x_2 are two points lying on a line perpendicular to thickness. T_1 and T_2, respectively, are the temperatures at these two points. For heat flow along the radius of a hollow tube, the heat flux is given by:

$$q = -k(2\pi l)(T_2 - T_1)/(\ln D_2 - \ln D_1) \tag{16.9}$$

In this equation, l is the length of the tube, T_2 is the temperature at the end of an outer diameter, and T_1 is the temperature at the end of the inner diameter lying along the same outer diameter.

When the temperature changes with time at any point in a material, the rate of change of temperature with time is dependent on thermal diffusivity. Thermal diffusivity is the ratio of thermal conductivity to heat capacity per unit volume. The heat flux for a transient heat flow can then be written as Equation 16.10.

$$dT/dt = d[(k/\rho c_p)(dT/dx)]/dx \tag{16.10}$$

Here, ρ is the density of the material and c_p is the specific heat capacity at constant pressure, that is, heat capacity per unit mass at constant pressure. Now let us consider all three thermal properties in some detail.

16.2.2 Heat Capacity

The heat given to a material is utilized for various processes:

  a. For increasing the vibrational energy
  b. For increasing the rotational energy
  c. For increasing the energy level of electrons
  d. For giving rise to defects and thereby for disordering
  e. For changing magnetic orientation
  f. For altering a structure

Atoms vibrate about their mean positions in solids. An increase in vibrational energy increases their amplitude and the frequency of their vibrations. Defects such as Schottky and Frenkel are produced by the movement of atoms from their original positions. All the aforementioned processes increase the internal energy of the material and the configurational entropy.

Each atom possesses an average energy of kT obtained as a sum of ½kT each of potential and kinetic energies for each degree of freedom. Because there are three degrees of freedom along three axes, the total energy per atom becomes 3kT. A g-atom of the substance possesses 3NkT energy. The specific heat at constant volume per g-atom is given by:

$$c_v = (dE/dT)_v = [d(3NkT)/dT]_v = 3Nk = 5.96 \text{ cal/g-atom } °C \tag{16.11}$$

At higher temperatures, materials are found to acquire this constant value of heat capacity.

$$C_v/3Nk = f[(hv_{max}/k)/T] = f(\theta_D/T) \qquad (16.12)$$

According to Debye theory [1], for a maximum frequency of lattice vibration, v_{max}, hv_{max}/k is called the Debye temperature, θ_D. At low temperatures, heat capacity is found to be proportional to $(T/\theta_D)^3$.

Bond strength, elastic constant, and melting point of a material decide the minimum temperature at which c_v becomes a constant. The variation of heat capacity with temperature for four common ceramics is shown in Figure 16.1. It can be seen from this figure that the minimum temperature is in the range of one-fifth to one-half the melting temperature of the material in K.

The constant value of heat capacity beyond θ_D temperature is contributed by only the vibrational energy. And most of the heat supplied goes to increase this energy. The further slight increase seen beyond θ_D temperature is due to the contribution to other forms of energy given earlier. When we consider the heat capacity at constant pressure, there is a larger increase in its value beyond this critical temperature. This increase is found to be linear with temperature.

Porosity decreases the heat capacity. This is advantageous for furnaces that are to be heated and cooled rapidly, as well as for those that are used frequently. Furnaces use insulators as their constructing material. These are ceramics with high porosity. By providing a large amount of porosity, their heat capacity is low. Therefore, for a fixed amount of heat amount, they can be heated faster. They also have less heat content, so they can be cooled faster. Because of less heat intake, those furnaces are more economical.

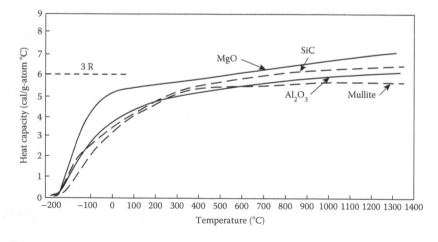

FIGURE 16.1
Variation of heat capacity with temperature.

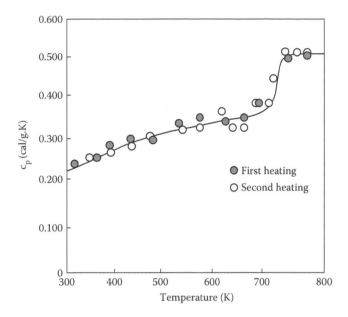

FIGURE 16.2
Variation of heat capacity with temperature for $0.15Na_2O-0.85B_2O_3$ glass. (From D. R. Uhlmann, A. G. Kolbeck, and D. L. de Witte, *J. Non Cryst. Solids*, 5, 426, 1971.)

Figure 16.2 [2] shows the variation of heat capacity with temperature for $0.15Na_2O-0.85B_2O_3$ glass. The sudden increase in heat capacity marks the transition to the liquid state. In the liquid state, configurational entropy comes into play, which results in higher heat capacity.

16.2.3 Thermal Expansion

Figure 16.3 shows the variation of lattice energy with interatomic distance. Lattice energy is the sum of two energies, namely the attractive energy and the repulsive energy. In the case of an ionic solid, the attractive energy arises from the coulombic attraction among unlikely charged ions, and the repulsive energy arises from the repulsion among likely charged ones. The repulsive energy falls very rapidly as interionic distance increases. This unequal variation of both the energies gives rise to a nonsymmetrical trough for the variation of the lattice energy. The minimum in this trough marks the equilibrium separation.

Temperature increases the equilibrium interionic separation. This is because the amplitude of vibration increases with temperature. But it brings about a more symmetrical arrangement. In other words, the entropy decreases as temperature is increased. The increase in thermal expansion and heat capacity is parallel with an increase in temperature, as shown in Figure 16.4. Both result from an increase in the internal energy of the material.

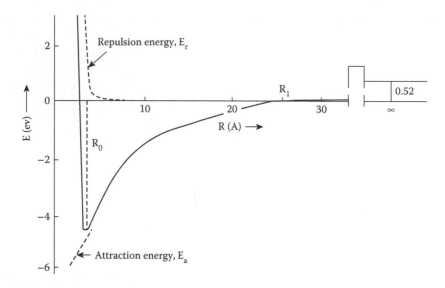

FIGURE 16.3
Variation of lattice energy with interatomic separation.

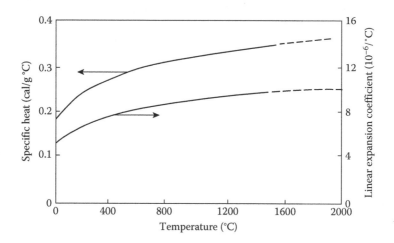

FIGURE 16.4
Parallel changes in specific heat and thermal expansion coefficient of alumina.

In the case of isotropic materials, such as those having cubic crystal structure, the thermal expansion is given by:

$$1 + \alpha \Delta T = 1 + 3a\Delta T + 3a^2 (\Delta T)^2 + a^3(\Delta T)^3$$
$$\alpha = 3a + 3a^2 (\Delta T) + a^3(\Delta T)^2 \qquad (16.13)$$

Here, $\bar{\alpha}$ is the average volume thermal expansion coefficient and a is the average linear expansion coefficient for a small temperature change ΔT. Because a is small for a small temperature change, higher orders of a than 1 can be neglected. Hence, the volume expansion coefficient for an isotropic material can be written as:

$$\bar{\alpha} = 3a \qquad (16.14)$$

For a nonisotropic crystal, such as a hexagonal one, the thermal expansion varies with the axis. For example, in the case of alumina, its average linear thermal expansion coefficient normal to the c-axis is 8.3×10^{-6} per °C, whereas it is 9.0×10^{-6} per °C parallel to the same axis. Another interesting phenomenon taking place in these crystals is that both c/a and a_c/a_a decrease as temperature is increased. That means, as stated earlier, the structure becomes more symmetrical. Some materials, such as aluminum titanate, show contraction with an increase in temperature in a direction normal to the c-axis. These materials exhibit resistance to thermal shock as their volume expansion with temperature becomes low.

Apart from crystal structure, the bond strength also decides the thermal expansion. Materials such as tungsten possess high bond strength. Consequently, their thermal coefficient is low. Oxide structures have dense crystal structures. Their linear thermal expansion coefficients at room temperature are in the range of 6 to 8×10^{-6} per °C.

16.2.3.1 Thermal Expansion of Glasses

The network structure in glasses has a low density. When modifiers are added, their density increases, even though the structure is loosened. Figure 16.5 [3] shows the effect of adding lead dioxide to silica glass.

The density of a glass depends on the rate of cooling of a glass melt. As the rate of cooling increases, its density decreases. Its thermal expansion coefficient varies with temperature. This is shown in Figure 16.6. There is a sudden increase seen in the expansion coefficient in the temperature range of 550°C to 600°C. At this temperature range, a structural rearrangement takes place in which a glassy solid transforms to a supercooled liquid. Above this temperature range, viscous flow takes place, and the sample contracts.

In the case of $Na_2O-B_2O_3$ systems, there is a minimum for the thermal expansion coefficient. This behavior is called the boric oxide anomaly. This happens because with low concentrations of the alkali oxide, the BO_3 triangles converts to BO_4 tetrahedra. This conversion reduces the expansion. For larger alkali concentrations, singly bonded oxygens are formed and the expansion increases. Figure 16.7 [4] shows the effect of various alkali oxides on the thermal expansion coefficent of borate glass. The larger the size of the cation, the larger the increase in the expansion cofficient.

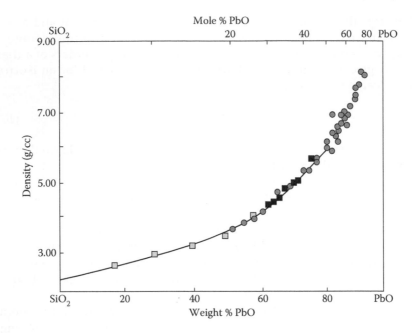

FIGURE 16.5
Effect of addition of PbO_2 to SiO_2 glass. (From D. R. Uhlmann, *J. Non Cryst. Solids*, 1, 474, 1969.)

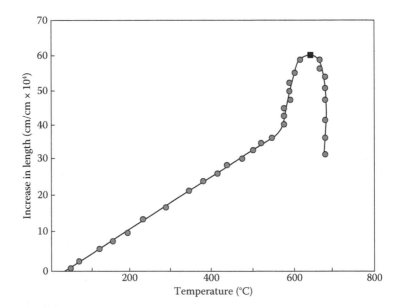

FIGURE 16.6
Variation of thermal expansion coefficient of a silicate glass.

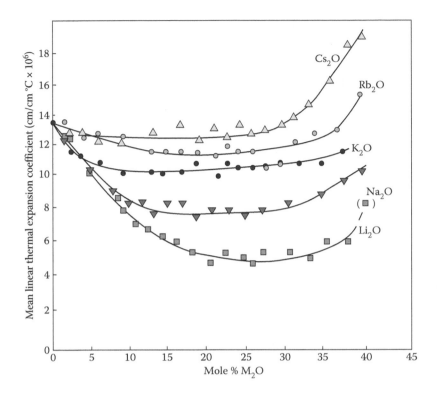

FIGURE 16.7
Thermal expansion coefficients of alkali borate glasses as a function of composition. (From R. R. Shaw and D. R. Uhlmann, *J. Non Cryst. Solids*, 1, 347, 1969.)

16.2.3.2 Thermal Expansion of Composites

In a composite, different phases have different expansion coefficients. But there will be restriction from neighboring grains for expansion and contraction. Because of this, microstresses are developed. The stresses on each grain are given by Equation 16.15.

$$\sigma_i = K(\alpha_r - \alpha_i)\Delta T \tag{16.15}$$

Here, σ_i and α_i are the stress and volume expansion coefficient, respectively, of the grain i; K is the bulk modulus, α_r is the average volume expansion coefficient of the composite, and ΔT is the temperature difference. The bulk modulus is given by:

$$K = -P/(\Delta V/V) = E/3(1-2\mu) \tag{16.16}$$

Here, P is the isotropic pressure, V is the initial volume, ΔV is the change in volume due to the pressure P, E is the elastic modulus, and μ is the Poisson's ratio. The summation of stresses in a volume is zero. That is,

$$K_1(\alpha_r - \alpha_i)V_1\Delta T + K_2(\alpha_r - \alpha_i)V_2\Delta T + \cdots\cdots = 0 \qquad (16.17)$$

and

$$V_1 + V_2 + \cdots = V_r \qquad (16.18)$$

Also,

$$V_i = F_i\rho_r V_r/\rho_i \qquad (16.19)$$

where V_i is volume, F_i, weight fraction, ρ_i, density of the ith phase, and ρ_r and V_r are the average density and volume, respectively, of the composite. From Equation 16.17, since $\Delta T \neq 0$,

$$K_1(\alpha_r - \alpha_1)V_1 + K_2(\alpha_r - \alpha_2)V_2 + \cdots\cdots = 0 \qquad (16.20)$$

Substituting the values of V_i from Equation 16.19 in Equation 16.20, and finding an expression for α_r, we get Equation 16.21 [5].

$$K_1(\alpha_r - \alpha_1)F_1\rho_r V_r/\rho_1 + K_2(\alpha_r - \alpha_2)F_2\rho_r V_r/\rho_2 = 0$$

$$K_1(\alpha_r - \alpha_1)F_1/\rho_1 + K_1(\alpha_r - \alpha_2)F_2/\rho_2 = 0$$

$$K_1\alpha_r F_1/\rho_1 - K_1\alpha_1 F_1/\rho_1 + K_2\alpha_r F_2/\rho_2 - K_2\alpha_2 F_2/\rho_2 = 0 \qquad (16.21)$$

$$\alpha_r(K_1 F_1/\rho_1 + K_2 F_2/\rho_2) = (K_1\alpha_1 F_1/\rho_1 + K_2\alpha_2 F_2/\rho_2)$$

$$\alpha_r = (K_1\alpha_1 F_1/\rho_1 + K_2\alpha_2 F_2/\rho_2)/(K_1 F_1/\rho_1 + K_2 F_2/\rho_2)$$

This equation is plotted for two composites in Figure 16.8 [6]. It shows good agreement between predicted and experimental values.

16.2.3.3 Polymorphic Transformations

Polymorphic transformations are accompanied by volume changes. Therefore, these transformations affect the thermal expansion coefficient. The values for the average volume expansion coefficient can be calculated using Equation 16.21 itself; α for each phase here will have to be substituted by $\Delta V/V_0\Delta T$. Figure 16.9 shows the variation of the linear thermal expansion of a silica containing quartz plus crystobalite phases and only a crystobalite phase. The material containing the two phases shows an increase in the value of the expansion coefficient at the transformation temperature.

16.2.3.4 Microstresses

In a polyphase material, when there is a large difference in thermal expansion coefficients among the different phases, microstresses develop during

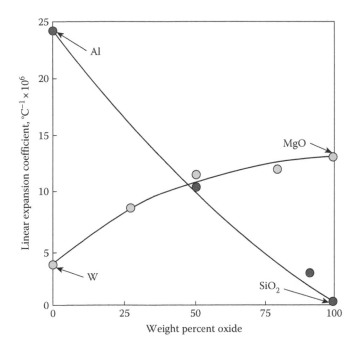

FIGURE 16.8
Variation of thermal expansion coefficients of two composites. The circles represent the experimental values. (From W. D. Kingery, *J. Am. Ceram. Soc.*, 40, 351, 1957.)

heating and cooling of the material. These stresses can also develop in a single-phase material having large differences in thermal expansion coefficients along the different crystallographic directions. Sometimes, these stresses can lead to microfissures in the material. The existence of such fissures can be detected by thermal cycling. The plot of change in length versus temperature will give a hysteresis loop in such materials. Figure 16.10 shows such a loop. In large grain materials, formation of expansion hysteresis is more predominant.

16.2.3.5 Glaze Stresses

These are the stresses that develop in glazes. This happens because of the unequal expansion coefficients of the glaze and the substrate. If the Young's modulus (E) of the glaze and the substrate is the same, then the stresses developed in a thin glaze and in an infinite slab substrate are given by Equations 16.22 and 16.23, respectively.

$$\sigma_{gl} = E(T_0 - T')(a_{gl} - a_b)(1 - 3j + 6j^2) \tag{16.22}$$

$$\sigma_b = E(T_0 - T')(a_b - a_{gl})j(1 - 3j + 6j^2) \tag{16.23}$$

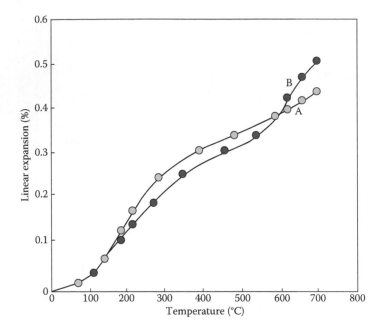

FIGURE 16.9
Effect of polymorphic transformation on linear thermal expansion coefficient. A: silica containing crystobalite phase. B: silica containing crystobalite and quartz phases.

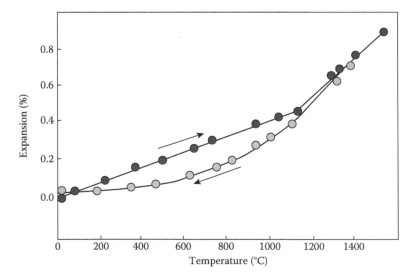

FIGURE 16.10
Formation of hysteresis loop in TiO_2 because of the existence of microfissures.

These equations give the stresses developed by heating the glazed material from a stress-free state at T_0 to a temperature T′. The letter "j" is the ratio of the thickness of the glaze to that of the substrate.

On cooling, the glaze should develop a compressive stress. Otherwise, it may develop crack-like features. The development of such features is known as crazing. Crazing is shown in Figure 16.11. Crazing develops under tensile stress. These crazes can lead to cracks. Failure under compressive stress requires substantial stress. Such a failure is called *shivering*.

Even if the glazing is under compressive stress, crazing can occur in service. The reason is that glazes can absorb moisture and expand. Expansion brings about the tensile stress. The solutions for this problem are as follows:

a. Increase the compressive stress to such an extent that, even after expansion during service, the glaze is still under compression

b. Select a composition that does not absorb moisture

The composition can be made more vitreous, or a glaze that is alkali-free can be used. Both ways prevent moisture absorption.

Crazes have been used for many centuries for decorative purposes. Lately, they have been used to protect ceramic body surfaces against corrosion and mechanical forces. Corrosion protection is brought about by making the crazes impermeable to fluids. By incorporating compressive stresses, they become resistant to mechanical failure. For example, moderate-expansion glazes are used to strengthen high-expansion glass-ceramic bodies such as $Na_2O-BaO-Al_2O_3-SiO_2$ systems. The compressive strengths obtained are on the order of 170 MPa, and the overall strengths are 260 MPa [7].

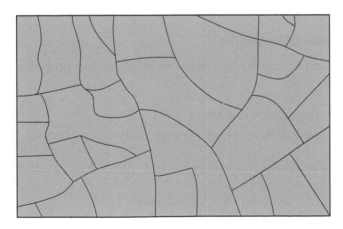

FIGURE 16.11
Crazed glaze.

16.2.4 Thermal Conduction

Thermal conduction takes place when thermal energy is transferred. Thermal energy transfer takes place under the influence of a temperature gradient. It depends on the energy concentration present per unit volume, its velocity of movement, and its rate of dissipation to the surroundings. In gases, the energy per unit volume is the same as the heat capacity per unit volume. The velocity of energy movement is the same as the velocity of the movement of the atoms and molecules making up the gases. This velocity can be obtained from the kinetic theory of gases. Supposing N molecules move with an average velocity v, then the velocity in one direction is Nv/3. If at (0,0) the mean energy of the molecules is E_0, then their energy at (0,l) is given by $E_0 + l\partial E/\partial x$. The distance l is the mean free path of the molecules. Now, we can write the energy flux as:

$$q/A = k(\partial T/\partial x) = Nvl(\partial E/\partial x) \tag{16.24}$$

In this equation, k is the thermal conductivity. The term $N(\partial E/\partial x)$ can be written as Equation 16.25.

$$N(\partial E/\partial x) = N(\partial E/\partial T)(\partial T/\partial x) = c(\partial T/\partial x) \tag{16.25}$$

Here, c is the heat capacity per unit volume.
Substituting the value of $N(\partial E/\partial x)$ in Equation 16.24, we get an expression for k as:

$$k = cvl/3 \tag{16.26}$$

This relationship is strictly applicable for ideal gases.

16.2.4.1 Phonon Conductivity

The conduction of heat through dielectric solids such as ceramics can happen in two ways:

1. By the propagation of elastic waves
2. By the movement of thermal energy units called phonons

In the first case, the conductivity can be expressed as Equation 16.27.

$$k = [\int c(\omega)vl(\omega)d\omega]/3 \tag{16.27}$$

In this equation, $c(\omega)$ is the specific heat per frequency and $l(\omega)$ is the attenuation length for the waves. It is the phonon–phonon interaction that contributes more to the value of thermal conductivity. This type of interaction results in phonon scattering called the *Umklapp process*. Apart from this interaction, there will also be scattering of phonons in the lattice. This decreases the mean free path of the phonons and thereby decreases the conductivity.

16.2.4.2 Photon Conductivity

Photon conductivity also adds to the total thermal conductivity. It comes from the electromagnetic radiation. This energy per unit volume is given by:

$$E_T = 4\sigma n^3 T^4/c \qquad (16.28)$$

Here, σ is the Stefan–Boltzmann constant (1.37×10^{-12} cal/cm^2 s K^4), n is the refractive index of the material, and c is the velocity of light (3×10^{10} cm/s). The heat capacity contributed by the radiation is given by Equation 16.29.

$$C_R = \partial E_T/\partial T = 16\sigma n^3 T^3/c \qquad (16.29)$$

The velocity of movement of the radiation within the material is given by Equation 16.30.

$$v = c/n \qquad (16.30)$$

Substituting the values of C_R and v in Equation 16.26, we get the radiant energy conductivity as:

$$k_r = 16\sigma n^2 T^3 l_r \qquad (16.31)$$

In this equation, l_r is the mean free path of the photons. The radiant energy transfer is negligible in opaque materials because $l_r \approx 0$. It is significant in the case of silicates and single crystals at moderate temperature levels, where l_r reaches macroscopic dimensions.

16.2.4.3 Phonon Conductivity in Single-Phase Crystalline Ceramics

The mean free path of phonons is affected by a variety of processes. All of these processes therefore determine the thermal conductivity of ceramics. One major process that influences the mean free path is the Umklapp process. At low temperatures, the mean free path corresponding to this process becomes large.

The dependence of thermal conductivity on temperature is illustrated for a dielectric material such as alumina in Figure 16.12. At very low temperatures, even though the mean free path becomes large, the boundary

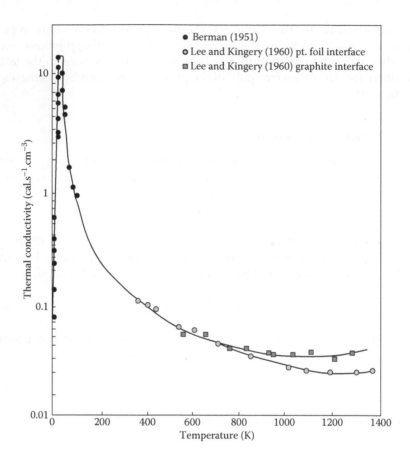

FIGURE 16.12
Variation of thermal conductivity of single-crystal alumina with temperature.

effects predominate. Because of this, the thermal conductivity decreases and finally drops to zero at 0 K. From 0 K, the temperature increases rapidly and reaches a maximum. The variation of k in this range corresponds approximately to exp $(-\theta/\alpha T)$. Above the Debye temperature, the decrease in k follows approximately 1/T relation. At high temperatures, the mean free path corresponds to the crystal lattice dimensions. At this stage, it does not depend on temperature, and k value remains at a minimum value.

Both *structure* and *composition* affect phonon conductivity in single-phase crystalline ceramics. Complex structures scatter lattice waves to a greater extent. Hence, thermal conductivity is lower in those structures. For example, magnesium aluminate spinel, consisting of two oxides, has a lower thermal conductivity than the single oxides alumina or magnesia.

Thermal conductivity is also dependent on crystallographic directions. To illustrate this point, let us take the example of titania. At 200°C, the thermal conductivity of this oxide is not the same along the directions normal and parallel to the c-axis. The ratio of the values normal to parallel is 1.52. As temperature is increased, the ratio decreases. That is because the structure becomes more symmetrical at higher temperatures.

As the difference among the atomic weights of the constituents increases, the thermal conductivity decreases. For example, TiC has less conductivity than SiC.

Another factor affecting the phonon conductivity in single-phase crystalline ceramics is the *grain boundary*. The thermal conductivity of single crystalline and polycrystalline materials is the same at low temperatures. At high temperatures, single crystals possess greater conductivity. This is because photon conductivity, which is predominant at high temperatures, is greater in single crystals because of the absence of grain boundaries. When the material contains fissures at grain boundaries or is made up of more than one phase having different thermal expansion coefficients, the thermal conductivity decreases in such material. This is because of the greater disruption of heat flow.

The effects of *impurities* and *solid solutions* on phonon conductivity in single-phase crystalline ceramics are discussed next. Impurities and solute atoms tend to decrease thermal conductivity. These increase the phonon scattering by way of differences in mass, binding force, and elastic strain field. As the temperature is raised, the scattering increases at low temperatures. At temperatures greater than about half the Debye temperature, it becomes independent of temperature. This is because the average wavelength at these temperatures becomes comparable with or less than the point imperfection.

The inverse of the total mean free path is the sum of the inverses of different mean free paths, as given in Equation 16.32.

$$(1/l_{total}) = (1/l_{thermal}) + (1/l_{impurity}) + \cdots\cdots \qquad (16.32)$$

For a given total mean free path, when one type of scattering decreases, the other type increases. Thus, the effect of solute scattering is greatest in simple lattices and at low temperatures at which thermal scattering mean free path is large. The scattering for low concentrations of solute is directly proportional to the volume fraction added. This is shown in Figure 16.13. In this figure, it can be seen that the slopes are the same at all temperatures. This means the effect of solutes on resistivity is independent of temperature. This in turn shows that the mean free path for impurity scattering is independent of temperature. The effectiveness of solutes in decreasing thermal conductivity depends on the mass difference, size difference, and the binding energy difference of the solute. The effect of

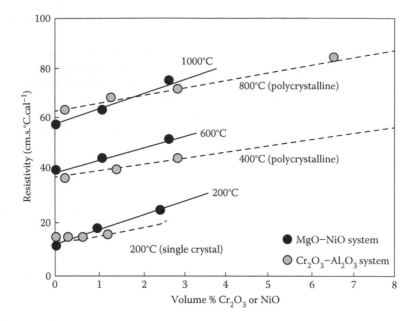

FIGURE 16.13
Effect of solute concentration on resistivity.

solutes on the conductivity is greatly dependent on temperature. This is also evident from Figure 16.13. As temperature increases, the resistivity also increases.

In the case of solid solutions, when they become non stoichiometric in composition, variation in composition can have large effect. For example, when UO_2 is made into UO_{2+x}, the conductivity decreases by 75%. When it is made in a solid solution with ThO_2, conductivity is further reduced. The lowest observed conductivity for an O-deficient Th-U composition has a value of about 0.003 cal/cm^3 °C.

Neutron irradiation gives rise to displaced atoms and lattice strains in crystalline ceramics. Therefore, it decreases the thermal conductivity.

16.2.4.4 Phonon Conductivity in Single-Phase Glasses

Glass structure being amorphous, the mean free path of the phonons is limited to interatomic distances. Hence, conductivity is limited as compared to crystalline materials.

There are three stages of the *effect of temperature* on the thermal conductivity of glass. From a very low value at 0 K, it increases rather rapidly as temperature is increased. After a certain increase, the value remains almost constant. At very high temperatures, it increases again rather sharply. This sharp increase is due to photon conductivity.

Composition also has an effect on glass conductivity. The value of conductivity for fused silica is about 0.0033 cal/s °C cm² at room temperature. It decreases to 0.0013 for silica containing 80% PbO. Silica glasses have higher conductivity than borosilicate glasses. Borosilicate glasses have higher value than glasses containing heavy metal ions. The mean free path decreases as the structure becomes more complex. This leads to a decrease in conductivity in complex structures as compared to simple ones.

16.2.4.5 Photon Conductivity

The photon conductivity equation was given by Equation 16.31. The mean free path for photons (l_r) is the inverse of the absorption coefficient (a). Absorption coefficient and refractive index are independent of temperature. But they depend on frequency.

Single crystals of dielectric materials are transparent in the visible region of the electromagnetic spectrum. In the ultraviolet (uv) region, they are opaque. This is because uv radiation excites electrons. Infrared radiation is absorbed. When the material contains transitional metal ions, they absorb even visible radiation because of electronic transitions. As the temperature is raised in the range in which photon conductivity is predominant, the black body emission spectrum shifts to lower wavelengths. This increases the mean free path, correspondingly increasing photon conductivity.

The *mean free path for photon conductivity* is fixed by absorption and scattering of photons. When materials have a low absorption coefficient, photon conduction becomes important at temperatures of the order of a few hundred degrees centigrade. The absorption coefficient depends on the wavelength. For a wavelength of 2 to 4 microns, the absorption coefficient is low. For higher values, it increases rapidly.

The sum of all types of mean free paths is called the integrated mean free path. The variation of this with temperature is greater for materials with good transmission. Examples of these materials are glasses and single crystals. Temperature affects the mean free path in two ways. The distribution of radiant energy and the absorption edge shift to lower wavelengths as the temperature is raised. The net effect is that for both glasses and single crystals, the mean free path increases as the temperature is increased.

The major form of photon attenuation in most ceramics results from light scattering. Pores act as the main scattering centers. There is a large difference in the index of refraction between the pores and the solid. And the pore sizes are usually small. These two factors act together to reduce the transmission greatly, even with 0.5% porosity. Most ceramics contain a certain amount of porosity. Because of this, the effective mean free path is substantially less in these materials than that in glasses and single crystals. As a result, the photon energy transfer only becomes important for sintered materials and at quite high temperatures on the order of 1500°C.

The *temperature dependence of photon conductivity* depends on the integrated mean free path. This increase with temperature raised the power in the range of 3.5 to 5.

For photon conductivity, *boundary effects* are found to predominate when the sample size is similar to the photon mean free path. Energy transfer between two boundaries can be identified by three separate processes:

1. Direct transfer from one boundary to the other
2. Energy exchange between the boundary and the material
3. Photon processes within the material

In the first process, the intervening material tends to only alter the photon velocity. In turn, the rate of heat transfer is altered. In the second process, the transfer is limited to a region having a thickness of the order of magnitude of the mean free path of the photons. The rate of energy exchange is determined by the temperature of the boundary and the temperature gradients in the material. The third process is independent of the boundary.

The measured *apparent photon conductivity* includes the boundary effect because photon mean free paths range from 0.1 to 10 cm. Common sample sizes are also in this range. Let us consider a situation where the electromagnetic radiation does not interact with the material, and photon conduction is the only energy transfer process. In this situation, the temperature gradient in the material is independent of the rate of heat transfer. Increasing the ratio of the distance between the boundaries and the photon mean free path (d/l_r) increases the apparent conductivity.

16.2.4.6 Conductivity of Polyphase Ceramics

Polyphase ceramics are more common than single-phase ones. They contain mixtures of one or more phases and pores. The conductivity of such ceramics depends on:

a. Amount of each phase
b. Arrangement of each phase
c. Conductivity of each phase

Three idealized microstructures are shown in Figure 16.14 to represent the amount and arrangement phases in a two-phase structure. Of these, the parallel slab arrangement is uncommon. The structure shown in Figure 16.14(b) is that of a continuous major phase with a dispersed minor phase. This kind of structure is typical of many ceramics. Porosity is also commonly seen dispersed in a continuous matrix. In glass ceramics, sometimes the glass phase becomes the continuous phase and is the minor phase. This kind of structure is shown in Figure 16.14(c). This structure is

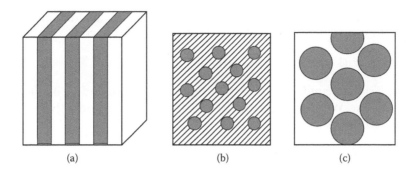

FIGURE 16.14
Idealized two-phase structures: (a) parallel slabs; (b) continuous major phase; (c) continuous minor phase.

also typical of metal-bonded ceramics such as cemented carbides. In insulating powders, each particle will have this structure, where the pores are the discontinuous major phase.

In order to ascertain the effect of the amount of phases and their distribution on the conductivity of ceramics, let us consider Figure 16.14(a). When heat flows parallel to the slabs, it is like electricity passing through a parallel circuit. The thermal conductivity for such an arrangement is given by:

$$k_m = v_1 k_1 + v_2 k_2 \tag{16.33}$$

Here, all the slabs are assumed to have the same dimensions so that their volumes are proportional to their vertical and horizontal cross-sectional areas. In the case of $k_1 \gg k_2$, the thermal conductivity for this structure is given by Equation 16.34.

$$k_m \approx v_1 k_1 \tag{16.34}$$

This means that the heat conduction is dominated by the better conductor. When the conduction takes place across the slabs, the relevant equation for the conductivity is given by:

$$(1/k_m) = (v_1/k_1) + (v_2/k_2) \tag{16.35}$$

The heat flow in this case can be likened to electricity flow through a circuit connected in a series. Again, considering $k_1 \gg k_2$, we get the conductivity of the system from Equation 16.36.

$$k_m \approx k_2/v_2 \tag{16.36}$$

In this case, the conduction is dominated by the poorer conductor.

For the more realistic picture depicted in Figure 16.14(b) and (c), the resultant thermal conductivity has been derived by Maxwell:

$$k_m = k_c[1 + 2v_d(1 - k_c/k_d)/(2k_c/k_d + 1)]/[1 - v_d(1 - k_c/k_d)/(k_c/k_d + 1)] \quad (16.37)$$

Here, k_c is the conductivity of the continuous phase and k_d is that of the dispersed one. When k_c is greater than k_d, the resultant conductivity is given by Equation 16.38.

$$k_m \approx k_c\,[(1 - v_d)/(1 + v_d)] \quad (16.38)$$

In case, $k_d > k_c$, then the resultant conductivity is:

$$k_m \approx k_c\,[(1 + 2v_d)/(1 - v_d)] \quad (16.39)$$

Alumina porcelains contain corundum and glass phases. Both phases are continuous at a composition corresponding to about 9 vol% glass. In such a case, the thermal conductivity value will be in between that of the two phases. Generally, the glass is continuous in vitreous ceramics. Therefore, the conductivity of these materials is closer to that of the glass phase.

The effect of porosity is different at low and high temperatures. At low temperatures, pores have lesser conductivity than at the solid phase. Therefore, conductivity decreases as porosity increases. At high temperatures, radiation across the pores also contributes to the conductivity.

Supposing the temperatures of the opposite surfaces of a pore are T_1 and T_2, the radiant energy transfer between these surfaces is given by Equation 16.40.

$$q = n^2 \sigma e_{eff} A\left(T_1^4 - T_2^4\right) \quad (16.40)$$

In this equation, e_{eff} is the effective emissivity. On factoring $(T_1^4 - T_2^4)$, we get:

$$q = 4n^2 \sigma e_{eff} A T_m^3 \Delta T \quad (16.41)$$

Here, T_m is the mean temperature between T_1 and T_2. If the pores have parallel surfaces separated by a distance of d_p, the effective conductivity (k_{eff}) is obtained from Equation 16.42.

$$q = -k_{eff} A \Delta T/d_p \quad (16.42)$$

Substituting the value of q from Equation 16.41 in Equation 16.42 and rearranging, we get an expression for k_{eff} as:

$$k_{eff} = 4d_p n^2 \sigma e_{eff} T_m^3 \quad (16.43)$$

$$5TiO_2 + 5.715MoSi_2 = 1.143Mo_5Si_3 + Ti_5Si_3 + 5SiO_2 \quad (16.44)$$

$$2.5Ti_3O_2 + 5MoSi_2 = Mo_5Si_3 + 1.5Ti_5Si_3 + 2.5SiO_2 \qquad (16.45)$$

$$W_p = W_t - W_e \qquad (16.46)$$

$$\varepsilon_{gb} = \Delta\alpha\Delta T/(1-v) \qquad (16.47)$$

$$\varepsilon_{gbc} = \left(24\gamma_{gb} / El_b\right)^{1/2} \qquad (16.48)$$

Thus, it can be seen from Equation 16.43 that pore size and temperature decide the conductivity through the pores. Pores of a larger size increase the conductivity at high temperatures. Equation 16.43 is applicable for opaque materials containing pores. The conductivity is also increased if the emissivity is greater, as is evident from the same equation.

In the case of translucent materials, another factor will have to be added to Equation 16.43. This factor is called the optical thickness. It is a product of the thickness of the material and the extinction coefficient. For opaque materials, the optical thickness is one. It is greater than unity for translucent materials. Hence, conductivity is increased for translucent materials. For translucent materials the efficiency of small pores as radiation shields decreases.

Powders and fibers have much less conductivity compared to solids. Even though their porosity is low, the pores are continuous. Pores have less conductivity than solids.

In two-phase systems such as Al_2O_3–ZrO_2, MgO–$MgAl_2O_4$, and alumina–mullite, thermal conductivity is found to be less than that of individual materials. This is attributed to the formation of microcracks along boundaries due to unequal expansion and contraction along different crystallographic directions during their manufacture or use. Because cracks enclose pores, they act in a similar manner and decrease conductivity. Though the volume fraction enclosed is negligible, they are more or less continuous.

Heat treatment given to ceramics may result in the creation of new cracks or in the closing of old cracks. Creation of new cracks decreases the conductivity and the closing of cracks acts in the opposite manner.

Materials with large thermal conductivity at low temperatures generally have a negative temperature coefficient—and vice versa for those with low conductivity at low temperatures.

16.3 Thermo-Mechanical Properties

16.3.1 Introduction

Thermo-mechanical properties are required for high-temperature materials. The last few decades have witnessed the development of various boride-based materials. This development has occurred due to their combination of properties, which include high hardness, elastic modulus, abrasion resistance, and superior thermal and electrical conductivity. The targeted applications include high-temperature structural materials, cutting tools, armor material, electrode materials in metal smelting, and wear-resistant parts.

The boride-based structural ceramics, because of their refractoriness and high-temperature strength, are well-suited for applications at high temperatures [8]. Among the structural ceramics, titanium diboride (TiB_2) is considered as the base material for a range of high-technology applications [9,10]. TiB_2 is a refractory material with a combination of attractive properties, including exceptional hardness ($\approx$25–35 GPa at room temperature, more than three times harder than fully hardened structural steel), which is retained to high temperatures. It has a high melting point (>3000°C), good creep resistance, good thermal conductivity (~65 W/m K), high electrical conductivity, and considerable chemical stability. TiB_2 also has properties similar to TiC, an important base material for cermets and many of those properties, namely hardness, thermal conductivity, electrical conductivity, and oxidation resistance, are even better than those of TiC [11–14]. This unique combination of properties makes TiB_2 a cadidate l for heavy-duty wear applications, particularly at elevated temperatures.

With respect to chemical stability, relevant in high-temperature machining applications, TiB_2 is more stable in contact with pure iron than are tungsten carbide (WC) and Si_3N_4; therefore, TiB_2-based materials may be preferred over WC-based materials for high-temperature applications. The good electrical conductivity of TiB_2 (electrical resistivity $\approx$13 $\times$ 10^{-8} Ω m) makes it an excellent candidate for special electrical applications such as cathodes used in aluminum electrosmelting or in vaporizing elements for vacuum metal deposition installations [15]. The relatively low fracture toughness of monolithic TiB_2 ($\approx$5 MPa $\cdot$ $m^{\frac{1}{2}}$) is the bottleneck for engineering applications [16].

In the case of nonoxide ceramics, such as titanium diboride (TiB_2), the type and amount of sinter-aid influence the microstructural development and, consequently, the mechanical properties. From the perspective of high-temperature applications, different transition metal borides with melting points higher than 3000°C have attracted wider attention for a range of technological applications [17–21]. Ceramics such as borides, carbides, and nitrides of Group IVB and VB elements are known as ultra-high-temperature ceramics (UHTCs) [22,23]. There has been a growing interest in UHTCs due to the increasing demand for hypersonic aerospace vehicles and reusable atmospheric reentry vehicles for space applications, which require temperature capabilities of more than 1800°C [17,22,23]. Also, diborides of transition metals, such as zirconium, titanium, and hafnium, are candidate materials for various structural applications, including furnace elements, high-temperature electrodes, refractory linings, microelectronics, and cutting tools, in addition to aerospace applications [17,18,24–27].

Among the borides, titanium diboride (TiB_2) is one of the materials of choice for cathodes in aluminum electrosmelting, armor components, cutting tools, wear-resistant parts, and various other high-temperature applications [8,23,28]. However, such a wider application of TiB_2 is limited because of the difficulties in obtaining full density. TiB_2 is difficult to sinter mainly because of the covalent bonding and low self-diffusion coefficient [17,22].

Generally, pressure-assisted sintering and high sintering temperatures ($\geq 1800°C$) are needed to attain full densification of monolithic borides (TiB_2, ZrB_2, etc.) [22]. Hence, the use of sintering aids is necessary to overcome the intrinsically low sinterability of ZrB_2 ceramics. Metallic additives and liquid-phase sintering (LPS) techniques have been utilized to enhance densification. However, the residual secondary phases degrade the properties of the borides. Since 2000, attempts have been made toward using various nonmetallic additives (such as SiC, ZrC, HfN, AlN, Y_2O_3, $TaSi_2$, Si_3N_4, and $MoSi_2$), often in combination with advanced techniques, such as spark plasma sintering (SPS) and reactive hot pressing (HP) [19,29–46].

Flexural strength and hot hardness measurements are necessary to evaluate the high-temperature mechanical behavior of ceramics.

16.3.2 High-Temperature Flexural Strength

In order to demonstrate the high-temperature flexural strength of ceramics, TiB_2 was taken as an example. It was hot pressed (HP) TiB_2 with and without the help of sintering aids. The sintering aids used were $MoSi_2$ and $TiSi_2$. $MoSi_2$ was used in two weight percentages, 2.5 and 10, whereas $TiSi_2$ was used in three weight percentages, 2.5, 5, and 10. The samples were hot pressed for 60 min at 30 MPa. The sample size was $3 \times 4 \times 40$ mm^3. After hot pressing, they were tested in air with a four-point flexural configuration. The results are shown in Table 16.1. At room temperature (RT), the four-point flexural strength values of monolithic TiB_2 and TiB_2–(2.5 wt%) $MoSi_2$ composite were measured to be approximately 390 MPa. Typically, the flexural strength of HP TiB_2, as reported by others [47], varied in the

TABLE 16.1

Effect of Temperature on Flexural Strength of TiB_2 and $TiSi_2$/$MoSi_2$ Ceramics

Material Composition (Wt%)	Hot Pressing Temp.	Relative Density (% ρth)	Grain Size (μm)	Flexural Strength (MPa)			Ref.
				RT	500°C	1000°C	
TiB_2–(0)$MoSi_2$	1800	98	1.5	387 ± 52	422 ± 29	546 ± 33	8
TiB_2–(2.5)$MoSi_2$	1700	99	1.2	391 ± 31	442 ± 34	503 ± 27	33
TiB_2–(10.0)$MoSi_2$	1700	97	1.3	268 ± 70	312 ± 28	261 ± 30	33
TiB_2–(2.5)$TiSi_2$	1650	>98	2.3	381 ± 74	–	433 ± 17	37
TiB_2–(5.0)$TiSi_2$	1650	>99	3.0	426 ± 69	479 ± 33	314 ± 17	37
TiB_2–(10.0)$TiSi_2$	1650	>99	3.5	345 ± 60	375 ± 50	325 ± 25	37

range 300–400 MPa. The addition of 2.5 wt% $MoSi_2$ does not degrade the strength properties of TiB_2. However, the lowest strength (~268 MPa) was recorded with the TiB_2-based composite densified using 10 wt% $MoSi_2$ sinter-additive. The fracture strength of the TiB_2, irrespective of $MoSi_2$ content, increases with temperature up to 500°C. In an earlier study, Baumgartner and Steiger [48] reported an increase in strength of mono-lithic TiB_2 with temperature, and they attributed it to the relief of residual internal stresses. At 1000°C, the flexural strength of the TiB_2 composite with 10 wt% $MoSi_2$ sinter-additive is reduced (~261 MPa), compared with strength at 500°C (~312 MPa). Among all the TiB_2–$TiSi_2$ compositions, TiB_2–(5 wt%)$TiSi_2$ exhibited the highest room temperature strength (~426 MPa; see Table 16.1). Up to 500°C, the fracture strength increases for all the TiB_2 compositions, which were densified to more than 97% of the theoretical density (ρ_{th}). Both the baseline monolithic TiB_2 and TiB_2–(2.5 wt%) $MoSi_2$, or (2.5 wt%)$TiSi_2$ ceramics, could increase flexural strength to more than 400 MPa at 1000°C. When hot pressing was completed for $TiSi_2$, an addi-tional phase, Ti_5Si_3, was formed. At 1000°C, the flexural strength of refer-ence TiB_2 is observed to be higher than all the TiB_2–$TiSi_2$ compositions. It had been hot pressed at a higher temperature of 1800°C. It implies that at elevated temperature the grain boundary sliding at TiB_2/$TiSi_2$ and TiB_2/Ti_5Si_3 interfaces results in fracture at low loads for the TiB_2 composites. Because the brittle-to-ductile transition temperature of $TiSi_2$ is 800°C, this phase could exhibit plasticity at or above 800°C with the application of load [49–53]. Therefore, the plastic deformation of $TiSi_2$ at high temperature can lead to strength degradation in the TiB_2 composites.

The preceding discussion implies that it is advantageous to use $TiSi2$/$MoSi_2$ as a sintering aid to retain high-temperature strength and hardness prop-erties. To achieve a high density (>97% ρ_{th}) with monolithic TiB_2, a high hot pressing temperature of 1800°C is needed. One could achieve maximum den-sity of 99% ρ_{th} at a lower hot-pressing temperature of 1650°C with the use of $TiSi_2$ as a sintering aid. Also, the retention of strength at high temperatures is only possible with better density and a minimal amount of sinter-additive. The TiB_2–(2.5 wt%)$TiSi_2$ composite exhibited a better combination of density and strength values at high temperatures.

16.3.3 Hot Hardness

For the hot hardness measurements, the TiB_2–(2.5 wt%)$MoSi_2$ composite possessing the maximum sintered density among the composites was selected, along with monolithic TiB_2. They were hot pressed at 1700°C and 1800°C, respectively. The measured hot hardness values are provided in Table 16.2. The hardness of TiB_2 decreased from approximately 26 GPa at RT to approximately 8.5 GPa at 900°C, while the measured hardness reduced from approximately 28 GPa at RT to approximately 10.5 GPa at 900°C for the TiB_2–(2.5 wt%)$MoSi_2$ composite. Brittle materials can be

TABLE 16.2

Hot Hardness Values of TiB_2–$MoSi_2$/$TiSi_2$ Ceramics

Material (wt%)	Hot Hardness (GPa) at Various Temperatures (°C)						Reference
	23	200	300	600	800	900	
TiB_2–(0)$MoSi_2$	25.6	–	15.1	11.5	–	8.5	8
TiB_2–(2.5)$MoSi_2$	27.6	–	18.9	13.6	–	10.5	8
TiB_2–(2.5)$TiSi_2$	27.0	–	15.1	11.5	–	8.9	29
TiB_2	25.0	14.7	12.8	–	5.3	–	37
TiB_2–(5)$TiSi_2$	27.0	–	13.3	11.0	–	7.0	37
TiB_2–(10)$TiSi_2$	24.0	–	13.0	10.0	–	5.0	37
TiB_2	28.0	24.0	18.0	14.0	8.0	7.0	54
ZrB_2	20.0	–	12.0	9.0	7.0	7.0	54
HfB_2	27.0	26.0	16.0	10.0	9.0	6.0	54

plastically deformed even at temperatures below 0.5 times the melting temperature [54].

However, the higher hardness of the TiB_2 composite can be attributed to better sintered density (~99% ρ_{th}). Even though $MoSi_2$ is softer ($H_v \sim 9$ GPa) compared with TiB_2, the addition of small amounts of $MoSi_2$ to TiB_2 did not have any negative effect on the hardness [55]. It has been reported in the literature that the hardness of TiB_2 cermets is much lower than both monolithic TiB_2 and TiB_2 reinforced with the ceramic additives. The hardness of TiB_2-based cermets varied between 7.3 GPa (for TiB_2 – (5 vol%)[Fe – Fe_2B]) and 4.8 GPa (for TiB_2–(20 vol%)[Fe–Cr–Ni–Fe_2B]) at 800°C [56]. Also, monolithic TiB_2 could retain a maximum hardness of approximately 5 GPa at 800°C [57]. Earlier reports indicated that TiB_2–$MoSi_2$ materials had better hardness values of more than 10 GPa at 900°C. Among all the TiB_2–$TiSi_2$ samples, the hardness varied from 27 GPa at RT to 8.9 GPa at 900°C for TiB_2–(2.5 wt%)$TiSi_2$ (see Table 16.2). At RT, the monolithic TiB_2 (HP at 1800°C) shows a little lower hardness than the TiB_2–(5 wt%)$TiSi_2$. However, the hardness of the monolithic TiB_2 is comparable with other TiB_2 samples (containing ≥5 wt% $TiSi_2$) at elevated temperatures. For example, the hardness of monolithic TiB_2 and TiB_2–(5 wt%) $TiSi_2$ were recorded as 7–8.5 and 7.0 GPa, respectively, at 900°C.

The RT hardness of TiB_2–$TiSi_2$ ceramics varies from 24 to 27 GPa (see Table 16.2). Such a hardness variation can be attributed to differences in sintered density and to the amount of sinter-additive. The hardness increased with $TiSi_2$ addition (up to 5 wt% $TiSi_2$), further increase in the $TiSi_2$ content to 10 wt% lowers hardness (24 GPa) of TiB_2 despite its full densification (99.6% ρ_{th}). The reason for this could be due to the lower hardness of the additive. Although the TiB_2 composites consist of relatively softer phases such as $TiSi_2$ (8.7 GPa) and Ti_5Si_3 (9.8 GPa) [58], the addition of small amounts of $TiSi_2$ (≤5 wt%) does not degrade the hardness of TiB_2 to any greater extent. Ti_5Si_3 is a sintered reaction product.

Problems

16.1　Calculate the effect on thermal conductivity of ThO_2 containing 10% porosity at 1500°C of varying pore size from 5 mm in diameter to 0.5 mm in diameter when the pores are isolated.

16.2　Which of the materials listed below would be best suited for an application in which a part experiences sudden and severe thermal fluctuations while in service?

Material	k_{th}, W/(m.K)	Modulus, GPa	K_{Ic}, MPa.m$^{1/2}$	α, K^{-1}
1	290	200	8	9×10^{-6}
2	50	150	4	4×10^{-6}
3	100	150	4	3×10^{-6}

16.3　Estimate the heat loss through a 0.5 cm thick 1000 cm^2 window made of soda lime if the inside temperature is 25°C and the outside temperature is 0°C; k_{th} of soda lime is 1.7 W/m.K.

16.4　Estimate the heat loss through a 0.5-cm-thick 1000 cm^2 furnace wall made of porous firebrick if the inside temperature is 1200°C, and the outside temperature is 0°C; k_{th} of porous firebrick is 1.3 W/m. K.

References

1.　C. Kittel, *Introduction to Solid State Physics*, 3rd Edn., John Wiley, New York, 1968.
2.　D. R. Uhlmann, A. G. Kolbeck, and D. L. de Witte, Heat capacities and thermal behavior of alkali borate glasses, *J. Non Cryst. Solids*, 5: 426, 1971.
3.　D. R. Uhlmann, Effect of phase separation on the properties of simple glasses, *J. Non Cryst. Solids*, 1: 474, 1969.
4.　R. R. Shaw and D. R. Uhlmann, The thermal expansion of alkali borate glasses and the boric oxide anomaly, *J. Non Cryst. Solids*, 1: 347, 1969.
5.　P. S. Turner, Thermal expansion stresses in reinforced plastics, *J. Res. NBS*, 37: 239, 1946.
6.　W. D. Kingery, Thermal Conductivity: XIII, Effect of Microstructure on Conductivity of Single-Phase Ceramics, *J. Am. Ceram. Soc.*, 40: 351, 1957.
7.　D. A. Duke, J. E. Megles, J. F. MacDowell, and H. F. Bopp, Strengthening glass-ceramics by application of compressive glazes, *J. Am. Ceram. Soc.*, 51: 98, 1968.
8.　B. Basu, G. B. Raju, and A. K. Suri, Processing and properties of TiB_2-based materials: A review. *Int. Mater. Rev.*, 51(6): 352–374, 2006.
9.　R. Telle, Boride and carbide ceramics, in *Materials Science and Technology*, Vol. 11, R. W. Cahn, P. Haasen, and E. J. Kramer (Eds.), Structure and Properties of Ceramics. Ed. M. Swain, VCH, Weinheim, Germany, 1994, 175 p.
10.　G. V. D. Goor, P. Sägesser, and K. Berroth, Electrically conductive ceramic composites, *Solid State Ionics*, 101–103: 1163–1170, 1997.

11. P. Ettmayer, H. Kolaska, W. Lengauer, and K. Dreyer, Ti(C,N) cermets—Metallurgy and properties, *Int. J. Refract. Metals Hard Mater.*, 13: 343–351, 1995.

12. G. E. D' Errico, S. Bugliosi, and E. Gugliemi, Tool-life reliability of cermet inserts in milling tests, *J. Mater. Proc. Technol.*, 77: 337–343, 1998.

13. D. Moskowitz and M. Humenik, Jr., Cemented titanium carbide cutting tools, *Mod. Dev. Powder Metall.*, 3: 83–94, 1966.

14. S. Put, J. Vleugels, G. Anne, and O. Van Der Biest, Functionally graded ceramic and ceramic-metal composites shaped by electrophoretic deposition, *Colloids Surf. A Physicochem. Eng. Asp.*, 222(1–3): 223–232, 2003.

15. Y. Liu, X. Liao, F. Tang, and Z. Chen, Observations on the operating of TiB$_2$-coated cathode reduction cells, in *Light Metals*, E. Cutshall (Ed.), The Minerals, Metals and Materials Society, Warrendale, PA, 1991, pp. 427–429.

16. F. De Mestral and F. Thevenot, Ceramic composites: TiB$_2$-TiC-SiC—Part I properties and microstructures in the ternary system, *J. Mater. Sci.*, 26: 5547–5560, 1991.

17. W. G. Fahrenholtz, G. E. Hilmas, I. G. Talmy, and J. A. Zaykoski, Refractory diborides of zirconium and hafnium, *J. Am. Ceram. Soc.*, 90(5): 1347–1364, 2007.

18. J. R. Ramberg and W. S. Williams, High temperature deformation of titanium diboride, *J. Mater. Sci.*, 22: 1815–1826, 1987.

19. J. J. Melendez-Martinez, A. Dominguez-Rodriguez, F. Monteverde, C. Melandri, and G. de Portu, Characterisation and high temperature mechanical properties of zirconium boride-based materials, *J. Eur. Ceram. Soc.*, 22: 2543–2549, 2002.

20. F. Peng and R. F. Speyer, Oxidation resistance of fully dense ZrB 2 with SiC, TaB$_2$ and TaSi$_2$ additives, *J. Am. Ceram. Soc.*, 91(5): 1489–1494, 2008.

21. R. A. Cutler, Engineering properties of borides, in *Engineering Materials Handbook, Ceramic and Glasses*, Vol. 4, S. J. Schneider Jr., (ed.) ASM International, Metals Park, OH, 1991, pp. 787–803.

22. F. Monteverde and L. Scatteia, Resistance to thermal shock and to oxidation of metal diborides–SiC ceramics for aerospace application, *J. Am. Ceram. Soc.*, 90(4): 1130–1138, 2007.

23. R. G. Munro, Material properties of titanium diboride, *J. Res. Natl. Inst. Stand. Technol.*, 105(5): 709–720, 2000.

24. A. K. Kuriakose and J. L. Margrave, The oxidation kinetics of zirconium diboride and zirconium carbide at high temperatures. *J. Electrochem. Soc.*, 111(7): 827–831, 1964.

25. F. Monteverde, S. Guicciardi, and A. Bellosi, Advances in microstructure and mechanical properties of zirconium diboride based ceramics, *Mater. Sci. Eng. A*, 346: 310–319, 2003.

26. Y. Yan, Z. Huang, S. Dong, and D. Jiang, Pressureless sintering of high-density ZrB$_2$-SiC ceramic composites, *J. Am. Ceram. Soc.*, 89(11): 3589–3592, 2006.

27. F. Monteverde, C. Melandri, and S. Guicciardi, Microstructure and mechanical properties of an HfB$_2$ + 30 vol% SiC composite consolidated by spark plasma sintering, *Mater. Chem. Phys.*, 100: 513–519, 2006.

28. W. Wang, Z. Fu, H. Wang, and R. Yuan, Influence of hot pressing sintering temperature and time on microstructure and mechanical properties of TiB$_2$ ceramics, *J. Eur. Ceram. Soc.*, 22: 1045–1049, 2002.

29. F. Monteverde and A. Bellosi, Effect of the addition of silicon nitride on sintering behavior and microstructure of zirconium diboride, *Scr. Mater.*, 46: 223–228, 2002.

30. F. Monteverde and A. Bellosi, Beneficial effects of AIN as sintering aid on microstructure and mechanical properties of hot pressed ZrB_2, *Adv. Eng. Mater.*, 5: 508–512, 2003.

31. F. Monteverde and A. Bellosi, Efficacy of HfN as sintering aid in the manufacturing of ultra-high temperature metal diboride-matrix ceramics, *J. Mater. Res.*, 19: 3576–3585, 2004.

32. F. Monteverde and A. Bellosi, Development and characterization of metal-diboride-based composites toughened with ultra-fine SiC particulates, *Solid State Sci.*, 7: 622–630, 2005.

33. S. S. Hwang, A. L. Vasiliev, and N. P. Padture, Improved processing and oxidation resistance of ZrB_2 ultra-high temperature ceramics containing SiC nanodispersoids, *Mater. Sci. Eng. A*, 464: 216–224, 2007.

34. S. Zhu, W. G. Fahrenholtz, and G. E. Hilmas, Influence of silicon carbide particles size on the microstructure and mechanical properties of zirconium diboride-silicon carbide ceramics, *J. Eur. Ceram. Soc.*, 27: 2077–2083, 2007.

35. L. Rangaraj, C. Divakar, and V. Jayaram, Fabrication and mechanisms of densification of ZrB_2-based ultra-high temperature ceramics by reactive hot pressing, *J. Eur. Ceram. Soc.*, 30: 129–138, 2010.

36. Z. Wang, S. Wang, X. Zhang, P. Hu, W. Han, and C. Hong, Effect of graphite flake on microstructure as well as mechanical properties and thermal shock resistance of ZrB 2-SiC matrix ultrahigh temperature ceramics, *J. Alloys Comp.*, 484: 390–394, 2009.

37. A. L. Chamberlain, W. G. Fahrenholtz, and G. E. Hilmas, Pressureless sintering of zirconium diboride, *J. Am. Ceram. Soc.*, 89(2): 450–456, 2006.

38. D. Sciti, S. Guicciardi, A. Bellosi, and G. Pezzotti, Properties of a pressureless-sintered ZrB_2-$MoSi_2$ ceramic composite, *J. Am. Ceram. Soc.*, 89(7): 2320–2322, 2006.

39. A. L. Chamberlain, W. G. Fahrenholtz, and G. E. Hilmas, Low-temperature densification of zirconium diboride ceramics by reactive hot pressing, *J. Am. Ceram. Soc.*, 89(12): 3638–3645, 2006.

40. D. Sciti, L. Silvestroni, G. Celotti, C. Melandri, and S. Guicciardi, Sintering and mechanical properties of ZrB_2-$TaSi_2$ and HfB_2-$TaSi_2$ ceramic composites, *J. Am. Ceram. Soc.*, 91(10): 3285–3291, 2008.

41. G. B. Raju and B. Basu, Thermal and electrical properties of TiB 2-MoSi 2, *Int. J. Refract. Metals Hard Mater.*, 28: 174–179, 2010.

42. G. B. Raju, K. Biswas, and B. Basu, Microstructural characterization and isothermal oxidation behaviour of hot-pressed TiB_2-10 wt% $TiSi_2$ composite, *Scr. Mater.*, 61: 674–677, 2009.

43. A. Mukhopadhyay, G. B. Raju, A. K. Suri, and B. Basu, Correlation between phase evolution, mechanical properties and instrumented indentation response of TiB_2-based ceramics, *J. Eur. Ceram. Soc.*, 29: 505–516, 2009.

44. G. B. Raju and B. Basu, Densification, sintering reactions, and properties of titanium diboride with titanium disilicide as a sintering aid, *J. Am. Ceram. Soc.*, 90(11): 3415–3423, 2007.

45. G. B. Raju, K. Biswas, A. Mukhopadhyay, and B. Basu, Densification and high temperature mechanical properties of hot pressed TiB_2-(0–10 wt.%) $MoSi_2$ composites, *Scr. Mater.*, 61: 674–677, 2009.

46. G. B. Raju, B. Basu, N. H. Tak, and S. J. Cho, Temperature dependent hardness and strength properties of TiB_2 with $TiSi_2$ sinter-aid, *J. Eur. Ceram. Soc.*, 29(10): 2119–2128, 2009.

47. K. Nakano, H. Matsubara, and T. Imura, High temperature hardness of titanium diboride single crystal, *Jpn. J. Appl. Phys.*, 13(6): 1005–1006, 1974.
48. H. R. Baumgartner and R. A. Steiger, Sintering and properties of TiB_2 made from powder synthesized in plasma-arc heater, *J. Am. Ceram. Soc.*, 67(3): 207–212, 1984.
49. R. Mitra, Mechanical behavior and oxidation resistance of structural silicides, *Int. Mater. Rev.*, 51(1): 13–64, 2006.
50. H. Inui, M. Moriwaki, N. Okamoto, and M. Yamaguchi, Plastic deformation of single crystals of $TiSi_2$ with the C54 structure, *Acta Mater.*, 51: 1409–1420, 2003.
51. J. Li, D. Jiang, and S. Tan., Microstructure and mechanical properties of in situ produced $SiC/TiSi_2$ nanocomposites, *J. Eur. Ceram. Soc.*, 20: 227–233, 2000.
52. J. Li, D. Jiang, and S. Tan, Microstructure and mechanical properties of in situ produced Ti_5Si_3/TiC nanocomposites, *J. Eur. Ceram. Soc.*, 22: 551–558, 2002.
53. R. Rosenkranz and G. Frommeyer, Microstructures and properties of high melting point intermetallic Ti_5Si_3 and TiSi 2 compounds, *Mater. Sci. Eng. A*, 152: 288–294, 1992.
54. H. L. Wang and M. H. Hon, Temperature dependence of ceramics hardness, *Ceram. Int.*, 25: 267–271, 1999.
55. D. S. Park, B. D. Hahn, B. C. Bae, and C. Park, Improved high-temperature strength of silicon nitride toughened with aligned whisker seeds, *J. Am. Ceram. Soc.*, 88(2): 383–389, 2005.
56. T. Jungling, L. S. Sigl, R. Oberacker, F. Thummler, and K. A. Schwetz, New hard-metals based on TiB_2, *Int. J. Refract. Metals Hard Mater.*, 12: 71–88, 1993.
57. X. Zhong and H. Zao, High temperature properties of refractory composites, *Am. Ceram. Soc. Bull.*, 60: 98–101, 1999.
58. G. Berg, C. Friedrich, E. Broszeit, and C. Berger, Data collection of properties of hard materials, in *Handbook of Ceramic Hard Materials*, Vol. 2, R. Riedel (Ed.), Wiley-VCH Verlag GmbH, Weinheim, Germany, 2000, pp. 965–990.

Bibliography

M. W. Barsoum, *Fundamentals of Ceramics*, Taylor & Francis Group, LLC, New York, 2003, pp. 462, 464.
B. Basu and K. Balani, *Advanced Structural Ceramics*, 1st Edn., The American Ceramic Society, John Wiley, New York, 2011, pp. 259–260, 286–287, 309–312.
W. D. Kingery, H. K. Bowen, and D. R. Uhlmann, *Introduction to Ceramics*, 2nd Edn., John Wiley, New York, 1976, pp. 583–645.

Section III

Refractories

17

Classification

17.1 Introduction

Most refractories are ceramics. Carbon, which is the basis of organic materials, is also considered a refractory material. Among metallic materials, tungsten and molybdenum are refractories. All refractory materials possess a high melting point on the order of 2000°C and higher. They have the ability to retain their physical shape and chemical identity when subjected to high temperatures. This property makes them useful at high temperatures.

Refractories can be classified in two ways. One way is based on chemical composition. The various refractories are silica (SiO_2), alumina (Al_2O_3), magnesia (MgO), chromia (Cr_2O_3), alumino-silicate, and magnesia-chromia. In the second method, the refractories are classified as acidic, basic, and neutral refractories. This classification is based on the behavior of refractories toward slags. Following is a brief discussion of each of these classes.

17.2 Acid Refractories

Acid refractories react with basic slags. Therefore, they are not useful in basic conditions but are used under acidic conditions. In these refractories, SiO_2 is the basic constituent. Examples are silica, fireclay series with 30%–42% Al_2O_3, sillimanite, and andalusite with about 60% Al_2O_3.

17.3 Basic Refractories

Basic refractories react with acidic slags. Therefore, they are not useful in acidic environments. They are used under basic conditions and are based on MgO. Examples are magnesite, dolomite, chrome-magnesite, magnesite-chrome, alumina, and mullite.

17.4 Neutral Refractories

Neutral refractories do not react with either acidic or basic slags. Hence, they are useful in acidic and basic conditions. Examples are carbon, chromite ($FeO.Cr_2O_3$), and fosterite ($2MgO.SiO_2$).

Certain refractories are grouped under *special refractories*. Examples are zirconia, thoria, and beryllia. They possess special properties that make them useful in special applications. For example, thoria is a nuclear fuel that can sustain radiation, damage, and high temperatures.

Bricks are commonly produced by refractories and come in standard and nonstandard shapes. Nonstandard shapes are costlier. Standard shapes include rectangular prisms, tapered bricks, and tubular sleeves. Some refractory materials are supplied in granular form. These are called *pea-sized* refractories and are used in furnaces. To make the top layer of a furnace hearth, these pea-sized refractories are thrown onto the hearth while hot. They also are mixed with hot tar as a binder and are used as a lining for furnace hearths. Running repairs are made from *castables*. These are plastic preparations that can be made in any shape; they are dried and fired in situ.

Another class of refractories is called *insulating refractories*. These are designed to have very low thermal conductivity. This is achieved mostly by incorporating a high proportion of air into the structure. Bricks made in this way are called porous bricks. Another example of an insulating refractory is the mineral wool. This is not self-supporting. Hence, it should be contained for use. Asbestos is a natural insulator but is not useful as a refractory.

The widely used representatives from all these five classes will be discussed in Section III.

Bibliography

J. H. Chesters, *Refractories-Production and Properties*, The Institute of Materials, Oxford, UK, 1973.

J. D. Gilchrist, *Fuels, Furnaces and Refractories*, Pergamon Press, Oxford, UK, 1977, pp. 237–239.

18

Refractory Thermodynamic Principles

18.1 Introduction

Refractories are meant to be exposed to high temperatures. These high temperatures cause many changes, which are the result of the refractories' reactions with the environment. The environment can be in the form of solid, liquid, or gas. Reactions take place where the refractories and the environment make contact. Thus, we have solid–solid contact, solid–liquid contact, and solid–gas contact depending upon the solid refractories in contact with the environment in any of the three states of matter.

Thermodynamic principles help in predicting which reactions are impossible. This prediction is based on changes in free energy. It is obtained by subtracting the free energy of the reactants from the free energy of the products. In cases where the free energy change is positive, it is impossible for such reactions to take place spontaneously. This happens because any system will try to acquire a state of lowest energy. Thus, free energy becomes an important term in the field of thermodynamics.

18.2 Free Energy

What is defined by free energy can be seen in the *Gibbs free energy function*, which consists of two terms, enthalpy (H) and entropy (S).

18.2.1 Enthalpy

Heat capacity is defined as:

$$c = dq/dT \qquad (18.1)$$

Here, dq is the heat absorbed and dT is the temperature rise because of the heat absorption. *Enthalpy* is defined in Equation 18.2.

$$H = E + PV \qquad (18.2)$$

Here, E is the internal energy, P is pressure, and V is volume. On differentiating Equation 18.2, we get:

$$dH = dE + PdV + VdP \tag{18.3}$$

From the first law of thermodynamics, we have Equation 18.4.

$$dE = dq - dw \tag{18.4}$$

Substituting the value of dE in Equation 18.3, we get:

$$dE = dq - dw + PdV + VdP \tag{18.5}$$

At constant pressure, we have Equation 18.6.

$$dP = 0 \tag{18.6}$$

Also, at constant pressure, the work is the product of pressure and change in volume. Expressing it mathematically, we get:

$$dw = PdV \tag{18.7}$$

Substituting the values of dP and dw, we get:

$$dH = dq \tag{18.8}$$

Thus, enthalpy of a substance can be defined as the heat absorbed or released by it at a constant pressure. Substituting for dq in Equation 18.1, we get:

$$c_p = (dH/dT)_p \tag{18.9}$$

The heat capacity at constant pressure is expressed as c_p. The enthalpy difference between two temperatures can be obtained by integrating c_p between the two temperatures. The expression for this is given in Equation 18.10 where 298 K is taken as the standard temperature.

$$H^T - H^{298} = \int_{298} Tc_p dT \tag{18.10}$$

The H^{298} for elements is assumed to be zero. The enthalpy of a compound at any temperature is its heat of formation from its elements at that temperature. There is an empirical equation connecting temperature and c_p:

$$c_p = A + BT + C/T^2 \tag{18.11}$$

Example 1

The empirical relation for the c_p of Al is:

$$c_p = 20.7 + 0.0124\,T \tag{18.12}$$

That of Al_2O_3 is given by Equation 18.13.

$$c_p = 106.6 + 0.0178T - 2,850,000\,T^{-2} \tag{18.13}$$

The enthalpy of formation of Al_2O_3 from its elements at 298 K is −1675.7 kJ/mol. Calculate the enthalpy of both Al and Al_2O_3 at 900 K.

Solution

The enthalpy of Al at 900 K $= H_{Al}^{900} - H_{Al}^{298}$

$$= \int_{298}^{900} (20.7 + 0.0124T)\,dT$$
$$= (20.7T + 0.0124T^2/2)_{298}^{900}$$
$$= [20.7(900 - 298) + 0.0124\,(900^2 - 298^2)/2]$$
$$= 20.7 \times 602 + 0.0062(81,0000 - 88,804)$$
$$= 124,61.4 + 0.0062 \times 721,196$$
$$= 16.933 \text{ kJ/mol} \tag{18.14}$$

$$H_{Al2O3}^{900} - H_{Al2O3}^{298} = \int_{298}^{900} (106.6 + 0.0178T - 2,850,000T^{-2})\,dT$$
$$= [106.6(900 - 298) + 0.0089(900^2 - 298^2)$$
$$+ 2,850,000\,(900^{-1} - 298^{-1})]$$
$$= [64,173.2 + 72,1196 - 6397.1]$$
$$= 778,972.1 \tag{18.15}$$

Therefore, $H_{Al2O3}^{900} = 778,972.1 - 1675.7$
$$= 777.296 \text{ kJ/mol} \tag{18.16}$$

18.2.2 Entropy

Entropy measures the degree of disorder in a system. The higher the degree of disorder, the higher will be the entropy. Degree of disorder increases with the amount of heat absorbed by the system. Thus, entropy at any temperature is given by the Equation 18.17.

$$S = q/T \tag{18.17}$$

Here, q is the heat absorbed reversibly by the system. In the microscopic domain, Boltzmann's equation holds well:

$$S = k \ln \Omega \tag{18.18}$$

Here, k is the Boltzmann's constant and Ω is a measure of the randomness. Entropy can arise by different means. Some of them are the following.

a. Configurational entropy: This is related to the different configurations of atoms in a crystalline lattice.

b. Thermal entropy: This arises from the thermal energy levels possessed by the atoms.

c. Electronic entropy: This is associated with the electrons.

d. Magnetic entropy: This is related to the magnetic moments.

e. Dielectric entropy: This concerns the dielectric moments.

18.2.2.1 Configurational Entropy

Configurational entropy arises from the disorder in the crystal lattice. This disorder is the result of the defects present in the lattice. Let us consider a lattice consisting of N atoms and n point defects. The number of configurations by which these two species can be arranged is given by Equation 18.19 [1].

$$\Omega = (n + N)!/n!N! \tag{18.19}$$

Substituting this equation in Equation 18.18, we get Equation 18.20.

$$S = k \ln\left[(n+N)!/n!N!\right] \tag{18.20}$$

Sterling's approximation is given by:

$$\ln x! \approx x \ln x - x + 1/2 \ln (2\pi x) \tag{18.21}$$

For large values of x, the second term on the RHS can be neglected. Hence, Equation 18.18 becomes Equation 18.22.

$$\ln x! \approx x \ln x - x \tag{18.22}$$

Expanding Equation 18.20 using Equation 18.22, we get Equation 18.23 for configurational entropy.

$$S_c = -k\left\{N \ln\left[N/(N+n)\right] + n \ln\left[n/(n+N)\right]\right\} \tag{18.23}$$

Example 2

Calculate the total number of possible configurations for 15 atoms and one vacancy in a 2-d rectangular lattice. Draw the various configurations.

Solution

Substituting N = 15, and n = 1 in Equation 18.16, we get:

$$\Omega = (1 + 15)!/1! \ 15!$$
$$= 16 \times 15 \times 14 \times 13 \times 12 \times 11 \times 10 \times 9 \times 8 \times 7 \times 6 \times 5 \times 4 \times 3 \times 2 \times 1/1$$
$$\times 15 \times 14 \times 13 \times 12 \times 11 \times 10 \times 9 \times 8 \times 7 \times 6 \times 5 \times 4 \times 3 \times 2 \times 1$$
$$= 16 \tag{18.24}$$

The 16 possible configurations are shown in Figure 18.1.

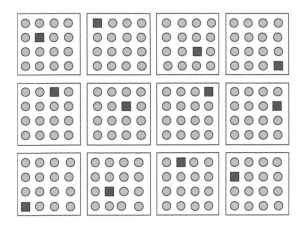

FIGURE 18.1
Sixteen possible configurations.

Example 3

Calculate the increase in entropy when 10^{20} vacancies are introduced to a part of a perfect crystal weighing 1 g atom.

Solution

$$S_c = -k \{N \ln [N/(N + n)] + n \ln [n/(n + N)]\}$$
$$= -1.38 \times 10^{-23}\{6.02 \times 10^{23}\ln [6.02 \times 10^{23}/(6.02 \times 10^{23} + 10^{20})] + 10^{20}\ln [10^{20}/(10^{20} + 6.02 \times 10^{23})]\}$$
$$= -1.38 \times 10^{-23}\{6.02 \times 10^{23}\ln [6.02 \times 10^{23}/10^{20}(6020 + 1)] + 10^{20}\ln [10^{20}/10^{20}(1 + 6020)]\}$$
$$= -1.38 \times 10^{-23}\{6.02 \times 10^{23}\ln[6020/6021] + 10^{20}\ln[1/6021]\}$$
$$= -1.38 \times 10^{-23} \times 10^{20} \{6.02 \times 10^3\ln [6020/6021] + \ln [1/6021]\}$$
$$= -1.38 \times 10^{-3} \{6020 \ln [6020/6021] + \ln [1/6021]\}$$
$$= - 0.00138 \{6020 \times [-0.00007] + [-3.7797]\}$$
$$= - 0.00138 \{-0.4343 - 3.7797\}$$
$$= 0.00138 \times 4.2139$$
$$= 0.0058 \text{ J/K} \tag{18.25}$$

18.2.2.2 Thermal Entropy

Atoms, ions, and molecules vibrate about their mean position. This vibration is a function of temperature. The entropy associated with this vibration is called thermal entropy, S_T. From Equation 18.17, it follows that:

$$dS = dq/T \tag{18.26}$$

Substituting for dq from Equation 18.1, we get, for change of thermal entropy, Equation 18.27.

$$dS_T = cdT/T \tag{18.27}$$

The change in thermal entropy at constant pressure can be written as:

$$\Delta S_T = \int_0^T \left(c_p/T\right)dT \tag{18.28}$$

Atoms, ions, or molecules can vibrate about their mean position. Einstein considered them as simple harmonic oscillators. Let us consider a mole of such oscillators. Let the frequency of oscillation be v_e. The thermal entropy per mole is given by:

$$S_T = 3N_{AV}k\left[hv_e/kT\left(e^{hv_e/kT}-1\right)-\ln\left(1-e^{-hv_e/kT}\right)\right] \tag{18.29}$$

Here, N_{AV} is the Avogadro's number. For $kT \gg hv_e$, $e^{-hv_e/kT}$ can be written as in Equation 18.30.

$$e^{-hve/kT} = 1 - hv_e/kT \tag{18.30}$$

Similarly, $e^{hve/kT}$ is given by Equation 18.31.

$$e^{hve/kT} = 1 + hv_e/kT \tag{18.31}$$

Substituting the values of $e^{-hve/kT}$ and $e^{hve/kT}$ in Equation 18.29, we get:

$$S_T = 3R\left[1 - \ln\left(hv_e/kT\right)\right] \tag{18.32}$$

The following conclusions can be drawn from Equation 18.32.

a. S_T increases as temperature increases. As temperature increases, atoms possess greater energy levels.
b. S_T decreases with an increasing frequency of vibration.

If the frequency changes from say v to v', then the change in thermal entropy is given by Equation 18.33.

$$\Delta S_T = 3R\left[1 - \ln\left(hv'/kT\right) - 1 + \ln\left(hv/kT\right)\right]$$
$$= 3R\left\{\ln\left[\left(hv/kT\right)/\left(hv'/kT\right)\right]\right\} \tag{18.33}$$
$$= 3R\left[\ln\left(v/v'\right)\right]$$

18.2.2.3 Electronic Entropy

Electronic entropy is associated with electrons and holes. It indicates the uncertainty in fixing their energy levels. This uncertainty will increase with temperature as the electrons and holes are excited to higher energy levels. These higher energy levels are greater in number with very small differences in energy. Hence, electrons or holes can jump to any of these closely spaced levels.

18.2.2.4 Magnetic and Dielectric Entropies

Magnetic and dielectric entropies are associated with materials possessing magnetism and dielectricity, respectively. In the case of magnetic materials, their magnetic moments may be oriented with order or disorder. When there is perfect order, then magnetic entropy is zero. Magnetic entropy increases with increases in disorder. This can also happen with increases in temperature. The same is the case with dielectric entropy.

18.2.2.5 Total Entropy

The total entropy of a system can be obtained by adding the different entropy values it possesses.

Thus,

$$S_{tot} = S_c + S_T + S_e + S_m + S_d \tag{18.34}$$

Here, S_e is electronic entropy, S_m is magnetic entropy, and S_d is dielectric entropy.

18.2.3 Free Energy

Free energy determines the stability of any system. Stability is associated with *equilibrium*. At equilibrium, the system will be at its lowest free energy value. Free energy is defined by the following equation:

$$G = H - TS \tag{18.35}$$

Here, S represents the total entropy of the system. If any change takes place in free energy at constant temperature, it is given by Equation 18.36.

$$\Delta G = \Delta H - T\Delta S \tag{18.36}$$

At equilibrium,

$$\Delta G = 0 \tag{18.37}$$

This is shown schematically in Figure 18.2. In this figure, the reaction coordinate represents the extent of a reversible reaction. At equilibrium,

$$\Delta G / \Delta \xi = 0 \tag{18.38}$$

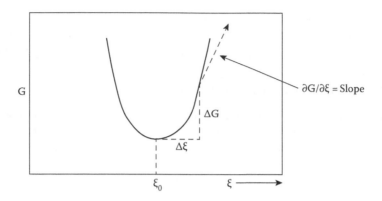

FIGURE 18.2
Schematic of free-energy versus reaction coordinate curve.

Free energy change is an extensive property. That means the value depends on the size of the system. The rate of change of free energy with the number of moles of a species, n_i, is called the *chemical potential*, μ_i, of that species. Expressed mathematically,

$$\mu_i = \partial G / \partial n_i |_{P, T, J} \qquad (18.39)$$

Thus, the chemical potential of a species in a system of J components is the work that would be required to remove it from the system to an infinite distance at a constant pressure and temperature.

Because there is no absolute value for the chemical potential in accordance with other thermodynamic properties, it is specified relative to a standard state. The value in the standard state is specified as μ_i°. When a species is not in its standard state, the deviation from its standard state is expressed in terms of its activity, a_i. Then the expression for its chemical potential is:

$$\mu_i = \mu_i^\circ + RT \ln a_i \qquad (18.40)$$

The activity for a species i is given by Equation 18.41.

$$a_i = \gamma_i X_i \qquad (18.41)$$

Here, γ_i is its activity coefficient, and X_i is its mole fraction in the system. In the standard state because μ_i is equal to μ_i°, a_i becomes equal to 1. Also, for an ideal solution,

$$a_i = \gamma_i^\circ X_i \qquad (18.42)$$

Here, γ_i° is called the *Raoultian activity coefficient*, and its value is 1.

18.3 Chemical Equilibrium

Chemical equilibrium is associated with chemical equations. To explain it, let us consider the following hypothetical chemical reaction:

$$M(s) + \tfrac{1}{2} X_2(g) = MX(s) \qquad (18.43)$$

In this equation, "s" represents a solid and "g" a gas; ΔG of this reaction is the driving force. It can be shown that:

$$\Delta G = \Delta G° + RT \ln K \qquad (18.44)$$

Here, $\Delta G°$ is the free energy change when the reactants and the products are in their standard states. K, the *equilibrium constant* for the reaction, is given by Equation 18.45.

$$K = a_{MX}/a_M \, (P_{X2})^{1/2} \qquad (18.45)$$

Here, P represents partial pressure. This equation is based on *mass action expression*. At equilibrium, there is no change in free energy. Therefore, Equation 18.44 reduces to:

$$\Delta G° = -RT \ln K \qquad (18.46)$$

And K is given in terms of standard free energy change as:

$$K = \exp - (\Delta G°/RT) \qquad (18.47)$$

18.4 Chemical Stability

Solid materials are chemically stable when their activities are within certain ranges of values. For example, refractories possess activities such that they are stable in oxidizing conditions. The range of activity values in which a material is chemically stable is the chemical stability domain for that material. With the help of thermodynamics, we can calculate the stability domains.

Let us consider the oxidation of a hypothetical metal M. Let us say that it forms two types of oxide. They are MO_z and MO_y. Let us say that $y > z$. Let us now consider the oxidation reaction of M to form its lower oxide. It is shown in Equation 18.48.

$$M + (z/2) \, O_2 = MO_z \qquad (18.48)$$

Let the standard free energy change for this reaction be $\Delta G_1°$. Using Equation 18.46, we can write the partial pressure of oxygen as:

$$\ln P_{O_2} = 2\Delta G_1°/zRT \tag{18.49}$$

Further oxidation of MO_z gives MO_y. This reaction is given in Equation 18.45.

$$[2/(y-z)]\, MO_z + O_2 = [2/(y-z)]\, MO_y \tag{18.50}$$

The corresponding partial pressure of oxygen is:

$$\ln P_{O_2} = \Delta G_2°/RT \tag{18.51}$$

Here, the $\Delta G_2°$ value is given by Equation 18.52.

$$\Delta G_2° = \left[2/(y-z)\right]\left[\Delta G_{MOy}° - \Delta G_1°\right] \tag{18.52}$$

In this equation, $\Delta G_{MOy}°$ is the standard free energy change for the formation of MO_y. Thus, the chemical stability domain for MO_z at a temperature T is between the partial pressures of oxygen corresponding to $\Delta G_1°$ and $\Delta G_2°$.

Example 4

The following standard free energies of formation of different iron oxides at 1000 K are:

$$\Delta G_{FeO} = -206.95 \text{ kJ/mol} \tag{18.53}$$

$$\Delta G_{Fe_3O_4} = -792.60 \text{ kJ/mol} \tag{18.54}$$

$$\Delta G_{Fe_2O_3} = -561.80 \text{ kJ/mol} \tag{18.55}$$

Calculate chemical stability domains for these oxides.

Solution

The reaction for the formation of FeO is given by Equation 18.56.

$$Fe + \tfrac{1}{2} O_2 = FeO \tag{18.56}$$

This equation is of the same form as Equation 18.48. Here, z is equal to 1. Therefore, Equation 18.49 can be written as:

$$\ln P_{O_2} = -2 \times (-206.95 \times 1000)/8.3146 \times 1000$$
$$= 49.7799 \tag{18.57}$$

Therefore, the oxygen partial pressure for the formation of FeO at 1000 K is given by:

$$P_{O_2} = \text{anti ln } (49.7799)$$
$$= 4.1 \times 10^{15} \text{ MPa} \tag{18.58}$$

When the oxygen partial pressure over FeO is increased, at a certain value, FeO gets converted to Fe_3O_4. The reaction is given by:

$$3FeO + \tfrac{1}{2} O_2 = Fe_3O_4 \tag{18.59}$$

The standard free energy change for this reaction is:

$$\Delta G^\circ = \Delta G_{Fe_3O_4} - 3\Delta G_{FeO}$$
$$= -792.6 - 3 \times (-206.95) \tag{18.60}$$
$$= -171.75 \text{ kJ/mol}$$

The partial pressure of oxygen required for the reaction given in Equation 18.59 is obtained from Equation 18.61.

$$\ln P_{O_2} = 2 \times 171.75 \times 1000 / 8.3146 \times 1000$$
$$= 41.3129 \tag{18.61}$$

The value of this pressure is given in Equation 18.62.

$$P_{O_2} = \text{anti ln } (41.3129) = 8.7 \times 10^{11} \text{ MPa} \tag{18.62}$$

Fe_3O_4 will be stable up to a certain oxygen partial pressure at which it will react with oxygen to produce Fe_2O_3. This reaction is given by:

$$2Fe_3O_4 + \tfrac{1}{2} O_2 = 3Fe_2O_3 \tag{18.63}$$

The standard free energy change for this reaction is given by Equation 18.64.

$$\Delta G^\circ = 3\Delta G_{Fe_2O_3} - 2\Delta G_{Fe_3O_4}$$
$$= 3 \times (-561.80) - 2 \times (-792.6) \tag{18.64}$$
$$= -100.20 \text{ kJ/mol}$$

The partial pressure of oxygen can be obtained from:

$$\ln P_{O_2} = -2 \times (-100.20) / 8.3146$$
$$= 24.1022 \tag{18.65}$$

The value of this is given in Equation 18.66.

$$P_{O_2} = \text{anti} \ln 24.1022$$
$$= 2.9 \times 10^4 \text{ MPa} \tag{18.66}$$

18.5 Gibbs–Duhem Relation

The Gibbs–Duhem relation gives the interrelationship of the chemical potentials of the components in a binary compound.

A binary oxide has the general formula MO_ξ. Its equilibrium with its ions can be represented as:

$$MO_\xi \leftrightarrow M^{\xi(2+)} + \xi O^{2-} \tag{18.67}$$

In terms of the *electrochemical potential*, we can write Equation 18.68 for this system.

$$\eta_{MO_\xi} = \eta_{\xi(2+)} + \xi \eta_{O2-} \tag{18.68}$$

In this equation, η is the electrochemical potential, given by:

$$\eta_i = \mu_i + z_i e \varphi \tag{18.69}$$

Here, z_i is the number of charges on the ion i, e is the electronic charge, and φ is the electric potential. The electrochemical potential of the oxide is constant. Therefore,

$$d\eta_{MO_\xi} = 0 \tag{18.70}$$

From Equation 18.68, therefore we get Equation 18.71.

$$d\eta_{M\xi(2+)} = -\xi d\eta_{O2-} \tag{18.71}$$

Substituting the values of η following Equation 18.69 in Equation 18.71, we get:

$$d\mu_{M\xi(2+)} + 2\xi e d\varphi = -\xi \left(d\mu_{O2-} + 2e d\varphi \right)$$
$$d\mu_{M\xi(2+)} = -\xi d\mu_{O2-} \tag{18.72}$$

Equation 18.72 is known as the *Gibbs–Duhem relation*.

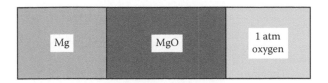

FIGURE 18.3
A MgO crystal in equilibrium with Mg and O_2 separately on its sides.

Example 5

18.5.1 A MgO crystal is placed between Mg metal on one side and pure oxygen on the other side (Figure 18.3). Calculate the activity of Mg and the partial pressure of O_2 at each interface at 1000 K. $\Delta G_{MgO}°$ is −492.95 kJ/mol.

18.5.2 Show that the Gibbs–Duhem relationship holds well.

Solution

18.5.1 The formation reaction for MgO is given by:

$$Mg\ (s) + ½\ O_2\ (g) = MgO\ (s) \tag{18.73}$$

The expression for the standard free energy change for this reaction is given by Equation 18.69.

$$\Delta G° = -RT\ \ln\ (a_{MgO}/a_{Mg}P_{O2}^{1/2}) \tag{18.74}$$

Substituting the values applicable on the Mg side, we get for the partial pressure of oxygen at the interface:

$$\Delta G° = -RT \ln\left(a_{MgO}/a_{Mg}P_{O2}^{1/2}\right)$$

$$-492.95 = -8.31467 \ln\left(1/1 \times P_{O2}^{1/2}\right)$$

$$\text{Therefore, } \ln\left(1/1 \times P_{O2}^{1/2}\right) = 59.2868$$

$$\text{Therefore, } \left(1/1 \times P_{O2}^{1/2}\right) = \text{anti}\ \ln 59.2868 = 5.6 \times 10^{25} \tag{18.75}$$

$$\text{Therefore, } P_{O2}^{1/2} = 1/5.6 \times 10^{25} = 1.8 \times 10^{-26}$$

$$\text{Taking logarithm, we get: } \tfrac{1}{2} \log P_{O2} = \log 1.8 - 26$$

$$\text{Therefore, } \log P_{O2} = -51.4895$$

$$\text{Therefore, } P_{O2} = 3.2 \times 10^{-52}\ \text{MPa}$$

Now, considering the oxygen side, the unknown is the activity of Mg at the interface. Substituting the values in Equation 18.74, we get this value from Equation 18.76.

$$-492.95 = -8.31467 \ln\left(1/a_{Mg} \times 1\right)$$

$$\ln\left(1/a_{Mg} \times 1\right) = 59.2868$$

$$\text{Therefore, } 1/a_{Mg} = \text{anti ln } 59.2867 = 5.5969 \times 10^{25} \tag{18.76}$$

$$\text{Therefore, } a_{Mg} = 1.8 \times 10^{-26}$$

18.5.2 The Gibbs–Duhem relationship will hold well if the chemical potential of magnesium oxide at the metal side is equal to that at the oxygen side. That is:

$$\mu_{MgO}|_{oxygen} = \mu_{MgO}|_{metal} \tag{18.77}$$

Writing the chemical potential of magnesium oxide in terms of its constituents, we get Equation 18.78.

$$\mu_{O_2}^o + (RT/2)\ln P_{O_2} + \mu_{Mg}^o + RT \ln a_{Mg}|_{oxygen}$$
$$= \mu_{O_2}^o + (RT/2)\ln P_{O_2}|_{metal} + \mu_{Mg}^o + RT \ln a_{Mg}|_{metal} \tag{18.78}$$

Substituting the values, we find from Equation 18.79 that LHS = RHS.

$$\ln a_{Mg}|_{oxygen} = \tfrac{1}{2}\ln P_{O_2}|_{metal} = -59.2867 \tag{18.79}$$

Hence, the Gibbs–Duhem relationship holds well in this case.

18.6 Interfaces

Interfaces determine many properties and processes. At interfaces, the structure changes from one type to the next. Thus, we have interfaces between solids at solid–solid contact, between solid and liquid at solid–liquid contact, and between solid and gas at solid–gas contact. In order to know the mechanical properties, chemical phenomena, and electrical properties, we should understand the structure, composition, and properties of these contacts.

18.6.1 Solid–Solid Contact

Interfaces between two solids can form during the transformation of one solid to another. For example, we find a number of allotropic forms for silica. So, at different temperatures, there will be solid state transformations among these allotropic forms. Transformation of α-quartz to β-quartz is one such transformation in the solid state.

The Clausius–Clapeyron equation is of great importance for calculating the effect of change of pressure (P) on the equilibrium transformation temperature (T) of a pure substance. The equation is represented as Equation 18.80.

$$dP/dT = \Delta H/T\Delta V \tag{18.80}$$

Here, ΔH is the heat of transformation, and ΔV is the volume change associated with the transformation. In the case of transformation from α-quartz to β-quartz, we can write Equation 18.80 as:

$$dP/dT = \Delta H_{tr}/T(V_\beta - V_\alpha) \tag{18.81}$$

Here, ΔH_{tr} is the molar heat of transformation, and Vα and Vβ are the molar volumes of α-quartz and β-quartz, respectively. Equation 18.81 can also be written as Equation 18.82.

$$(dT/dP) = T(V_\beta - V_\alpha)/\Delta H_{tr} \tag{18.82}$$

This equation may be used to calculate the change in the transition temperature with a change of pressure.

18.6.2 Solid–Liquid Contact

18.6.2.1 Surface Tension

A force is required to extend a liquid surface. Surface tension, γ, is defined as the reversible work, w, required to increase the surface of a liquid by a unit area:

$$dw = \gamma dA \tag{18.83}$$

Knowing the conditions under which this work is done, we can relate surface tension to other thermodynamic properties.

Consider a two-phase system, as shown in Figure 18.4. Following the first law, we get

$$dE = TdS - PdV + \gamma dA + \sum \mu_i dn_i \tag{18.84}$$

for the interface between two phases, which is planar. Here, dE is the change in internal energy, and dA is the change in area of the interface. For a constant entropy, volume, and composition, we get:

$$\gamma = (\partial E/\partial A)_{S,V,ni} \tag{18.85}$$

Similarly, surface tension can also be written in terms of Gibbs free energy change, dG. From the second law of thermodynamics, we have:

$$dG = -SdT + VdP + \gamma dA + \sum \mu_i dn_i \tag{18.86}$$

Therefore,

$$\gamma = (\partial G / \partial A)_{P,T,ni} \tag{18.87}$$

when pressure, temperature, and composition are constant. The increase in free energy when the surface area is increased from A_1 to A_2 is given by Equation 18.88.

$$\Delta G = \int_{A1}^{A2} \gamma dA = \gamma (A_2 - A_1) \tag{18.88}$$

18.6.2.2 Thermodynamics

Let us consider Equation 18.84 again for the interface between the two phases shown in Figure 18.4. In this equation, dS is the excess entropy due to the presence of the interface, dV is the volume of the interface, and dn_i represents the excess moles of species I within the interface. For a constant composition, we get the internal energy of the interface by integrating Equation 18.84. The result is shown in Equation 18.89.

$$E = TS + \gamma A - PV + \sum \mu_i n_i \tag{18.89}$$

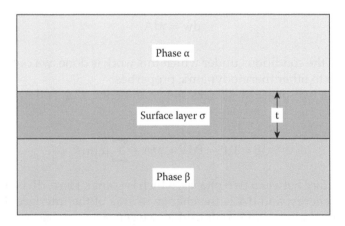

FIGURE 18.4
Interface layer between two phases.

An expression for surface tension can be obtained from this equation as given in Equation 18.90. The first three terms together inside the second bracket are the Gibbs free energy.

$$\gamma = (1/A)\left(E - TS + PV - \sum \mu_i n_i\right) \tag{18.90}$$

Therefore, surface tension of a flat surface is the excess free energy per unit area. By differentiating Equation 18.89, we get:

$$dE = TdS + SdT + \gamma dA + Ad\gamma - PdV - VdP + \sum \mu_i dn_i \tag{18.91}$$

Writing the RHS of Equation 18.84 for dE, we get:

$$0 = SdT + Ad\gamma - VdP + \sum n_i d\mu_i$$
$$Ad\gamma = VdP - SdT - \sum n_i d\mu_i \tag{18.92}$$

For a unit area, Equation 18.92 becomes Equation 18.93.

$$d\gamma = vdP - sdT - \sum \Gamma_i d\mu_i \tag{18.93}$$

Here, v, s and Γ_i are the volume, excess entropy, and excess moles of species i per unit area, respectively. At constant temperature and pressure, Equation 18.93 becomes:

$$d_\gamma = -\sum \Gamma_i d\mu_i \tag{18.94}$$

For a two-component system, we have:

$$d\gamma = -\Gamma_1 d\mu_1 - \Gamma_2 d\mu_2 \tag{18.95}$$

From the Gibbs–Duhem equation given in Equation 18.72, we can write Equation 18.96.

$$x_1\, d\mu_1 + x_2\, d\mu_2 = 0 \tag{18.96}$$

In this equation, x_1 and x_2 are the mole fractions of the two components in the phase. From this equation, we get:

$$d\mu_1 = -(x_2/x_1)\, d\mu_2 \tag{18.97}$$

Substituting the value of $d\mu_1$ in Equation 18.96, we get Equation 18.98.

$$d\gamma = \Gamma_1(x_2/x_1)d\mu_2 - \Gamma_2 d\mu_2$$

$$= \left[\Gamma_1(x_2/x_1) - \Gamma_2\right]d\mu_2 \qquad (18.98)$$

$$O_r - d_\gamma = \left[\Gamma_2 - \Gamma_1(x_2/x_1)\right]d\mu_2$$

This equation can be written as Equation 18.99.

$$-d\gamma = \Gamma_{2(1)}\, d\mu_2 \approx \Gamma_{2(1)}\, RT\, d\ln c_2 \qquad (18.99)$$

Here, $\Gamma_{2(1)}$ is given in Equation 18.100; c_2 is the concentration of the component 2.

$$\Gamma_{2(1)} = [\Gamma_2(x_2/x_1) - \Gamma_1] \qquad (18.100)$$

Its value is assumed to be low so that its activity coefficient is nearly constant. Equation 18.99 is the Gibbs adsorption isotherm. Using this equation, we can write:

$$\Gamma_{2(1)} = -d\gamma/d\mu_2 = -d\gamma/RT\, d\ln a_2 \approx -d\gamma/RT\, d\ln c_2 \qquad (18.101)$$

Equation 18.101 shows that the excess of component 2 in the interface is related to the variation of surface tension.

18.6.2.3 Curved Surfaces

18.6.2.3.1 Pressure Difference

Curved surfaces are characterized by a pressure difference existing across them. This can be illustrated by taking the example of a capillary inserted in a liquid. This is shown in Figure 18.5.

A bubble is blown through the capillary. The resistance to the expansion of the bubble is surface area created. At equilibrium, the work of expansion must equal the increased surface area. That is,

$$\Delta P\, dv = \gamma\, dA \qquad (18.102)$$

In this equation, ΔP is the excess pressure required to create the area dA. The change in volume is given by Equation 18.103.

$$dv = 4\pi r^2\, dr \qquad (18.103)$$

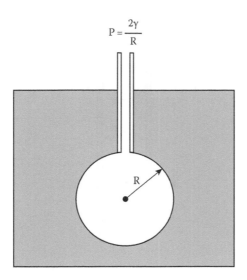

$$P = \frac{2\gamma}{R}$$

FIGURE 18.5
A capillary immersed in liquid with a bubble at its end.

Here r is the original radius of the bubble and dr is the increase in the radius because of the increased area dA. The increased area is given by:

$$dA = 8\,\pi r\,dr \qquad (18.104)$$

Using Equations 18.101 through 18.103, ΔP can be written as:

$$\Delta P = \gamma\, dA/dv = \gamma\,(8\,\pi r\,dr/4\pi r^2\,dr) = \gamma(2/r) \qquad (18.105)$$

For a general shape having two principal radii, r_1 and r_2, Equation 18.105 modifies to Equation 18.106.

$$\Delta P = \gamma[(1/r_1) + (1/r_2)] \qquad (18.106)$$

The capillary rise observed is the result of the pressure difference due to the surface tension. When this difference is balanced by the hydrostatic pressure of the liquid column, the capillary rise takes place. Capillary rise is shown in Figure 18.6. As shown in this figure, the meniscus inside the capillary has a concave surface. The contact angle for this meniscus is θ. Thus, the radius of the meniscus becomes R/cos θ. Therefore, we can write the pressure difference across the meniscus as:

$$\Delta P = \gamma(2/r) = \gamma(2\cos\theta/R) = \rho gh \qquad (18.107)$$

In this equation, ρgh is the hydrostatic pressure of the liquid column. The various terms in it are the density of the liquid, ρ; the gravitation constant, g;

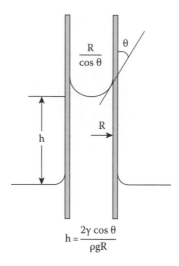

FIGURE 18.6
Capillary rise.

and the height of the liquid column inside the capillary, h. Thus, Equation 18.107 gives another way of determining the surface tension of a liquid from the knowledge of the contact angle. The pressure difference across a curved surface is important in the sintering and vitrification of ceramics and refractories, as well as in grain growth.

18.6.2.4 Solid–Liquid Equilibria

Solid–liquid equilibria can be observed when a refractory melts. Applying the Clausius–Clapeyron equation to such equilibria, we get:

$$dP/dT = \Delta H_f/T(V_{liq} - V_{solid}) \qquad (18.108)$$

or

$$dT/dP = T(V_{liq} - V_{solid})/\Delta H_f \qquad (18.109)$$

Here, ΔH_f is the molar heat of fusion, and V_{liq} and V_{solid} are the molar volumes of liquid and solid, respectively. Equation 18.109 may be used to calculate the change in the melting point of a refractory with a change of pressure.

18.6.3 Solid–Gas Contact

18.6.3.1 Vapor Pressure of a Curved Surface

The pressure difference across a curved surface also affects the vapor pressure over such a surface. This pressure over a convex surface happens to be

TABLE 18.1

Effect of Particle Radius on Pressure Difference and Relative Vapor Pressure

Material	Surface Diameter (µ)	Pressure Difference, MPa	Relative Vapor Pressure, (p/p₀)
SiO₂ glass	0.1	12.07	1.02
(1700°C)	1.0	1.21	1.002
γ = 0.30 J/m²	10.0	0.12	1.0002
Solid Al₂O₃	0.1	36.21	1.02
(1850°C)	1.0	3.62	1.002
γ = 0.91 J/m²	10.0	0.36	1.0002

higher than that over a planar surface. The pressure difference across such a surface can be written as:

$$V\Delta P = RT \ln(p / p_o) = V_\gamma \left[(1/r_1) + (1/r_2)\right] \quad (18.110)$$

Here, V is the molar volume of the material having the convex surface, p is the vapor pressure over the convex surface, and p_o is the vapor pressure over the planar surface. Therefore,

$$\ln(p / p_o) = (V_\gamma / RT)\left[(1/r_1) + (1/r_2)\right] = (M_\gamma / \rho RT)\left[(1/r_1) + (1/r_2)\right] \quad (18.111)$$

In this equation, M is the molecular weight and ρ is the density.

Equation 18.106 tells us that the difference in vapor pressure increases with a decrease in particle size. Table 18.1 illustrates this point. Clays are widely used in the manufacture of products out of ceramics and refractories. Clays are characterized by fine particle sizes. Because fine particles have a large surface area per unit volume, the greater surface energy forces bring about good sinterability in clay-mixed green products.

Problems

18.1 Derive $'S_c = -k \{N \ln [N/(N + n)] + n \ln [n/(n + N)]\}'$.

18.2 Show that $'\Delta G° = -RT \ln K'$.

18.3 For the Si–SiO₂ system:

18.3.1 Find the equilibrium partial pressure of oxygen at 1000 K.

18.3.2 If oxidation is occurring with water vapor, calculate the equilibrium constant and the H₂/H₂O ratio in equilibrium with Si and SiO₂ at 1000 K.

18.3.3 For the same H₂/H₂O ratio obtained earlier, compute the equilibrium partial pressure of oxygen in the gas mixture.

18.4 Find whether Al can reduce Fe_2O_3 at 1200°C.

18.5 Is it possible to oxidize Ni in a CO/CO_2 atmosphere with a ratio of 0.1 at 1200°C? Substantiate your answer.

18.6 Will SiO_2 oxidize Zn at 700°C? Give a reason for your answer.

Reference

1. C. Newey and G. Weaver (Eds.), *Materials Principles and Practice*, Butterworth, London, 1990, p. 212.

Bibliography

M. W. Barsoum, *Fundamentals of Ceramics*, Taylor & Francis Group, LLC, New York, pp. 110–126, 129–131, 133–135.

W. D. Kingery, H. K. Bowen, and D. R. Uhlmann, *Introduction to Ceramics*, 2nd Edn., John Wiley, New York, 1960, pp. 177–181, 185–188.

G. S. Upadhyaya and R. K. Dube, *Problems in Metallurgical Thermodynamics and Kinetics*, Pergamon Press, Oxford, UK, 1977, pp. 77, 81–84.

19

Properties and Testing

19.1 Introduction

Refractory materials, in addition to their resistance to heat, should also behave appropriately with other environmental factors that are encountered in their service life. These factors are mechanical and thermal stresses; corrosion from solids, liquids, and gases; and gaseous diffusion. Different refractory products are designed and manufactured to meet the required properties useful for a particular application. After their manufacture, the products are tested to ascertain their properties. In this chapter, the various properties required for refractories are discussed along with the tests conducted to determine those properties.

Most of the refractories are used in iron and steel works. Iron is produced as pig iron in a blast furnace. Here, iron ore is reduced with the help of coke in the presence of limestone. A blast furnace comes under the class of furnaces called shaft furnaces. It is the tallest among all shaft furnaces. The raw material—which is a mixture of iron ore, coke, and limestone—is dropped from the top of the blast furnace. Therefore, the refractories forming the furnace top should be resistant to abrasion. At the furnace hearth, two liquid layers are formed. The bottom layer consists of liquid metal over which liquid slag stands. Both layers are tapped separately and run to a cast house. The tap hole material should be such that the hole can be conveniently drilled for tapping molten iron and slag out, and then it should be possible to plug the hole so the process can be repeated. In the cast house, the molten iron flows into a trough and the slag to a slag pit. The properties of the refractory making up the trough should be such that it resists the impact and splashing of the molten iron.

One of the common steelmaking methods is called the basic oxygen furnace (BOF) process. The raw material for this furnace is a mixture of molten iron derived from a blast furnace and scrap steel. Steelmaking essentially reduces the impurity content of molten iron. These elements mainly are carbon, sulfur, phosphorus, and manganese. These excess elements are removed by oxidizing them with the help of oxygen gas. The required oxygen is supplied through a top blown or bottom blown process. In the top blown process, an oxygen lance is used; for the bottom blown process, oxygen is passed through pipes. In both processes, the molten metal experiences vigorous stirring and intense heating. The refractory used for the furnace construction, therefore, should resist not only the high temperature liquid metal but also the high temperature liquid slag that is developed. Basic slag is formed in this process. Therefore, a basic refractory should also be used for the construction of the furnace. A magnesia-based refractory containing carbon is primarily used for this purpose.

A second method of steelmaking uses an electric arc furnace (EAF). The raw material for this process is steel scrap. The refractory should be able to resist the mechanical impingement of this scrap. Also, *hot spots* will develop when the arc strikes the furnace lining. Hot spots are subjected to intense heat. Again, the slag developed is basic in nature. Because of these requirements, the refractory material should have sufficient mechanical strength, high refractoriness, and basic slag resistance.

In *ladle metallurgy,* alloying additions are done in the ladle itself. Such ladles are called *ladle metallurgical furnaces* (LMF). The refractories making up these furnaces should be able to resist the alloying operations. Sometimes, these ladles will have to be reheated.

In processing nonmetals, the property requirements of the refractories are entirely different from those of iron and steelmaking. For example, in the extraction of aluminum, the temperature requirements are much lower. But, there is a problem with penetration of the refractory by molten aluminum. Thus, the refractory needs to be nonwetting to liquid aluminum. This can be achieved by the addition of certain additives to the refractory material during its manufacture.

Another application of refractories is in hydrocarbon industries, such as in oil and gas industries. The property requirements for refractories here are different from those of metal industries. The temperature requirements are lower, but the refractories should be abrasion resistant to the flow of high velocity particles. The refractories also should conserve heat energy during conversion processes by having thermal conductivity within a certain limit.

Glass making is another important industry in which refractories are used. Here, liquid glass will be in constant contact with the refractory material. Hence, the refractory should be nonporous. Fused refractories are used for making glass tanks.

From this discussion, it must be obvious that the property requirements are different for different applications of refractories. The most important requirements are as follows:

a. Rigidity and maintenance of size, shape, and strength at the operating temperature

b. Ability to withstand thermal shock

c. Resistance to chemical attack

For insulating properties, cold crushing strength and thermal conductivity may become important. When one property is greater, the other property may become less. For example, high porosity, which is required for insulation, will lead to reduced slag resistance.

In today's competitive world, selecting the most suitable refractories to prolong service life and to make a process more cost effective has become all the more important. In order to do this, materials must be carefully assessed to make sure they are suitable for the environment and conditions in which they will operate. This is accomplished by testing the refractory materials.

Testing of refractories is carried out for the following reasons:

a. To characterize new materials

b. For quality control of processes and products [1]

c. To help rank and select materials for a particular application and to predict in-service behavior

d. To examine the effects of service conditions

e. For fault diagnosis

f. To establish properties required for mathematical modeling to compare current and new designs

There are a number of test methods available to check all the properties of refractories. The majority of these methods have been standardized. Important refractory properties that are tested are as follows:

a. Refractoriness

b. Hot strength

c. Thermal shock resistance

d. Resistance to chemical attack

e. Cold crushing strength

f. Thermal conductivity

19.2 Refractoriness

Refractoriness represents the melting point of a refractory. It is the ability to withstand high temperatures in an unstressed condition. In the case of refractories, measuring the melting point is not practicable. This is because they do not possess a sharp melting point. They melt over a range of temperatures. This range may extend over several hundreds of degrees of centigrade temperature. Also, the determination of solidus and liquidus temperatures would not be easy. And, they might not be very useful. A refractory material may be partly liquid at a high temperature but appears as a solid. At a still higher temperature, the proportion of the liquid may increase. At this point, the material will either collapse and appear liquid or pasty, or it will exude liquid and collapse slowly later. The temperature of collapse may become a function of the rate of heating.

Refractoriness is tested by a comparison method. For this comparison, standard cones (called Seger cones) are available. The refractory to be tested is made in the shape and dimension of these cones. If the material is available in solid form, then it is cut and shaped. If it is available as a powder, it is mixed with dextrin. Dextrin binds the powder particles; therefore, the powder can be shaped in the form of a cone. Each Seger cone is identified by a number. After making the cone, it is heated along with standard cones at a standard rate in an oxidizing temperature until the test cone bends over. This bent cone is matched with one of the Seger cones. The number of this matched cone is quoted as the pyrometric cone equivalent (PCE). Sometimes, PCE may be reported between two standard cone values if the softening point occurs between the two. The temperature corresponding to this number is the PCE temperature. This temperature is usually much higher than the working temperature of the refractory brick. Hence, it is not of much practical use. ISO 528:1983 gives the standard procedure for the test.

PCE is affected by the presence of impurities. Iron and alkali metals reduce the value. They act as fluxes. A refractory selected for an application should have its working temperature well below its PCE temperature.

19.3 Hot Strength

Hot strength is an important quality of a brick. It is also measured indirectly and is defined as the temperature at which deformation under a standard load is rapid. The test is called the refractoriness under load (RUL) test. It can be compared to a creep test in a short time. RUL is the ability of a material to withstand specific conditions of load, temperature, and time [1].

The standard load applied is either 45.36 kg or double this weight. The standard specimen can be a cylinder or a square prism. The dimensions of the cylindrical specimen are 50 mm in diameter and 50 mm in height. The dimensions of the prismatic specimen are 38 mm × 38 mm × 114 mm [2]. The sample is kept in a carbon granule furnace between refractory plates and in contact with a dilatometer. The dilatometer is in turn attached to an extensometer to measure deformation. A specified load is applied on top of the sample. The furnace is then heated at a standard rate of 10°C per minute. As it is heated, a record of the height of the sample is made. Figure 19.1 shows such a record.

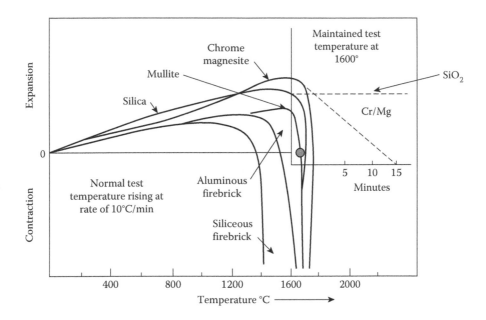

FIGURE 19.1
Plots for the RUL test for various bricks.

The test is continued until the test piece collapses or it contracts to 90% of its height. RUL may be quoted as the range of three temperatures:

1. *Initial softening.* This range corresponds to the one in which the curve remains horizontal.

2. *Rapid collapse.* This is the range in which the sample contracts at a rate of 1.27 mm per minute.

3. *Total collapse.* This range corresponds to a contraction beyond 10% shortening.

Heating can be stopped at a suitable temperature. This temperature may be relative to the working temperature. After stopping the heating, the creep in the specimen is measured as a function of time. This is shown in Figure 19.1 for silica and chrome-magnesite bricks at a constant temperature of 1600°C. The purpose of this check is to determine the maximum temperature at which the brick could be used under a compressive load. Creep strength of a refractory is dependent on the softening point and the amount of its glass phase. As the softening point lessens, the creep strength also becomes less. As the amount of glass phase increases, the creep strength decreases.

Where the material is to be selected for high temperatures to sustain a load, its creep strength has to be determined—such as for a glass-melting tank crown. To conduct the creep test separately, the same setup was used as for the RUL test. The test sample used is a cylindrical specimen 50 mm in diameter and 50 mm in height. It is heated to the test temperature under a stress of 0.2 MPa for dense refractories. The deformation of the sample with respect to time is noted. From this data, creep percent versus time is plotted. When selecting a material for an application where creep is important, its creep deformation should not exceed 0.2%. The standard for creep in compression is ISO 3187:1989.

When the bricks are used in furnaces, only one side faces the hot temperature inside the furnace. This is called the hot face of the brick, and the opposite one is the cold face. Even when the hot face reaches its softening temperature, the cold face will still be below that temperature. Hence, it is difficult to strictly apply the RUL test result in this case. The standard procedure is given in ISO 1893:1989.

Hot strength is a function of structure. Sometimes the strength may be retained beyond the solidus temperature. For example, this happens with silica and mullite. During heating, silica in quartz form transforms to tridymite. Tridymite has a hexagonal crystal structure, which is not symmetrical. Hence, once it forms, the strength is retained beyond the solidus temperature by the interlocking of the crystals. The same thing happens in the case of mullite, which has an orthorhombic crystal structure. This does not happen with cubic crystals, which are symmetrical. Therefore, the crystals of basic bricks do not interlock, possessing cubic crystal structure. Such is the case with glass, which is amorphous and isotropic.

19.4 Thermal Shock Resistance

Thermal shock resistance is the ability to withstand rapid changes in temperature with minimal cracking. This kind of situation will arise under the following circumstances:

a. When the material is heated and cooled suddenly, such as with nozzles
b. Under thermal cycling, such as the regenerators
c. With the severe thermal gradient that exists in sliding gate plates

Under these thermal shock conditions, if the material experiences a strain greater than its fracture strain, it will break into pieces.

Table 19.1 lists the properties required for good thermal shock resistance. It also lists how each property produces its beneficial effect. Table 19.2 lists three refractories possessing good thermal shock resistance [3]. The properties that provide this benefit are also indicated. It is difficult to exactly simulate service conditions in a laboratory test. Thermal shock resistance

TABLE 19.1

How Different Properties Bring About Good Thermal Shock Resistance

Sl. No.	Property	Thermal Shock Resistance
1.	Low thermal expansion coefficient	Reduces stress associated with temperature gradient
2.	High thermal conductivity	Reduces temperature gradient
3.	High toughness	Inhibits crack propagation
4.	High strain to failure	Accommodates thermal stress
5.	Low modulus of elasticity	Minimizes stress associated with thermal expansion

TABLE 19.2

Three Materials with Good Thermal Shock Resistance and Their Properties

Sl. No.	Material	Properties
1.	Carbon	High strain to failure, low thermal expansion, high thermal conductivity, high work to fracture
2.	Silicon carbide	High thermal conductivity, low thermal expansion
3.	Castable	High strain to failure, low thermal expansion

is highly dependent on specimen geometry, uniformity of heating, and the stresses imposed by service conditions. There are several practical methods used to determine thermal shock resistance:

1. Finding the degree of cracking
2. Number of thermal cycles to failure
3. Reduction in strength
4. Reduction in modulus of elasticity
5. Calculation from several property measurements [3]

The test in which the number of cycles to failure is determined is called the spalling test. This test can be performed in different ways. The selection of the method depends on the brick material and its working conditions. For example, silica refractory has excellent thermal shock or spalling resistance above 300°C but poor resistance below it. Hence, a spalling test on silica should avoid temperatures below 300°C.

One way of testing for spalling resistance is by heating the brick sample slowly up to its working temperature, then cooling it rapidly for 10 min in a cold air jet or on a steel plate, and then reheating it for 10 min in a furnace maintained at the working temperature. This cycle is repeated until a piece of the brick is detached. The number of cycles to failure is noted. A brick surviving 30 cycles has good thermal shock resistance. The spalling resistance of such a brick is said to be +30 cycles.

In some tests, only one end of the brick is heated and cooled. This almost corresponds to the condition experienced in furnaces. In another kind of simulated test, a small wall of bricks is heated by gas burners and cooled with an air blast. The heating and cooling is restricted to one face of the wall.

The American Society for Testing of Materials (ASTM) has given a standard test for measuring the thermal shock resistance. It is numbered ASTM C1100-88 [2]. Test samples are supported 100 mm above a segmented line gas burner. They are then subjected to cyclic rapid heating and cooling for five cycles. Cooling is effected by forced air. Heating and cooling are for 15 min each. The hot face temperature can be in the range of 815°C–1093°C. The modulus of elasticity is determined before and after the cycle of heating and cooling. The reduction in the value is a measure of the vulnerability of the material to thermal shock. All samples are made to have a constant length of 22.86 cm.

In yet another method of determining thermal shock resistance, material properties are established first. Thermal shock resistance is calculated from the values of these properties, which are called R values. The higher the R value, the greater the thermal shock resistance. Hasselman [4] derived a series of parameters to rank materials according to their thermal shock resistance.

Spalling resistance is a function of four properties of the brick:

1. *Thermal conductivity.* This determines the temperature gradient set up during the test. The higher the conductivity, the lower the temperature gradient will be. In this case, the brick will be more resistant to spalling.

2. *Coefficient of thermal expansion.* This property determines the extent of the strain induced by the temperature gradient—the higher the coefficient, the greater the strain. The more the strain, the less resistance there will be to spalling.

3. *Modulus of elasticity.* The higher the modulus of elasticity, the greater the brick's resistance will be to spalling.

4. *Shear strength.* Modulus of elasticity and shear strength together determine the intensity of the stress. When the stress exceeds the shear strength of the material, a crack will develop to relieve the stress.

Once cracks develop, subsequent stress cycling will modify the temperature gradient and worsen the situation; the cracks will start propagating—further relieving the stress. These kinds of mutually dependent processes will continue until a break out from the brick takes place. That marks the end of the test.

Spalling resistance for a given material having a particular chemistry, crystallographic condition, and distribution of grains can be improved by selecting a good bonding material, decreasing porosity, and by having a thorough firing and a uniform distribution of grog particles. Grog particles tend to arrest cracks.

19.5 Resistance to Chemical Attack

Refractory materials are exposed to a wide variety of environments in their applications. Many of these environments can attack the materials. An environment could be in a molten state such as with molten metal and slag, both of which are highly corrosive. Hence, it is necessary to characterize the corrosion resistance of refractories in the particular environment in which they are going to be used.

Available methods can only compare different materials on the basis of their corrosion resistance. Their actual behavior cannot be determined because the service conditions cannot be simulated in laboratory tests. For example, service temperature tends to vary with time. Laboratory tests are carried out at a constant temperature. Hence, the effect of thermal gradients is not considered in laboratory tests.

Static or dynamic methods are used to find the attack of molten media. The test for *slag resistance* is a static test. There are three types of tests for slag resistance.

The most common among the three tests involves drilling holes in the brick and packing these with the slag samples against which the resistance is to be found out. This assembly is then heated to the working temperature at which the brick will be used. This temperature is maintained for an hour. Then, the assembly is cooled. Afterward, the brick is sectioned at the holes. The depth of penetration by the slag is measured. This depth is compared with that in other bricks, and the brick having the least depth of slag penetration is selected because it will have the best slag resistance.

In the second method, ground brick and slag are mixed in various proportions. Refractoriness of each of these mixtures is determined. A composition versus refractoriness graph is drawn for each brick for different slag as well as for each slag and different brick combination. Based on these graphs, proper brick–slag combination is selected.

In the third type of test, individual bricks are dipped in slag or hot slag is sprayed onto walls built into specially designed furnaces. Field trials can also be performed by incorporating test bricks into working equipment.

The extent to which a brick will be attacked by a slag depends on the following factors:

a. The chemical reactions likely to take place between them

b. Whether any of the products of reaction are liquid at the working temperature

c. The speed with which the reactions are likely to proceed

Generally, acid slags react with basic bricks and basic slags with acid bricks. But there are exceptions to this general rule. For instance, if iron oxide is present in a slag, both acid and basic bricks will be attacked. The attack by a slag can be predicted from the equilibrium diagram of the brick and slag constituents. But the actual situation will have to be ascertained by conducting a slag resistance test. For example, the equilibrium diagram of a FeO-SiO_2 system predicts that iron oxide reacts with silica to form a slag. But in practice, silica is resistant to iron oxide attack. This is because the iron silicate slag formed is very viscous; therefore, this viscous slag reduces further attack on silica.

The rotating panel test is a dynamic test that is described in ASTM C874-99. In this test, samples are cut into shapes and are assembled in a rotating test furnace. These samples form the hot face lining. This kind of arrangement allows the corroding medium to flow over the surface of the material. This prevents the formation of a protective coating [5,6]. Therefore, corrosion will be more severe compared to static tests.

There are a number of factors that decide the corrosion resistance of a particular refractory in a medium, such as its standard electrode potential in that medium as well as its microstructure, bulk density, apparent porosity, pore size distribution, permeability, and wetting angle. As the potential of a material in a medium is more positive, its corrosion resistance in that medium is greater. A single phase structure will have less corrosion than a multiphase structure. As bulk density increases, corrosion resistance increases because porosity decreases. Apparent porosity refers to open pores. Open pores increase the surface area exposed to the medium. Hence, an increase in apparent porosity will increase corrosion. For a given pore volume, larger-sized pores will have more surface area compared to smaller-sized ones. Hence, if a pore-size distribution has more larger-sized pores, the corrosion will increase. Permeability admits the corrosive medium; hence, the material will be corroded to a greater extent. As the wetting increases, the medium's spreading on the surface of the material decreases. Therefore, corrosion is decreased.

19.5.1 Resistance to Alkali Attack

In many applications, refractories come into contact with alkali vapors or melts. One example happens in glass melting furnaces. The crown of these furnaces comes into contact with alkali vapors. Of the alkali vapors, sodium and potassium vapors are the most common. ASTM C987-00a is the standard for testing the resistance of refractories against vapor attack. The attack is assessed by suspending the refractory sample over a crucible containing alkali salts. After a period of time, the samples are observed—first visually, and then they are weighed to see if there is any weight loss. The results are compared and the most resistant refractory material is selected.

19.5.2 Resistance to Acid Attack

Refractories used in chimney linings, incinerators, and kiln linings should have sufficient resistance to acid attack. ISO 8890:1988 gives the standard procedure for checking the acid resistance of a refractory. The attack is determined under the boiling/condensing conditions of the particular acid involved.

19.5.3 Resistance to Carbon Monoxide

Refractories tend to deposit carbon in a carbon monoxiderich atmosphere. The reaction is given in Equation 19.1 [7].

$$2CO = CO_2 + C \qquad (19.1)$$

Any free iron in the refractory acts as the nucleation site for this deposition. The free iron gets converted to iron carbide in the process. When iron carbide grows, it leads to the disintegration of the refractory.

For this reason, refractories need to be tested for their resistance to carbon monoxide attack. The temperature range in which carbon deposition leads to carbon growth, cracking, and refractory disintegration is 400°C–600°C. ISO 12676:2000 prescribes a test temperature of 500°C. The test involves keeping refractory prisms in a sealed furnace. A stream of carbon monoxide is admitted to the furnace. The samples are visually checked at regular intervals throughout the test, which continues for 200 h or until disintegration takes place, whichever occurs earlier. After the test, the appearance of the samples is noted.

19.5.4 Resistance to Hydration

Magnesia and dolomite refractories are susceptible to hydration. To some extent, dead burning helps to reduce this problem. Dead burning is firing at temperatures high enough for grain growth to take place. This reduces the surface area of the grains. When the surface area is reduced, the surface available for carbon deposition decreases, thereby reducing the carbon deposition. Hydration of magnesia and dolomite takes place by the following reactions:

$$MgO + H_2O = Mg(OH)_2 \qquad (19.2)$$

$$MgO \cdot CaO + H_2O = Mg(OH)_2 \cdot Ca(OH)_2 \qquad (19.3)$$

These reactions lead to the breaking down of the refractories concerned. Factors controlling the hydration reaction are temperature, exposure time, and degree of product susceptibility. An increase in each of these factors increases the hydration.

Tests for hydration resistance can be done under pressure or without pressure application. Under pressure, tests can be conducted in an autoclave or in a pressure cooker. A humidity chamber can be used for testing without pressure.

The specimens are kept in the equipment for a set time or until disintegration occurs [2,8].

If the sample is in the form of magnesia granules, the test is done in an autoclave or in a flask. A sample of a particular sieve size is selected and then is dried and weighed. Next, the sample is kept in an autoclave or flask for a set duration. If a flask is used, it is fitted with a condenser. The sample is suspended in the flask, which contains boiling water. The hydration tendency is checked by finding the percentage of the sample passing through the initial sieve [2,8].

In the case of dolomite grains, the graded and dried sample is kept in a Petri dish. This dish is then kept in a steam humidity cabinet for a longer duration than for that of magnesia. (It could be 24 h.) After the test, the sample is weighed and then dried again. This redried sample is passed through the initial sieve. The amount passing through this sieve is expressed as a percentage of the hydrated sample weight [2,8].

The standard procedure for the hydration tendency of dolomite is given in ASTM C492-92 and for that of magnesite in ASTM C544-92.

19.5.5 Resistance to Molten Glass

Testing for resistance to molten glass is called exudation testing. Alumina-zirconia-silica (ANS) refractories form the side walls and superstructure of a glass tank. During the firing of these refractories, depending on the temperature, glassy exudate can form. As the temperature increases, the amount of exudate becomes larger and larger. When refractories are used for the construction of a glass tank, these exudates contaminate the molten glass. Contaminants will appear in solid glass as vitreous and crystalline faults.

The exudation potential can be checked by firing discs or bars of the material. The change in volume indicates the exudation. The glass phase will come onto the surface of the sample as exudate and will increase the volume. The standard for this test is ASTM C1223-98 [9]. When the increase in volume is beyond 2%, those bricks will cause defects in the glass.

19.6 Cold Crushing Strength

A cold crushing strength test is used to measure the cold strength of a brick. It is used to show whether or not the brick has been properly fired. This test, generally a quality control check, also indicates whether the brick will damage to corners and edges in transport. This test does not give any indication of the material's in-service behavior. Within the same batch of bricks, there could be variations in the cold crushing strength with respect to their firing position in the kiln. It is better to specify a minimum value for the strength below which the bricks will be rejected. Cold crushing strength is the maximum load at failure per unit of cross-sectional area when compressed at ambient temperature. Expressed in the form of an equation:

$$\text{Cold crushing strength} = \text{Load at failure/Area} \tag{19.4}$$

The unit for cold crushing strength is MPa. ISO 8995:1986 gives the standard test method. The specimen is prepared as a 50 mm cube or 50 mm diameter cylinder. A load is increasingly applied at a specified rate. The load at which it breaks is used to calculate its cold crushing strength. Typically a 60% alumina brick (the rest being silica) has a cold crushing strength of 39 MPa. The strength increases to 55 MPa for a 50% alumina brick.

19.7 Thermal Conductivity

Thermal conductivity is measured as the amount of heat flow through a material in unit time per unit temperature gradient along the direction of flow and per unit cross-sectional area [6]. The application decides the required thermal conductivity value. To conserve heat, a material with low conductivity value is required. In cases where the furnace has to be cooled, highly conducting refractories are required. Thermal conductivity of a material depends on temperature, density, porosity, and atmosphere. Table 19.3 gives the thermal conductivity values for a few refractories. Thermal conductivity can be measured by many methods, but all are grouped into two categories: *steady-state* and *transient*.

TABLE 19.3

Thermal Conductivity Values for a Few Refractories

Sl. No.	Material	Thermal Conductivity, W/mK
1.	Insulation panel	0.02
2.	Insulation brick	0.15
3.	Firebrick	0.80
4.	High-alumina brick	1.50
5.	Magnesia brick	5.00
6.	Carbon brick	13.50
7.	Semigraphite carbon brick	30.00
8.	Graphite brick	107.00

19.7.1 Steady-State Methods

Steady-state methods involve the measurement of heat flux between the hot and cold face of the sample that are each kept at constant temperatures. Following are examples of this method.

19.7.1.1 Calorimeter Method

The standard procedure is illustrated in ASTM C201-98. This method is useful for materials having a conductivity of up to 20 W/mK. A calorimeter is fitted at the cold face of the sample. The calorimeter obtains the mass, specific heat, and temperature rise of the heat flux.

19.7.1.2 Split Column Method

Whole brick is used for measuring thermal conductivity by this method. This brick is kept on a hot plate with its largest face touching the plate's surface. The brick's sides are insulated and a brass plate is kept on top of it. This brass plate distributes the temperature evenly. A blackened copper disc is inserted in the center of this plate. This disc is thermally insulated from the brass plate. The hot plate is then heated to the test temperature, which is maintained during the testing period. The copper disc is allowed to attain a constant temperature. The temperatures of the hotplate and the copper disc are measured using fine thermocouples. Draught screens are erected round about the brick, and the temperature of still air over the assembly is measured. From these measurements, the heat loss from the disc by radiation and convention is obtained using empirical formulae. The thermal conductivity is calculated using the heat throughput and temperature gradient. The arrangement is shown in Figure 19.2.

19.7.2 Transient Methods

In these methods, measurements are taken after a certain amount of heat is input to the sample. The sample may initially be at room temperature.

19.7.2.1 Cross Array Hot Wire Method

This method, which uses a cross-wire welded at the center, is given in ISO 8894-1:1987. One of the two wires is connected to a power source used as a heating element. The other wire is a thermocouple wire. This cross-wire is sandwiched between two blocks of the refractory material. Power is fed into the heating element for a short time. The temperature rise in the blocks is measured. This temperature rise is related to the thermal conductivity of the material. Thermal conductivity values up to 2 W/mK can be measured by this method.

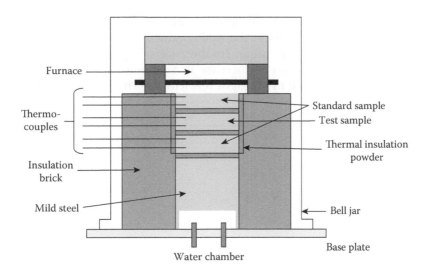

FIGURE 19.2
Split column setup.

19.7.2.2 Parallel Hot Wire Method

In this method, which is illustrated in the standard ISO 8894-2:1990, the heating element and the thermocouple wires are arranged in a parallel manner. This type of arrangement is useful in measuring the thermal conductivity up to 25 W/mK.

Eschner et al. [10] suggested that the differences in the transient and steady-state methods were about 10%–15%. The transient methods give higher values. Later investigations [11] suggested that if sufficient care was taken with methods and specimen preparation, then only a scatter of maximum 5% may occur. Possible reasons for the differing results are as follows:

a. In steady-state methods, the temperature at which the conductivity is measured is the average temperature of the hot and cold surfaces. There is a vast difference between these two temperatures. There is only narrow temperature variation in transient methods.

b. In steady-state methods, the heat flow is unidirectional. In transient methods, it is bidirectional. In the case of anisotropic materials, therefore, the measured value will be the average conductivities in the two directions.

As temperature increases, thermal conductivity decreases. This is because the vibrations of phonons become greater as the temperature increases. This makes the heat transfer more random rather directional. Hence, heat transfer in any direction decreases. This is true in dense refractories, where the conduction mode of heat transfer is greater than the other two modes. Figure 19.3 shows variations in thermal conductivity with the temperature for magnesite and 90% alumina. In the case of lightweight refractories, conductivity increases with an increase in temperature. Figure 19.4 shows such a variation for some lightweight refractories. This is because the pores

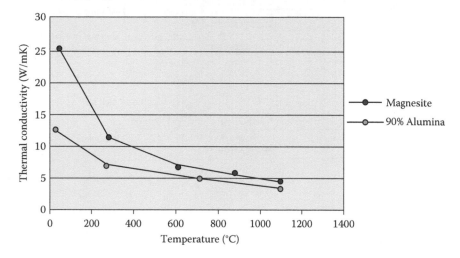

FIGURE 19.3
Variation of thermal conductivity with temperature for two dense refractories (CERAM data).

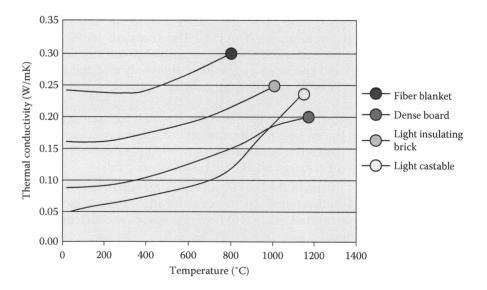

FIGURE 19.4
Thermal conductivities of some lightweight refractories (CERAM data).

become dominant in conducting the heat at higher temperatures. At lower temperatures, conduction of heat takes place mainly through the solid—the gas inside the pores. As temperature increases, the gas inside the pores is heated and then it conducts heat by convection. An additional increase in the temperature radiation mode of heat transfer also takes place through the pores, as this mode is proportional to the fourth power of temperature.

References

1. D. Herrel and F. T. Palin, *Refractories—Testing and General Properties*, 11th Edn., Mechanical Engineers Reference Book, New York, pp. 172–179.
2. ASTM C 1161, Test method for flexural strength of advanced ceramics at ambient temperature, in Annual Book of ASTM Standards, 2003, vol. 15.01, sect. 15.
3. D. A. Bell, *Thermal Shock*, Sterling Publications, London, 1993, pp. 106–108.
4. D. P. H. Hasselman, Thermal stress parameters for brittle refractory ceramics—A compendium, *Bull. Am. Ceram. Soc.*, 49(12): 1033–1037, 1970.
5. G. Evans and S. Baxendale, Refractory testing service, *Glass*, 77(2): 21–22, 2000.
6. S. Baxendale and S. Farn, Glass making refractories—The importance of corrosion assessment, *Refract. Eng.*, 15–18, 2003.
7. A. E. Dodd and D. Murfin, *Dictionary of Ceramics*, 3rd Ed., The Institute of Materials, Minerals and Mining, London, 1994.
8. British Standard Institute. BS 1902-1A:1966 Methods of testing refractory materials: Sampling and physical tests, London.
9. M. D. M. Innocenti, A. R. Studert, R. G. Pileggi, and V. C. Pandolfelli, How PSD affects permeability of castables, *Am. Ceram. Soc. Bull.*, 80(5): 31–36, 2001.
10. A. Eschner, B. Grosskpopf, and P. Jeschke, Experiences with the hot wire method of measurement of thermal conductivity of refractories, *Tonind-Ztg*, 98(9): 212, 1974.
11. W. R. Davis and A. Downs, The hot wire test—A critical review and comparison with the BS 1902 panel test, *Trans Brit Ceram Soc.*, 79: 44–52, 1980.

Bibliography

J. D. Gilchrist, *Fuels, Furnaces and Refractories*, Pergamon Press, Oxford, UK, 1977, pp. 240–251.
C. A. Schacht (Ed.), *Refractories Handbook*, CRC Press, New York, 2004, pp. 1–3, 435, 441–442, 445–446, 448–462, 468–473.

20

Production

Production

20.1 Introduction

The production of refractory products starts with raw materials. Each kind of brick is made from a different raw material. The treatments for preparing each type of brick are also different. But there are some common features in the production of refractory products. These are dealt with in this chapter.

Only a relatively few chemical elements that have high fusing compounds are found in the Earth's crust in sufficient abundance to be useful for heavy refractories. Compounds of silicon, aluminum, magnesium, calcium, chromium, and zirconium are refractory and are found in abundance in the Earth's crust. The refractory materials formed from these elements are alumina-silicates, magnesia, dolomite, chromite, and zirconia. These refractory materials are found as mineral deposits that are in the form of clays, sands, ores, and rocks. These deposits are mined or quarried. The big lumps obtained are then crushed and the resulting products are then subjected to mineral-dressing. Mineral-dressing comprises several different purification processes. It is a beneficiation process, in which impurities are segregated. Some of the important mineral-dressing operations are magnetic separation, flotation, and electrostatic separation. The method selected depends on the type of mineral selected. For example, electrostatic separation is used for a mineral that is magnetic. After beneficiation, the minerals are graded. Later, they are stored separately with respect to particle size.

20.2 Blending

The powder of a given refractory material will contain a range of particle sizes. The porosity of the final product is determined by controlling the fractional portions of the different sizes. Porosity can be decreased by blending coarse and fine-sized particles. The fine particles fill the interstices between the coarse particles. If uniform pores are required, same sized particles are used. For such a situation, blending is not required. But mixing will be necessary for uniform distribution of all ingredients required for processing. The ingredients other than the refractory powder are flux, bond material, and grog. Flux material reduces the melting point of the refractory. This will be needed in melting glass as well as to form a glass phase in the refractory brick. Bonding material binds the hard refractory particles and gives strength to the fired product. Grog is prefired material. Scrap brick is crushed and used as grog. Sometimes, grog is separately made by hard firing the refractory material. For instance, synthetic mullite grog is produced by firing mullite at 1750°C. This is used to produce high-duty mullite refractory. This refractory possesses high abrasion resistance and low permeability to gases.

Blending is carried out in a paddle mill with a kneading action. Along with the raw material and additives, some water is also added. This brings plasticity to the product—the bonding agent clay absorbs water and become plastic. The particles of the refractory raw material become embedded in the plastic clay.

20.3 Mixing

As mentioned in the previous section, mixing results in uniformity of the brick's various ingredients, which range from a mixture of same-sized refractory material powder and additives to different refractory materials' powders and additives. The additives, as mentioned earlier, may be flux, bond material, and grog. The water content varies from less than 5% up to 20%. To obtain a mix with less than 5%, a fine spray of water is used. The water content depends on the type of molding, the next stage of operation.

20.4 Molding

Molding involves giving shape to the product obtained after blending or mixing. After molding, the part should be able to be handled without damage.

There are different ways of molding. Two are hand molding and machine molding. Hand molding uses a plastic mix containing about 14%–20% water. This is hand-filled into a wooden box type of mold. The advantage of this method is that there is no need for an expensive molding machine. The disadvantage is that the process is very slow. This disadvantage is overcome in machine molding, which is used for mass production. Pressure brings the shape of the mold to the product; a 10%–12% water-mixed semi-plastic compound is used here. The pressure requirement is moderate and usually involves a two-stage process. The first stage is extrusion, which gives a rough shape to the product. Later, pressing gives the product its final shape.

Nonplastic mixtures as well as clays containing not more than 5% water are molded by dry pressing. Here, pressures applied are in the range of 35–140 MPa. These high pressures can cause air entrapment that, in turn, can cause lamination in the brick. To avoid air entrapment, the mixture is de-aired before molding. Otherwise, vacuum blocks are used in the mold wall. The green strength of dry pressed bricks is low. Hence, it is difficult to maintain the edges and corners of the bricks. For the same reason, handling should be mechanized as much as possible.

The other important molding process is slip casting. Fine refractory particles are suspended in water. This suspension is called slip. It is poured into a mold prepared from plaster of Paris. When a completely solid part is required, the water is allowed to be fully absorbed by the mold wall. The powder particles agglomerate inside the mold forming the shape of the inside cavity. The mold is then dried along with the solid in it. Later, the mold is cut open to remove the form. During the absorption of water, the solid thickness increases starting from the inside wall of the mold. This phenomenon is used to make a hollow portion by slip casting. When a sufficient thickness is built up, the remaining slip is poured out of the mold. Then, the process is continued as before. Slip casting is advantageous for making complicated shapes. But the product will be highly porous.

20.5 Drying

After shaping the refractory powders, the next step in the production process is drying. All of the molding processes use water, which then has to be removed from the product. This is done in the drying process.

Two methods of drying are used. In the first method involves drying floors. The refractory bricks are laid out in an open tray. Then these trays are kept on the drying floor, which is then heated by waste heat coming from kilns. In the second method, tunnel kilns are used for drying and for producing regular shapes. This method gives greater throughput because the process can be made continuous. Installed in a horizontal position, the refractory parts are stacked on bogeys. The bogeys are admitted at one end of the tunnel kiln, which is kept heated by a stream of hot air passed through the inside of the tunnel from the bottom. As the bogeys come down the kiln counter current to the hot air, the drying takes place. The speed of the bogeys is adjusted in such a way that when they exit the kiln, drying is complete.

20.6 Firing

Firing is the final stage in brick making. Formerly, beehive kilns were used for firing. A number of newly developed furnaces are used today. In one method, a number of furnaces are arranged side by side. The furnace used for the firing stage is heated by burning coal or coke. The oxygen required for burning is supplied by air. This air is preheated before it is used by passing through those furnaces that are in the cooling stage after firing. Thus, waste heat utilization takes place. This also minimizes the total energy requirement. After passing through the firing furnace, the high temperature air is passed through those furnaces that are to be heated for firing. This way, further waste heat utilization takes place. The heat is utilized to preheat the bricks before their temperature is raised to that of firing, and this conserves some more energy. After giving much of the heat, the combustion products of the burning coal or coke are passed through the stack to the atmosphere.

The second type of furnace used for firing is similar to the tunnel kiln used for drying (see Figure 20.1). As before, the stock is loaded onto the bogeys. But here, in contrast to the previous method of firing, the stock moves. Tunnel kilns are thermodynamically very efficient counterflow heat exchangers.

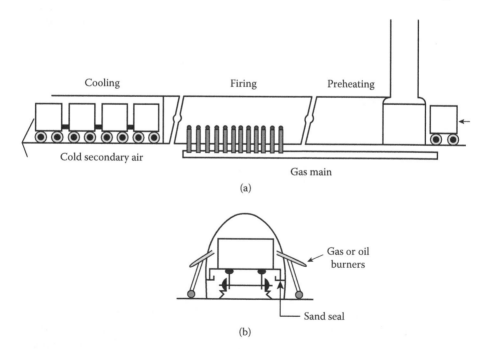

FIGURE 20.1
(a) Tunnel kiln; (b) cross-section in the firing zone.

The firing zone is in the middle of the kiln's length. Secondary air enters from the exit end. As it passes the cooling zone, this air absorbs heat from the bricks and is preheated. At the same time, the fired bricks are cooled. At the firing zone, the oil burners operate with the help of the oxygen in the preheated air. From the firing zone, the combustion products carry heat and pass through the preheating zone. As the green bricks pass through this zone, they absorb heat from the combustion products and are preheated. The firing temperature is maintained at least as high as the working temperature for sufficiently long enough to ensure that all possible reactions are completed and that dimensional stability is attained.

In some cases, the cooling rate is critical. For example, silica undergoes phase changes at low temperatures. As it is cooled, the tridymite form of silica is converted to quartz, causing extensive contraction. If the cooling rate is fast, thermal stresses are set up inside that can result in cracking.

In situ firing has been used in recent years. This means that bricks are not separately fired. Green bricks are made and equipment, such as a furnace, is built. During the operation of these furnaces, the bricks get fired. An example is chemically bonded basic brick where $MgSO_4$ is incorporated as a temporary bond material. The green bricks containing this compound will have enough strength to be built into furnaces. During the operation of these furnaces, $MgSO_4$ decomposes, supplying MgO to the brick. MgO reacts with other ingredients in a way that is similar to the reactions taking place during separate firing. These bricks can also have steel reinforcements and steel casings. During usage, iron in steel becomes oxidized to iron oxide. After this oxidation, iron oxide becomes an ingredient in the refractory or serves as a bonding agent between adjacent bricks.

For making monolithic furnace walls, steel tubes are packed with magnesite, and furnace walls are built with this mixture. On heating, iron oxide is formed and a permanent monolithic structure results. Other examples of avoiding a separate firing stage are in the case of making Linz-Donawitz converters and ladles. An L-D converter is made from green, tar-bonded blocks. Ladles are lined with sand bricks. These bricks are made from clay-bonded sands. These green bricks are built into the ladles. Another way of making the ladles is by first shaping a sheet steel in the form of a ladle and then gunning the sand mixture into position round the sheet steel. Drying and firing take place during the usage of the ladle.

Another common use of refractories without utilizing a separate firing stage is that of refractory cements. Originally, these cements were used to patch worn parts. Today, these have replaced the bricks in certain applications. For example, it is preferable for the throat of a blast furnace to not have any joints. One type of composition for making the cement for furnace throat is a mixture of sillimanite with calcium trisilicate. This mixture is then tempered with water, laid in position, and allowed to set like concrete. On heating, water molecules will be detached because of the breakage of hydrogen bonds. They leave behind a fired refractory concrete.

20.7 Unshaped Refractories

20.7.1 Introduction

Unshaped refractories are not shaped like bricks. In the United States, these originated as *specialties*. The year 2014 marked the centenary year of the origin of these specialties, which were actually monolithic refractories. They were used as unshaped and unburnt refractory materials [1]. New and improved monolithic were developed during the 1920s and 1930s. Wars bring large demands for materials—as well as for monolithic refractories. The production of these refractories increased during World War II, especially for the plastic and ramming types. Monolithic refractories found much use in dehydration plants, which produced hundreds of thousands of kilograms of powdered eggs for shipping to the military. There was about a 35% increase in the production of monolithic refractories in the form of ramming mixes, castables, mortars, and coatings during this period [1].

After the invention of *plastic refractory* specialties in 1914, the *castable refractory* was invented in 1923. *Gunning technology* was established in Europe in the 1950s. Then, in the 1970s, *prefabricated shapes* were used to make soaking pits and the walls of pusher-type furnaces. In the same period, *deflocculated castables* were invented. This was a big leap in the development of monolithic refractories because it gave rise to two different types of these materials [2]. In 1980s, one type of deflocculated castable, called a *self-flowing castable* (SFC), was developed. SFCs filled any space by itself. The 1990s saw the development of *shotcreting castables* (SCCs). The right equipment for SCCs came from the building industry.

20.7.2 Definitions, Classifications, and Standardization

Initially, when unshaped refractories were first used, the property requirement was kept the same as for shaped refractories, in that they mainly replaced linings. Later on, when the use of these refractories became widespread, new standards had to be developed. The first standards on unshaped refractories by the American Society for Testing of Metals (ASTM) were published in 1943 [3]. They were numbered as C 179 and C 181; C 179 covered "drying and firing linear change of plastics and rammings" and C 181 dealt with "workability index of plastic refractories." As noted earlier, World War II saw an increase of about 35% in unshaped refractory production. Specifications were required to obtain the required items from many manufacturers.

There was no uniform standard in Europe until the 1960s. In 1953, the Federation Européenne des Fabricants de Produit Refractaires (PRE) was established to provide technical recommendations. To avoid any divergence of ideas between the refractories manufacturers belonging to PRE and the refractory users, mixed groups were set up in 1969 under the name Siderurgie et PRE (SIPRE). In February 1964, PRE Bulletin No. 65 was published, which contained the definitions and classifications of unshaped refractories and refractory mortars. The first PRE recommendations were published in the early 1970s.

To achieve easier communication and trading within the European community, there was a need to create European standards that was recognized in the mid-1980s. Refractories were also included in these standards. The standard for unshaped refractory products was numbered EN 1402.

20.7.2.1 Definitions

Various terms such as *mixes, refractory mixes, ramming mixes,* and *monolithics* are included in the definition of *unshaped refractory products.* These products are distinguished from bricks, which have a definite shape, are prefired, and are laid. Unshaped refractories are placed in larger sections. For example, a furnace lining consists of unshaped refractories. They have few joints.

The definition for an unshaped refractory is given in ISO 1927 (5) and EN 1402-1: It is a mixture, which consists of an aggregate and a bond or bonds, prepared ready for use either directly in the condition in which it is supplied or after the addition of a liquid that satisfies the requirements on refractoriness given in ISO R 836. They may contain metallic, organic, or ceramic fiber material. This mixture is either dense or insulating. An insulating mixture is one whose true porosity is not less than 45% when determined in accordance with EN 1094-4, using a test piece prefired to specified conditions.

There are four types of bond used for making unshaped refractory products:

1. A hydraulic bond with setting and hydraulic hardening at ambient temperature
2. A ceramic bond with hardening by sintering during firing
3. A chemical bond with hardening by chemical reaction at ambient temperature or at a temperature lower than that of a ceramic bond
4. An organic bond with hardening at ambient temperature or at higher temperatures

Many bonds are mixed. For instance, in the case of phosphate-boned ramming materials, a combination of chemical and ceramic bonds are used. Also, in the case of materials that are repaired with a hydraulic bond, the bond is made up of calcium aluminate cements and considerable amounts of clays. This means that the total bond is a mixture of hydraulic and ceramic bonds. When it is a mixed bond, the bond is named after the material that plays the major part during hardening.

The *maximum grain size* is the mesh width through which at least 95% by weight of the material can pass. Another definition relevant to an unshaped refractory is the *yield by volume,* which is the mass of the material as delivered that is necessary to place 1 m [3] of the material, expressed in tons, to the nearest 1%.

20.7.2.2 Subdivision of Unshaped Refractory Materials

At present, there is an extensive range of unshaped refractory materials. Various subdivisions are discussed in the sections that follow.

20.7.2.2.1 Refractory Castables

Out of all the unshaped refractory material subdivisions, refractory castables are the most important. These are produced as dense or insulating materials. Formerly, these were made as mixtures of aggregates and calcium aluminate cements. After continuous improvement, they are produced on a very high technical level today. Because of their varieties, they are further classified as follows.

20.7.2.2.1.1 Regular Castables Regular castables are hydraulically bonded and contain a cement.

20.7.2.2.1.2 Deflocculated Castables Deflocculating castables contain at least one deflocculating agent in addition to a cement. They should also contain a minimum of 2% by weight of ultrafine particles. These are particles less than one micron in size.

Examples of ultrafine particles in the deflocculated castables are microsilica and reactive alumina. These are further divided according to CaO content; CaO is used as the cement. The divisions are shown in Table 20.1.

More heat is required as the cement content decreases to achieve sufficient green strength.

20.7.2.2.1.3 Chemically Bonded Castables Chemically bonded castables contain one or more chemical bonds. Because they contain chemicals, they are set by chemical reactions.

20.7.2.2.2 Refractory Gunning Materials

Refractory gunning materials are also subdivided into three groups, which are as follows.

TABLE 20.1

Classification of Deflocculated Castables

Sl. No.	Class	% CaO	
		Minimum	Maximum
1.	Medium-cement castable (MCC)	>2.5	–
2.	Low-cement castable (LCC)	>1.0	<2.5
3.	Ultralow-cement castable (ULCC)	>0.2	<1.0
4.	No-cement castable (NCC)	0	<0.2

20.7.2.2.2.1 Castables Castables can also be dense or insulating. They are supplied dry and used after adding water. The water is added during and/or before gunning.

20.7.2.2.2.2 Deflocculated Castables As the name suggests, deflocculated castables contain deflocculating agents. They are used in the shotcreting process.

20.7.2.2.2.3 Plastics for Gunning Plastics for gunning are refractories that are made for gunning under high pressure. They can be used in the delivered condition.

20.7.2.2.3 Refractory Moldable Materials
Refractory moldable materials are also termed as plastic or ramming refractory materials. These are again subdivided into two as outlined in the following sections.

20.7.2.2.3.1 Plastic Refractory Materials Plastic refractory materials are the oldest monolithic material. They were developed as *pliable firebrick* in the United States in 1914 and were used a long time as the front part of many furnaces [3]. These are supplied in moldable, preformed blocks and placed by ramming. They harden under heat.

20.7.2.2.3.2 Refractory Ramming Mixes Refractory ramming mixes are incoherent. They are supplied in a dry or moistened condition. If they are supplied in a dry condition, a liquid binder is also provided. The refractory is mixed with the binder onsite using a paddle mixer, and then is applied. Refractory ramming mixes are filled behind a shuttering. They are installed by ramming or vibrating. Then they are hardened by heating.

20.7.2.2.4 Refractory Jointing Materials
Refractory jointing materials are used for jointing refractory bricks, preformed blocks, or insulating products. They consist of fine aggregates and bonds. These are also supplied either in a dry or in a ready-to-use state, similar to refractory ramming mixes. There are two types of jointing materials. These are explained in the following sections.

20.7.2.2.4.1 Heat Setting Jointing Materials These get hardened at elevated temperatures. The bonds formed are chemical or ceramic.

20.7.2.2.4.2 Air Setting Jointing Materials These harden at ambient temperature. The bonds are chemical or hydraulic.

20.7.2.2.5 Other Unshaped Refractory Materials

Apart from the four main subdivisions of unshaped refractory products, a few others are also used. They are enumerated in the following sections.

20.7.2.2.5.1 Dry Mixes
Dry mixes are used only in a dry state. They are placed in position by vibration or ramming. Their grain size distribution is such that they acquire a maximum tap density. They may contain a temporary bond. The materials for this bond may be organic additives or sintering agents. The permanent bond is ceramic, which is obtained by heating. The main uses of dry mixes are in the installation of induction furnaces, in lining of tundishes for wear resistance, or in ladles.

20.7.2.2.5.2 Injection Mixes
Injection mixes contain fine powders, which are injected by double-piston pumps. The pressure employed for injection is in the range of 1–20 MPa. These mixes are available in two forms—one that is ready-to-mix or one that must be mixed on site. Grouting mix is an example of an injection mix. It is injected through the steel jacket of a blast furnace and fills the gap between the jacket and the adjacent brick work to increase the cooling efficiency.

20.7.2.2.5.3 Coatings
Coatings are mixtures of fine refractory aggregates and bonds. They contain higher water content than mortars and are supplied in a ready-to-use form. The bonds may be ceramic, hydraulic, chemical, or organic. They can be applied manually, pneumatically, by mechanical projection, or by spraying. Manual application involves the use of a brush or trowel.

20.7.2.2.5.4 Tap Hole Mixes
Tap hole mixes are specifically used for filling and sealing of blast furnace tap holes. They are supplied as extruded preformed blocks. They are moldable because of their plastic consistency. They contain refractory aggregates and organic and ceramic bonds. Tap hole mixes develop a carbon bond on heating.

Table 20.2 shows the various divisions of unshaped refractories and their installation in tabular form.

20.7.2.3 Subdivision According to Chemical Composition

ISO R 1927 gives the subdivision of unshaped refractory products according to chemical composition. The various classes of these materials are listed in the following sections.

20.7.2.3.1 Al_2O_3-SiO_2 Products
Al_2O_3-SiO_2 products consist of alumina, silica, and alumino-silicates.

TABLE 20.2

Various Divisions of Unshaped Refractories and Their Installation

Primary Divisions	Secondary Divisions	Tertiary Divisions	Installation
Castables (dense or insulating)	Regular	–	Casting, rodding, vibrating
	Deflocculated	MCC LCC ULCC NCC	Vibrating, self-flowing, shotcreting
	Chemically bonded	–	–
Gunning materials	Castables (dense or insulating)	–	Gunning
	Deflocculated castables	–	Shotcreting
	Plastics	–	Gunning under high air pressure
Moldable materials	Plastic	–	Ramming
	Ramming mix	–	Ramming or vibrating
Jointing materials	Heat setting	–	–
	Air setting		
Other materials	Dry mixes	–	Ramming or vibrating
	Injection mixes	–	By injection
	Coatings	–	Manual, pneumatic, mechanical projection, spraying
	Tap hole mixes	–	–

20.7.2.3.2 Basic Products

Basic products consist of magnesia, doloma, magnesia-chrome, chrome ore, and spinel.

20.7.2.3.3 Special Products

Special products consist of oxides, nonoxides, and others not previously mentioned here. Examples are silicon carbide, silicon nitride, zircon, and zirconia.

20.7.2.3.4 Carbon-Containing Products

Carbon-containing products are those unshaped refractory products containing more than 1% carbon.

20.7.3 Methods of Installation

For a given quality of unshaped refractory material, the method of placement will decide its performance. The following sections detail the different methods of installation as used in various applications.

20.7.3.1 Concrete Work

Concrete work uses refractory castables. These are installed by casting, rodding, vibrating, or self-flowing. Self-flowing refers to casting without any external force such as that used in vibration. Ramming is also sometimes used to obtain better properties. Table 20.3 gives the methods of mixing and installation of different refractory castables. A paddle mixer cannot be used for lightweight insulating castables (bulk density, or BD). These contain aggregates in the form of perlite and vermiculite. A paddle mixer would destroy the aggregates by abrasion. It is used when the BD and thermal conductivity are to be increased. It is necessary with deflocculated castables to obtain the correct consistency.

Vibrators are used for greater compaction. They are adjusted to obtain the required thickness. Cold weather affects the setting. As the temperature drops, the setting time increases. Also because the initial strength is lower at lower temperature, the strength obtained after heating also lessens. Cement content also affects setting. As the cement content decreases, setting time increases. In such cases where the castable contains less cement, accelerators are used during winter time. The setting of refractory castables is an exothermic reaction, as shown in Equation 20.1.

$$\text{Calcium aluminate} + \text{aggregates} + \text{water} = \text{Hydrated calcium aluminate} + \text{heat} \tag{20.1}$$

TABLE 20.3

Methods of Mixing and Installation of Different Refractory Castables

	Insulating Castables		Dense Castables				
	B.D.<1.2 g/cc	B.D.>1.2 g/cc	RC	MCC	LCC	ULCC	NCC
Concrete mixer	Yes	Possible	Possible	No	No	No	No
Paddle mixer	No	Possible	Yes	Yes	Yes	Yes	Yes
Mixing time after water addition (min)	2	2	2–3	>3	≥3	≥4	≥4
Casting	No	No	Possible	No	No	No	No
Rodding	Yes	Yes	No	No	No	No	No
Vibrating	No	Possible	Yes	Yes	Yes	Yes	Yes
Self-flowing	No	No	No	Possible	Possible	Possible	Possible

The maximum temperature attained will depend upon the following factors:

a. Quantity of castable
b. Ratio of the area of the surface between the mold and the castable to which it is exposed
c. The type of the material

The greater the quantity of the castable, the higher the temperature attained. As the ratio of the areas increases, the temperature will increase. This is because the exposed surface releases more heat to the atmosphere than the mold. Whichever material absorbs more water will give out more heat. Correspondingly, there will be a greater increase in temperature. Because of the temperature rise during setting, there will be more moisture loss. This has to be prevented because setting requires water. Hence, during setting, the surface is covered or water sprinkling is adopted. It may take about 48 h to reach room temperature. Only then should the drying or heat-up procedure begin.

The cement content in deflocculated castables is low. Hence, the amount of heat they produce during setting is low. The time of installation can therefore be limited to 24 h. There are special ULCC castables, which are installed in blast furnace trough systems. They contain metallic additives and catalysts. Because of these, an intensive exothermal reaction takes place after the addition of water. Hydrogen is produced out of this reaction. It creates capillaries as it escapes from the structure. These capillaries help in the self-drying process because they increase the surface area from which heat is removed. Because of the fast reaction and drying, the furnace can be put into operation in a short time.

20.7.3.2 Ramming Process

Ramming is performed on plastic refractory materials and on refractory ramming mixes.

20.7.3.2.1 Ramming Plastic Refractory Materials

Plastic refractory materials are supplied in cardboard boxes in the form of preformed blocks or slices. These are rammed in position for application. The ramming may be done with a pneumatic rammer or a heavy rammer weighing about 1–1.5 kg. After the completion of the ramming process, the surface is leveled and scraped with a board containing nails. The nails produce narrow channels on the surface, thereby increasing the surface area. This increased surface area enhances drying. Drying produces shrinkage; this can result in shrinkage cracks. These are avoided by hitting shrinkage joints deep into the rammed material. These joints are approximately 1 mm thick steel plates. Better evaporation during drying is ensured by poking venting holes with a steel stud. The distance between two such holes is kept in the range of 150–200 mm. The thickness of the steel stud should be at least 3 mm.

20.7.3.2.2 Ramming Refractory Ramming Mixes

Refractory ramming mixes are installed with the help of shuttering. The ramming mixes are filled behind the shuttering to a height not exceeding 50 mm. After each layer is poured, it must be compacted with a pneumatic rammer. The frequency of this hammer, which weighs about 5 kg, is approximately 50 Hz. The stroke length is in the range of 120–180 mm.

20.7.3.3 Gunning Process

There are two types of gunning processes—dry gunning and shotcreting. Table 20.4 shows a comparison between the two.

20.7.4 Addition of Fibers

The definition of unshaped refractory products mentioned that these products may contain fibers. The addition of various fibers to these products will be discussed further in the following sections.

20.7.4.1 Organic Fibers

Organic fibers provide vaporization assistance during the drying and heat-up period. They are made from polypropylene. These fibers will dissolve as the temperature rises, and they permit easier evaporation. The efficiency of the fiber increases with increasing length and decreasing diameter. Standard lengths are in the range of 6–12 mm with a diameter of 10–30 mm.

TABLE 20.4

Comparison of Dry Gunning and Shotcreting

Particulars	Dry Gunning	Shotcreting
Type of product	Regular castables	Only deflocculated castables
Equipment	Rotor or double chamber guns	Paddle mixer and double piston pump
Transport	Through hose	Through hose (preferably metallic)
Pressure during transport	<0.6 MPa	>20 MPa
Nozzle technique	Adding water through a water ring	Injection of compressed air and accelerator
Type of setting	Normal	Quick hardening after hitting
Dust development	Big	Nil
Rebound	Big	No rebound
Properties	Those of regular castables	Those of cast or vibrated deflocculated castables

20.7.4.2 Metallic Fibers

Metallic fibers are made of steel. Much literature is available on steel fibers [4]. They are made using the following methods:

 a. Wiredrawing
 b. Centrifuging
 c. Milling

In centrifuging, steel fibers are centrifuged out of the melt. A block of steel is milled to obtain milled fibers. The number of fibers per kg varies from 25,000 to 48,000. This number depends on the following:

 a. Manufacturing method
 b. Shape of the fiber
 c. Alloy used

Fibers 25 mm long and 0.4 mm in diameter are most often used. The fibers can be straight, hooked, or corrugated. They constitute about 3% by weight in the castable.

Steel fibers improve the thermal shock resistance of the refractory. If a crack forms it is arrested. The thermal limit of the refractories will depend upon the particular alloy used. Under reducing atmospheres, some of the alloys can be effective above 1600°C. Under normal or oxidizing atmospheres, castables containing Cr-Ni-alloyed steel fibers should not be exposed to greater than 1200°C. There could be reactions among the oxides formed by the elements in steel and by those in the refractory. The effect of steel fibers on the properties of refractory castables is summarized in Table 20.5.

TABLE 20.5

Effect of Steel Fibers on Properties of Refractory Castables

Sl. No.	Property	Effect
1.	Water addition	Increases by about 0.10%–0.15%
2.	Consistency and flow behavior	More difficult
3.	Bulk density	Insignificant
4.	Porosity	-Do-
5.	Modulus of rupture	Increases
6.	Compressive strength	Insignificant
7.	Erosion resistance	-Do-
8.	Thermal shock resistance	Increases
9.	Thermal conductivity	Insignificant
10.	Permanent linear change	Increases
11.	Thermal expansion	-Do-
12.	CO resistance	No change for Cr-Ni-steel, decreases for carbon steel

20.7.4.3 Ceramic Fibers

Because of the limitations in temperature with respect to metallic fibers, ceramic fibers were developed for use at higher temperatures. These are made with a diameter of 2–5 µ. There are problems associated with the use of ceramic fibers because they are:

a. Difficult to disperse
b. Undergo self-crushing
c. React with the surrounding matrix

The reaction occurs at high temperatures; fibers become a part of the matrix due to the formation of a ceramic bond.

20.7.5 Preformed Shapes

Preformed shapes are made from refractory castables and ramming mixes. After shapes are formed, they are dried and tempered at the manufacturer's plant. Therefore, they can be put into service immediately. At the site, expenditures involving molds, installation, and heat-up can be reduced. Preformed shapes, such as burners, injection lances for pig iron or steel treatment, gas purging systems, or impact blocks for the bottoms of steel ladles, are examples of functional refractory products. Some of the advantages in using the preformed shapes are as follows:

a. Problems can be handed over to the manufacturer
b. Optimum processing
c. No in situ installation, removal of molds, and shuttering
d. Setting times are not required
e. Immediate start of operation

20.7.6 Drying and Heating Up

Drying and heating up are started once the setting of the unshaped refractory is over. If these processes are started after long periods, then the following rules should be observed:

a. The furnace, in which the lining of the unshaped product is placed, should be subjected to draft conditions. This reduces the humidity. Otherwise, hydrothermal reactions can occur during heating. In the case of refractory castables, alkali hydrolysis is possible. This will cause carbonation and disintegration of the lining [5].
b. With a phosphate-bonded ramming mix, rehydration will take place if the installed mix is kept idle for a long time. It can be restored by heating it up.

Drying and heating up should take place 48 h after installation. The amount of water retained on storing depends on the initial density of the product. As this density decreases, the amount of water also decreases. This is because as the density decreases porosity increases. As the porosity increases, the surface open to the atmosphere increases. This increases the release of water to the atmosphere. Thus, a dense castable lining retains most of the water initially present even after air drying for a long time. The amount of cement also plays a part in the drying process—the greater the amount, the easier the drying.

Deflocculated castables have a very low cement content. Also, the particles sizes are fine. Not only the amount of porosity but the size of the pores also matters in drying. As the pore size decreases, the drying becomes more difficult. Thus, in deflocculated castables, drying is a tedious process because of low cement content and pore size.

Clay-bonded products lose about 75% of their water content under air drying for about 15 days.

References

1. C. A. Krause, *Refractories: The Hidden Industry*, The American Ceramic Society, Westerville, OH, pp. 52, 76, 1987.
2. R. Krebs, Modern solution of refractory problems with unshaped refractories, in *Proceedings of the UNITECR 1999*, September 6–9, Berlin, 1999.
3. R. Krebs, State of the art of the standardisation of unshaped refractories in Europe, *CN Spec. Refract.*, 6(3), 72–76, 1999.
4. M. Braun, Reinforcing castables with steel fibres, *Interceram*, 29, 113–118, 1980.
5. L. D. Hogue and W. A. Jackson, Nature of carbonation of hydrated calcium aluminate cements in castable refractories, *Ind. Heat.*, 64(8): 45–49, 1997.

Bibliography

J. D. Gilchrist, *Fuels, Furnaces and Refractories*, Pergamon Press, Oxford, UK, 1977, pp. 207, 253–256.

F. H. Norton, *Refractories*, McGraw – Hill Book, 1968, p. 61.

C. A. Schacht (Ed.), *Refractories Handbook*, CRC Press, New York, 2004, pp. 287–305, 325–333.

21

Silica

21.1 Introduction

Silica bricks are made with the following composition: SiO_2 = <95%; Al_2O_3 <1%; alkalies = 0.3%. The raw materials, from which silica bricks are made, are ganister, quartzite, sand, and flint. Another raw material is silcrete. This is found in South Africa and is microcrystalline.

Ganister is clay-bonded sandstone. It has been used for furnace construction for centuries. The silica content in ganisters can go up to 98%. Most ganisters are expensive to mine. Quarzite is also sandstone, but it is bonded with colloidal silica. It is a hard material. Sand may contain quartzite.

The quality of a brick depends on the crystal size of the initial raw material, which is called rock. When the crystals are very fine, porosity becomes low after firing. Low porous bricks have good properties. Silcrete shows a conchoidal fracture and an extremely fine structure. High duty silica bricks are made from materials with high silica content and low alumina and alkali contents. They have a very fine grain size and low porosity.

21.1.1 Allotropy in Silica

There are 15 allotropes of silica. Most of them are metastable, and their transformations do not bring about much change in volume.

The naturally occurring form of silica is α-quartz. On heating, this transforms to β-quartz at 573°C. This transformation is rapid and reverse. The volume increases by 1.35%. Enough time should be allowed for this transformation to take place slowly. Otherwise, it can shatter the brick. β-quartz is stable up to 867°C. If the heating rate is faster, it may remain for a few hundred degrees more. Its conversion to tridymite is a slow process. A mineralizer such as calcium tungstate can catalyze the conversion. The conversion is associated with a considerable degree of atomic reorganization, which is the reason for the slowness of this conversion. Mineralizers cannot be added to any great extent because they act as a flux, bringing down the melting point of the refractory. The volume change for this transformation is 16.5%. Any damage due to large expansion can be avoided by slow heating.

At 1470°C, tridymite transforms to cristobalite. During this transformation, there is a contraction of 1.95%. Again this transformation is a slow process. Once transformed, cristobalite becomes stable up to the melting point, which is 1723°C.

If the heating rates are high, β-quartz and tridymite can directly transform to glass. The transformation from β-quartz to glass takes place at 1450°C and that from tridymite to glass at 1680°C. Liquid silica transforms to glass on fast cooling. It is also termed vitreous silica. If the temperature is maintained high but below the melting temperature, devitrification of glass takes place. The transformed product is β-cristobalite. Addition of a mineralizer enhances the transformation.

On cooling, β-cristobalite transforms to tridymite at 1470°C. This transformation is a slow process. Hence, unless the cooling is very slow, tridymite continues to remain in the same state up to 275°C. In the temperature range of 275°C–200°C, it transforms to α-cristobalite, which is a metastable state. This transformation is accompanied by a 3.15% decrease in volume. The highest temperature form of tridymite is β_4. Because its equilibrium transformation to β-quartz is a slow process, it readily undergoes metastable transformations when cooled. These are β_3 at 476°C, β_2 at 210°C, β_1 at 163°C, and α at 117°C. The contraction in volume for the β_2 to β_1 transformation is 0.18% and for β_1 to α is 0.45%. The other two metastable transformations have negligible volume changes.

Because of the many transformations involved in silica, as well as the volume changes, heating and cooling of silica bricks should be conducted very slowly. This ensures equilibrium transformations and will not result in any damage in the form of cracking and spalling. There are certain exceptions with respect to the actual behavior of silica bricks. Tridymite is less dense than vitreous silica. Stable tridymite is possible only with the help of impurities.

21.2 Manufacture

The general steps involved in the manufacture of silica bricks are as follows: Mined raw materials are crushed to obtain angular particles, which help in better packing during compaction. Bonding in the green compact among the particles is achieved by the addition of lime water, which is prepared such that it contains about 1.7% CaO. Lime water also aids compaction by imparting some plasticity. Dry pressing is carried out to get the green compact. Firing, at a temperature of 1500°C, is a slow process taking about two weeks. This long period of firing accommodates the various volume changes taking place during the many transformations. It also allows time for all the transformations—many of which are very slow. This slow firing process will then ensure that the bricks obtained are undamaged. Heating to beyond 573°C must be slow because the transformation from α-quartz to β-quartz is a fast process. Similarly, cooling from 300°C to 100°C must be slow to allow for the metastable transformations in cristobalite and tridymite. The soaking time at the firing temperature depends on the type of modification required in the final product. The time required for cristobalite is less than that required for tridymite.

Firing at a high temperature and for a long time is called hard firing. Hard firing for silica ensures that no quartz is left in the material. Once the quartz phase is eliminated, expansion during service becomes low. If the service temperature is high, there could be tridymite to cristobalite conversion in the initial service period. The density of the brick tells us the proportion of the phases. For single phases, the densities are 2650 kg/m³ for quartz, 2260 kg/m³ for tridymite, and 2320 kg/m³ for cristobalite. Two factors contribute to bonding in silica bricks:

1. Formation of calcium silicate glass
2. Interlocking action among tridymite needles

The most important property of silica bricks is its very high refractoriness under load (RUL). This is due to interlocking tridymite needles. These bricks are rigid even under loads very close to the melting point. The addition of CaO gives rise to a monotectic reaction in the SiO_2–CaO system. This reaction takes place at 1690°C. The single liquid remaining at the end of this reaction remains until 1440°C. With 1.7% CaO, this liquid is only about 5%. This amount of liquid cannot weaken the tridymite network to any great extent.

21.3 Properties

Silica bricks have refractoriness of 1710°C. At about 1630°C, initial softening occurs under load. Spalling resistance is very good above 300°C. They are acidic bricks but still have resistance to basic slags. The porosity is kept at about 15% in high duty bricks. Hence, permeability is very low. The apparent density of silica bricks is 1800 kg/m³. That means that they are fairly light.

There is another variety of silica bricks called *semi-silica bricks*. The raw materials for these bricks are low-grade ganisters or artificial sand/clay mixtures. The properties are in between those of silica and firebricks. They are inexpensive with low after-shrinkage—an advantage over firebricks. Semi-silica bricks have better spalling resistance than silica bricks. Once the surface is glazed, they are not attacked by slag. They exist in a metastable state. Their structure can be visualized as a matrix of firebrick in which silica particles are embedded.

21.4 Uses

The properties of silica bricks have made them suitable for two principal applications. One is as roofs of open hearth steelmaking furnaces. These roofs are subjected to compressive loads up to 1600°C. This compressive load comes from the arched shape of the roof. This load could be reduced because of the low density of silica bricks. Silica bricks also have a high RUL value. Another advantageous property of these bricks is their low thermal expansion coefficient, which makes them stable. Because of the high refractoriness of silica bricks, there are no drips from the hot face of the furnace roof. Their slag resistance is helpful in preventing attack by slag splash from the bath. In furnace roof construction, high-duty silica bricks are used alternatively with bricks of ordinary quality. In this construction, the silica bricks project more at the hot face. This protects the regular bricks from the severe conditions prevailing inside the furnace. This type of roof was called a *zebra* roof. Basic roofs developed almost simultaneously replaced zebra roofs.

The failure of silica roofs can come about in different ways. One is by the fluxes thrown up from the liquid bath in the furnace. These fluxes reduce the refractoriness of the bricks. The second way is due to the structural changes taking place in the roof. In due course, four roof layers develop. From the hot face up to a thickness where the temperature is 1470°C, a layer of cristobalite will form. This layer appears gray. A layer of tridymite forms where the temperature is 1470°C. This layer appears dark. The third layer is narrow. In this layer, fluxes diffuse from the earlier two layers. After this, the layer does not have any change of structure. Even though the refractoriness of the cristobalite layer increases because of the diffusion of fluxes, spalling takes place at the narrow zone—where the fluxes get collected.

The third reason for failure is thermal spalling. This happens during heating at a rate greater than 10° per hour between 200°C and 300°C. Overheating is the fourth reason for failure and happens when flame impinges on the roof. This overheating could cause the roof to melt.

The second principal application of silica bricks has been for making coke oven walls. The requirement here is for a thin wall that can withstand a temperature of 1400°C for long time. The thin wall is needed for efficient heat transfer. Today, mullite is also used for the same purpose. Both these materials can withstand the abrasion from the coke as it is being pushed out of the oven.

Another important use of silica bricks is in the making of the roof and upper checkerwork of Cowper stoves. These stoves send preheated air blasts to blast furnaces. They work as heat exchangers. Hence, the checkerwork material should have high thermal conductivity and heat capacity. In order to increase the productivity of blast furnaces, one of the techniques used is to increase the blast temperature. These stoves are now made to operate at temperatures of 1500°C. This high temperature exists at the upper checkerwork of the Cowper stoves. Here, the material should have sufficient RUL because it has to sustain its load at the high temperature in the form of a dome. Silica bricks are most suited for these conditions. Thermal conductivity and heat capacity are increased by making the bricks dense. The surface area of the checkerwork is increased by using bricks in the form of hexagonal prisms with hexagonal channels. The thickness of the bricks is fixed at about 3 cm. For the bottom part of the Cowper stoves, where the temperature is lower, mullite is the preferred material.

Silica bricks are also used in small electric arc furnaces, cupolas, and in soaking pits. Their resistance to slagging makes them useful in soaking pits. Furnaces such as acid open-hearth and Bessemer converters, where silica bricks were used, are now obsolete.

Semi-silica bricks are cheaper than silica bricks and are used as backing for silica bricks. In this position, conditions are less severe. Semi-silica bricks are used in coke ovens, kiln roofs, and in flues.

Today, *sand bricks* are used for lining furnaces such as L-D converters. These have properties similar to semi-silica bricks and are cheaper and quicker to install than normal bricks. They last as long as regular bricks and their thermal conductivity are low. This reduces the heat loss from the furnace. Small induction melting furnaces with up to a three-ton capacity are lined with pure silica sand, which is rammed against a steel former. Afterward, induction heating fires it.

Bibliography

J. D. Gilchrist, *Fuels, Furnaces and Refractories*, Pergamon Press, Oxford, UK, 1977, pp. 256, 273–283.

22

Alumina

22.1 Introduction

Alumina is a useful, abundant, and simple oxide refractory. It is substantially ionic in bond character. The powder is compacted and sintered in the temperature range of 1200°C–1800°C. Alumina parts are also made by the application of heat and pressure at the same time (hot pressing). Densification of powder compacts by these two methods brings about shrinkage. The void spaces between the particles decrease in size; on achieving 100% theoretical density, they are eliminated. Linear shrinkage ranges from 5% to 20%.

Because the sintering temperature is above the recrystallization temperature, recrystallization also takes place. In the case of ceramics, recrystallization results in large grain formation.

22.2 Manufacture

Pure alumina is extracted from bauxite ore. Bauxite is a mixture of two hydrates: gibbsite ($Al_2O_3 \cdot 3H_2O$), and diaspore ($Al_2O_3 \cdot H_2O$). The purification of bauxite ore is done by the Bayer process, which involves solution and precipitation. The precipitate obtained is aluminum hydroxide. Dehydration of aluminum hydroxide gives α-Al_2O_3 (corundum). This process gives rise to a purity of 99.9%. The major impurities are Na_2O, SiO_2, Fe_2O_3, TiO_2, CaO, Ga_2O_3, and B_2O_3.

Commercial alumina is also made from ammonium alum. This alumina is of a higher purity than with the Bayer process. Still higher purity alumina can be obtained from high-purity aluminum metal by chemical treatment. The purity can be as high as 99.99+.

22.2.1 Calcination

The aluminum hydroxide in the Bayer process is calcined to get alumina powder. On heating in the absence of air (calcination), aluminum hydroxide is decomposed to give aluminum oxyhydroxide. On further heating, aluminum oxyhydroxide decomposes to the cubic gamma alumina. On heating above 1150°C, gamma alumina is converted to the stable alpha alumina. Apart from aluminum hydroxide, other salts used to obtain alumina are aluminum chloride, ammonium aluminum alum, and aluminum sulfate.

All of the alpha alumina formed from gamma alumina is partially sintered. This gives rise to clusters of alpha alumina particles. Grinding is adopted to obtain individual powder particles. Grinding helps to get higher density compacts. The reason why this happens is that during grinding low-angle boundaries are destroyed. Thus, hindrance for volume diffusion during sintering is reduced. Faster diffusion enhances densification.

22.2.2 Grinding

A complete review on the grinding of aluminum oxide is available [1]. The first technique that was developed is wet grinding. This was done in a ball mill. It is still one of the most widely used techniques. The balls are made of high alumina. Alumina being second only to diamond in hardness, the use of steel balls gives rise to contamination of the obtained powder. Lately, vibratory mills are being used. These mills give higher productivity.

Alumina is also ground in a dry condition. In this technique, grinding aids such as long-chain fatty acids are added. The following are the advantages of dry milling:

1. Dewatering step is avoided.
2. Faster than wet milling.
3. Higher green densities are achieved to the extent of 75% [2].

The likely reasons for achieving higher densities in dry grinding are as follows:

1. Some compaction already takes place during dry grinding.
2. Agglomeration takes place during dewatering of the wet-ground powder.

Fluid energy mills are used as well. They are also called fluid impact mills because the fluid carrying the feedstock impacts on a hard surface.

The next method of producing powders is the sol-gel technique [3]. This has only a limited application for alumina because corundum cannot be used to form a stable sol. The other forms of alumina result in shrinkage and disruptive transformations of the alumina crystals during sintering.

Vapor decomposition techniques are the most promising for producing single-crystal particles. In the case of alumina, gamma alumina particles are produced rather than alpha alumina due to the low temperatures at which these techniques are carried out.

When activated powders are produced, the sintering temperature can be reduced. Many factors contribute to the enhanced sintering of compacts made from activated powders. The diffusion coefficient value determines initial sintering to a great extent. This value is decided mostly by vacancy concentration, especially when it is influenced by impurities. Activated processes produce activated powders. These processes introduce impurities into the powder mass, which pin down vacancies. The unpinned vacancies enhance the diffusion because they can move.

Milling is one way of producing activated powders. It introduces high dislocation density into the crystals of ductile materials. Dislocations again become mobile at the sintering temperature and reactivate sintering. Other characteristics responsible for activated sintering are particle shape and size. Fractured particles have sharp corners and a high surface area to volume ratio similar to small particles. This high ratio facilitates sintering. Milling also introduces impurities that enhance sintering.

22.2.3 Forming

Sintering, the step after forming, gives rise to shrinkage. This shrinkage can be reduced by reducing porosity in the formed part. The forming can be made easier by mixing the alumina powder with a certain amount of clay. If a pure product is required, then clay content should be avoided. Parts made of coarser powders shrink to a lesser extent than those made of finer powders.

The shape is normally produced by pressing. Bridging of powders during die filling and friction between the die wall and punches give rise to nonuniform density. Friction is reduced by the use of a die wall or admixed lubricant. In the case of admixed lubricants, a low temperature soaking is needed to remove the lubricant. Admixed powders are useful for automatic compaction.

High pressures tend to break down coarse particles. The resulting small particles occupy the interstices among the larger particles. This reduces the porosity of the compact. Spherical particles possess greater flowability than irregular particles. Hence, automatic compaction giving rise to greater productivity is facilitated with spherical powders. Flowability of fine particles can be increased by agglomeration.

Uniform density is better achieved in isostatic compaction than in uniaxial pressing. In isostatic compaction, generally oil is used to transmit pressure to the powder. This fluid transmits pressure from all directions equally. This equal pressure application results in uniform density. Isostatic compaction is less effective for automation than uniaxial pressing.

Another way of forming is powder injection molding (PIM). In this method, a sufficiently high amount of organic binder is added. This binder becomes plastic with heat. This plastic mixture containing the ceramic powder is injected through a nozzle into the mold. The injecting pressure fills the mold uniformly. This high amount of organic binder results in high porosity during sintering. The reduction of this high porosity at the sintering temperature causes large shrinkage of the sintered product.

Regular shapes are also made by extrusion. The plasticity required for extrusion is obtained by mixing the powder with varying amounts of organic binder, clay, and water. If sheets are to be produced, then this mixture is rolled. Clay is sometimes substituted with gels or colloids of alumina.

Slip casting is used when complicated shapes are needed. In order to distribute the particles in a liquid, a dispersant is used. For aluminum oxide, that dispersant is organic. The liquid may be acidic or basic. The problem with alumina is that its particles cast very fast. To slow down this process, organic gums are used. A better way of reducing the casting rate is by using colloidal alumina. This also strengthens the product. After the casting, the part has to be released from the mold. In order to facilitate the release, different techniques are used. In one method, the mold is precoated with an organic material such as oil or another aliphatic material. Using a salt, such as ammonium chloride, has been the most successful method.

In slip casting, dimensional tolerance is difficult to achieve. Hence, grinding is adopted. Aluminum oxide is ground with diamond abrasives because its hardness is second only to diamond on the Mohs scale. Depending on the strength of the product, the material may be machined before or after firing because strong bodies are difficult to machine after firing.

Both grain boundary and volume diffusion have been reported for the sintering of alumina [4–7]. The initial shrinkage associated with grain boundary diffusion is given by [8,9]:

$$L/L_0 = 1 - [2.14\gamma\Omega bD_g t/kTa^4]^{1/3} \tag{22.1}$$

Here, L is the height of the compact at any time t; L_0 is the original height; γ is the surface tension; Ω is the volume of the slowest moving species; b is the grain boundary width; D_g is the diffusion coefficient for grain boundary; k is the Boltzmann constant; T is the absolute sintering temperature; and a is the equivalent spherical radius of the particle. The same authors who have given the equation for grain boundary have also given the equation for volume diffusion (Equation 22.2):

$$L/L_0 = 1 - [5.34\gamma\Omega D_v t/kTa^3]^{1/2} \tag{22.2}$$

The volume diffusion coefficient is represented by D_v. Impurities control the diffusion mechanism. For example, it changed from grain boundary to volume diffusion when doped with manganese oxide.

When corundum is heated to >900°C it transforms to γ-Al_2O_3. Fused alumina may be cast to corundum, crushed, ground, graded, and formed into laboratory ware or prefired γ modification may be used. Ware should be fired at about 1700°C.

22.3 Properties

The properties of alumina are discussed under four headings—general, mechanical, electrical, and thermal.

22.3.1 General Properties

Al_2O_3 has a melting point of 2015°C. It can be used up to about 1900°C. Alumina refractory may be made impervious to gases. In that case, the resistance to spalling will be impaired. Porous ware is less susceptible to thermal shock. Slag resistance is poor, especially toward FeO and basic slags. It is unaffected by gases, except F_2. It is also resistant to fused alkalies.

22.3.2 Mechanical Properties

The strength of sintered alumina is an important property. Strength depends on density; the higher the density, the greater the strength. In high-strength materials, dislocation movement becomes difficult, causing failure by brittle fracture. In the case of brittle fracture, the Griffith criteria (Equation 22.3) hold well.

$$\sigma = [\gamma_f E/c]^{1/2} \tag{22.3}$$

In this equation, σ is the stress for the growth of a surface crack of size c, E is the elastic modulus of the material, and γ_f is the surface energy for fracture. The theoretical strength of a material corresponds to its strength when it contains no flaw. This is calculated to be one-tenth of its elastic modulus. That means, the theoretical strength of alumina should be greater than 7000 MPa. Alumina being a brittle material, its actual strength is determined by the maximum size of the cracks it contains—the greater the size, the lesser the strength, as given by Equation 22.3.

Whiskers have been found to possess the theoretical strength because they are single crystals containing no flaw. Large single crystals of alumina have been shown to attain its theoretical strength [10]. This high strength was obtained in sapphire rods by surface treatments such as flame polishing or chemical etching. These treatments eliminate any surface flaw. Alumina is brittle because the dislocations in it are immobile (a similar quality of ceramics in general).

Polycrystalline ceramics are weak because of the presence of grain boundaries. Grain boundaries are two-dimensional defects. A defective material tends to be weak. At high temperatures, there tends to be a transition from brittle to ductile behavior. Dislocations tend to become mobile as additional slip systems become operative. Also, grain boundary sliding can take place.

Surface damage to both single and polycrystals tends to decrease the strength of the ceramics. Surface annealing can repair surface damage [11,12].

The hardness of aluminum oxide is very high, as measured by Vickers hardness tester. High hardness shows that the material is less plastic. Only very high stresses can move the dislocations. One way of altering the dislocation mobility is by dissolving other oxides. The addition of iron oxide to alumina did not show much strengthening. When a solid solution was formed by the addition of chromia, the hardness of alumina was increased [13].

The elastic modulus of single and polycrystalline alumina is found to be the same [14]. This is to be expected because the elastic modulus depends only on the type of bond, which does not change for a given material. The value of elastic modulus for alumina is 4.09×10^5 MPa. The decrease of elastic modulus is linear up to 1000°C. The composition of sintered alumina determines its elastic modulus [15]. Alumina of 99.9% purity has a modulus value of 3.92×10^5 MPa; for a purity of 87.8%, the value is 2.57×10^5 MPa. There is no direct relation between strength and composition or strength and modulus. The modulus of a rupture test gives strength values between 345 and 193 MPa. In alumina, a higher modulus is associated with greater strength.

Porosity affects both modulus and strength. The empirical relationships between modulus and porosity and between strength and porosity are given as follows:

$$E = E_0 e^{-\alpha P} \tag{22.4}$$

$$\sigma = \sigma e^{-\beta P} \tag{22.5}$$

Here, E is the modulus of the material having P as the fractional porosity, and E_0 is the porosity at 0 porosity; α is a configurational constant. Similar parameters, corresponding to the strength, are given in Equation 22.5.

If the surface is made flawless, the σ_0 value equals $(1/3)$ σ_T (theoretical strength). The value of β in Equation 22.5 is unity for spherical pores. It varies between 1 and 10 for the other-shaped pores. Irregular shaped pores therefore reduce strength to a greater extent than spherically shaped pores. The reason for this is that the radius of curvature for such pores is less than that of the spherically shaped pores.

Another factor affecting the strength is the grain size. The Hall–Petch relation holds well in this case. The strength is inversely proportional to the square root of the grain size.

There are two methods by which alumina can be strengthened. In one method, surface damage is eliminated. This can be done by polishing or by chemical or thermal treatment. In the second method, the surface flaws are covered by a glaze.

22.3.3 Electrical Properties

Sintered alumina possesses excellent insulating properties. Therefore, it is used as an insulator in electrical and electronic industries. The electrical conductivity is some 20 orders of magnitude smaller than metallic conductors. As an insulator, sintered alumina is used at frequencies in the kilocycle to megacycle range. Impurities in the material tend to increase its conductivity.

Alumina can also be doped to use it as a semiconductor. At temperatures greater than 1500°C, alumina behaves as an intrinsic semiconductor. The activation energy then is found to be 5.5 eV [16]. Below this temperature, it behaves as an extrinsic semiconductor with an activation energy in the range of 2.5–5.5 eV.

22.3.4 Thermal Properties

The very high melting point of alumina makes it useful as a refractory. Even though alumina is soluble in glass, it is useful as a glass tank refractory because the rate of dissolution is small.

Creep is another thermal property. Single crystals creep by dislocation movement. During movement, crystals may encounter barriers. These barriers reduce the creep rate. Polycrystals creep by different mechanisms. Because of this, creep is greater in these crystals than it is in single crystals. In one creep mechanism, ions diffuse through the grains and grain boundaries. Other mechanisms are grain boundary sliding and dislocation mobility. Creep is dependent upon stress, grain size, porosity, and impurity content.

Porosity increases the creep rate. The stress dependence of creep in porous alumina was found to be between that of diffusion and dislocation creep mechanisms [17]. Strain rates were found to vary between the second and third power of the stress. With respect to the diffusion of ions, the diffusion of Al^{3+} ions was found to be the rate-determining step [18].

It was found that the diffusion coefficient decreased by a factor of 20 with the addition of 2000 parts per million of MgO to high-density sintered alumina [19]. Mg^{2+} ions can substitute Al^{3+} ions in the alumina lattice or a part of the ions can occupy interstices—the other part going to the lattice sites. When alumina was doped with $MgTiO_3$, it was found that the diffusion coefficient of Al^{3+} did not change. This is because there is charge compensation between the host cations and doped cations resulting in no point defects being produced.

If impurities do not dissolve in an oxide, at elevated temperatures the liquid phase may form by way of an eutectic reaction. Such liquids decrease the refractoriness drastically. Dense alumina finds use at high temperatures because of its low permeability toward gases. High purity is required to make it impervious.

22.4 Uses

The usual shapes in alumina are tubes and crucibles. Fused alumina is mixed with clay to produce *alundum* cement. This cement is tempered with water and can be hand-molded. It is useful in laboratories. On firing, it shrinks a little and gives rise to a porous refractory. It also spalls easily and is inexpensive. Depending on the grade, it is useful in the range of 1500°C–1700°C.

References

1. W. H. Gitzen, *Alumina Ceramics*, Technical Report, AFML-TR-66-13, Ohio State University, OH, 1966, p. 167.
2. R. L. Coble, Initial sintering of alumina and hematite, *J. Am. Ceram. Soc.*, 41: 55–62, 1958.
3. L. D. Hart and L. K. Hudson, Grinding low-soda alumina, *Bull. Am. Ceram. Soc.*, 47: 574, 1964.
4. R. D. Bagley, Effects of Impurities on the Sintering of Alumina, PhD Thesis, University of Utah, Salt Lake City, 1964.
5. D. L. Johnson and I. B. Cutler, Dihsion sitering: I, *J. Am. Ceram. Soc.*, 46: 541, 1963.
6. J. R. Keski, PhD Thesis, University of Utah, Salt Lake City, 1966, cited in A. M. Alper (Ed.), High temperature oxides, in *Refractory Materials*, Academic Press, New York, 1970.
7. G. C. Kuczynski, *J. Metals Trans. AIME*, 185: 169, 1949, cited in J. M. Blakely, Surface diffusion, *Prog. Mater. Sci.*, 10: 395–437, 1963.
8. D. L. Johnson, Magnesia, Alumina, Beryllia Ceramics: Fabrication, Characterization, in *Kinetics of Reactions in Ionic Systems*, T. J. Gray and V. D. Frechette (Eds.), Plenum Press, New York, 1969, p. 331.
9. S. D. Brown, O. E. Accountices, H. W. Carpenter, T. F. Schroder, and M. J. Serie, Technical Report, AFML-TR-67-194, 1967, cited in A. M. Alper (Ed.), High temperature oxides, in *Refractory Materials*, Academic Press, New York, 1970.
10. F. P. Mallinder and B. A. Proctor, *Proc. Br. Ceram. Soc.*, 6: 9, 1966, cited in A. M. Glaeser, Studies of interfacial behaviour in ceramics via micro-designed interfaces, *Mater. Sci. Monogr.*, 81: 33–70, 1995.
11. A. H. Heur and J. P. Roberts, *Proc. Br. Ceram. Soc.*, 6: 17, 1966, cited in A. M. Alper (Ed.), High temperature oxides, in *Refractory Materials*, Academic Press, New York, 1970.
12. L. M. Davies, *Proc. Br. Ceram. Soc.*, 6: 29, 1966, cited in S. M. Weiderhorn, B. J. Hockey, and D. E. Roberts, Effect of temperature on the fracture of sapphire, *Phil. Mag.*, 28: 783–796, 1973.
13. R. C. Bradt, *J. Am. Ceram. Soc.*, 50: 54, 1967, cited in A. M. Alper (Ed.), High temperature oxides, in *Refractory Materials*, Academic Press, New York, 1970.
14. S. C. Carniglia, Grain boundary and surface influence on mechanical behavior of refractory oxides-experimental and eductive evidence, in W. W. Kriegel and

H. Palmour (Eds.), *Materials Science Research*, Vol. 3, Plenum Press, New York, p. 425, cited in Martin H. Leipold, Impurity Distribution in MgO, *Journal of the American Ceramic Society*, 49(9): 498–502, 1966.

15. D. B. Binns and P. Popper, Mechanical properties of some commercial alumina ceramics, *Proc. Br. Ceram Soc.*, 6: 71–82, 1966.
16. P. J. Harrop, Intrinsic electrical conductivity in alumina, *Br. J. Appl. Phys.*, 16: 729, 1965.
17. G. M. Fryer and J. P. Roberts, *Proc. Br. Ceram. Soc.*, 6: 225, 1966, cited in H. Duong, and J. Wolfenstine, Creep behaviour of fine-grained two-phase $Al_2O_3.Y_3Al_5O_{12}$ materials, *Mater. Sci. Eng.*, 172: 173–179, 1993.
18. R. L. Coble and Y. H. Guerard, Creep of polycrystalline aluminum oxide, *J. Am. Ceram. Soc.*, 46: 353–354, 1963.
19. C. W. Hewson and W. D. Kingery, Effect of MgO and $MgTiO_3$ doping on diffusion-controlled creep of polycrystalline aluminium oxide, *J. Am. Ceram. Soc.*, 50(4): 218–219, 1967.

Bibliography

A. M. Alper (Ed.), High temperature oxides, in *Refractory Materials*, Academic Press, New York, 1970, pp. 129, 136, 142, 143, 160, 163–178.

J. D. Gilchrist, *Fuels, Furnaces and Refractories*, Pergamon Press, Oxford, 1977, pp. 303–304.

23

Alumino-Silicate

23.1 Al$_2$O$_3$–SiO$_2$ System

Alumino-silicate refractories are based on an Al$_2$O$_3$–SiO$_2$ system. The equilibrium diagram of this system is given in Figure 23.1. The composition of various alumino-silicate refractories are marked on the diagram, which includes the composition of semi-silica that was discussed in Chapter 21. Firebrick is one common aluminosilicate refractory. It contains 30 wt% Al$_2$O$_3$ and is a mixture of two phases—mullite and silica. At 1545°C, this mixture undergoes eutectic transformation. The eutectic composition is 5 wt% Al$_2$O$_3$. Immediately after this transformation, the firebrick will contain mullite and a liquid with eutectic composition. In the mixture, mullite is present as needle-shaped crystals. Similar to tridymite in silica brick, these needle-shaped crystals interlock among themselves. This interlocking is so rigid that firebrick retains its strength and rigidity up to 1810°C, even though it contains a liquid phase beyond the eutectic temperature. Of course, the strength depends on the percentage of mullite present. The percentage of alumina in firebrick varies from about 15%–40%. Therefore, applying the lever rule, mullite can vary from about 20%–55%. The higher the mullite percentage, the better the strength of the firebrick. Silica in firebrick can exist as tridymite, cristobalite, or as a mixture of both. If any flux is added during manufacture, then some silica can transform to silicate glass. Silicates are very viscous liquids. At 1810°C, firebrick undergoes incongruent melting.

Above the eutectic temperature, we get liquid in the interstices of mullite needles. With 25% Al$_2$O$_3$, applying the lever rule, we find that siliceous firebrick contains 70% liquid. Even though the liquid is viscous, it will be difficult for the remaining solid to sustain the original shape of the brick. Further, as the temperature is increased, the viscosity reduces. In those bricks containing fluxes, refractoriness is further reduced. In the case of bricks containing 45% Al$_2$O$_3$ (called high alumina brick), the amount of mullite is 60% immediately after the eutectic temperature. This brick can retain its shape beyond the eutectic temperature. As the temperature increases, the percentage of liquid slowly increases. At 1700°C, it is only 45%.

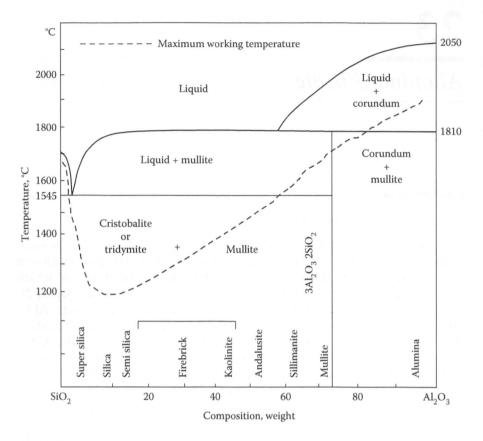

FIGURE 23.1
Al$_2$O$_3$-SiO$_2$.

At this temperature, siliceous brick would have slumped because it would have contained 80% liquid. High alumina brick can have refractoriness at 1700°C, but its refractoriness under load (RUL) value will be insufficient for any load-bearing application beyond the eutectic temperature.

Mullite contains 72% alumina. Therefore, it is a high alumina refractory. It melts at 1810°C incongruently. Mullite is useful beyond firebrick's useful range of temperature by about 400°C, as is evident from Figure 23.1, in which the working temperature line is drawn.

23.1.1 Firebricks

Firebrick is the regular brick of a furnace designer and is the most common refractory. This brick is made from fireclay, which occurs in association with coal measures. China clay is also used to make bricks. These bricks have more enhanced properties than firebricks.

23.1.1.1 Origin and Properties of Fireclays

Clays result from the decomposition of igneous rocks. This decomposition is caused by geological processes that take place over many years. The type of clay formed depends on the rock type and on the geological process. Fireclay comes from granite, which is acid rock that contains feldspar. Feldspar is any of a group of minerals, principally aluminosilicates of potassium, sodium, and calcium. In an environment containing H_2O, and CO_2, feldspar decomposes when high pressure and temperature prevail. The products of the decomposition reaction are K_2CO_3, $SiO_2 \cdot nH_2O$, and $Al_2O_3 \cdot SiO_2 \cdot 2H_2O$.

One type of hydrated fireclay is called kaolinte. The formula for this clay is that of the last product of decomposition. Kaolinite has a structure of alternate layers of SiO_4^{4-} and $Al(OH)_6^{3-}$. The SiO_4^{4-} layer is made up of tetrahedra, and the $Al(OH)_6^{3-}$ layer is made up of octahedra. Interlayer bonding gives rise to the formation of H_2O molecules. The charge balance is achieved by having common corners for the tetrahedra and octahedra. O^{2-} ions are located at these corners.

Two other types of fireclay are montmorillonite [$Al_2O_3 \cdot 5SiO_2 \cdot nH_2O$ $(CaMg)O$] and beidelte [$Al_2O_3 \cdot 3SiO_2 \cdot 4H_2O$]. These are also made up of tetrahedral and octahedral layers of $Al(OH)_6^{3-}$ and SiO_4^{4-}, respectively. The number of tetrahedral layers between two octahedral layers will be greater than one because there is more than one SiO_2 molecule per formula. In montmorillonite, half the Al^{3+} ions are replaced by Mg^{2+} ions. The charge balance is then achieved by incorporating Ca^{2+} ions. These ions are located in the interstices between two adjacent layers. The bond strength is less than that of the Si^{4+} and Al^{3+} ions. Incorporation of Ca^{2+} ions makes the clay more plastic, but the refractoriness is decreased.

Clays can get leached out of soluble matter by water flow. The resulting deposit is low in iron, alkalies, and alkaline earth and is called china clay. It is very pure, highly refractory, and fires white. The least pure clays are derived from basic rocks. They fire red and are useful as building bricks and for making drainage pipes. Between very pure and least pure clays are the intermediate pure ones. These include plastic ball clay, bond clay, flint fireclay, fireclay, and aluminous fireclay.

Plasticity in a clay is equivalent to fluidity in a liquid. In liquids, flow occurs when pressure is applied. The velocity of this flow is given by Equation 23.1.

$$V = K\varphi P \qquad (23.1)$$

Here, K is the proportionality constant, φ is fluidity, and P is the pressure applied. This equation is represented in Figure 23.2a. In the case of clays, the equation is:

$$V = K\mu(P - f) \qquad (23.2)$$

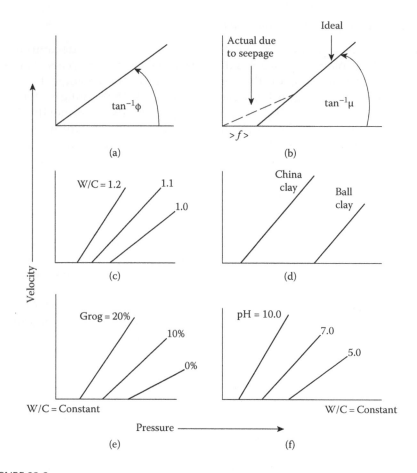

FIGURE 23.2
(a) Velocity versus pressure in a liquid; (b) velocity versus pressure for a clay; (c) dependence of water to clay ratio on mobility and yield stress of a plasticized clay; (d) dependence of plasticity on the type of clay; (e) dependence of plasticity on pH; (f) effect of grog on plasticity of clay.

In this equation, μ is the mobility and f is the yield stress below which there is no flow. This is shown in Figure 23.2b. Water to clay ratio decides both mobility and yield stress of a plasticized clay.

Figure 23.2c shows this dependence. As the ratio increases, mobility increases and yield stress decreases. As the yield stress decreases, the clay becomes more plastic. Plasticity differs based on clay materials. This is illustrated by Figure 23.2d. Plasticity also depends on pH as shown in Figure 23.2e.

Plasticity increases as pH increases. Grog addition also increases plasticity. This is shown in Figure 23.2f. In this case, the water to clay ratio is calculated by also considering grog as clay. Plasticity is important in molding. Higher plasticity is required to increase the clay's workability. At the same time, lower water content is desired to reduce the drying time.

23.2 Manufacture

23.2.1 Firebricks

A general method is adopted for the manufacture of firebricks. Slow drying is necessary for hand-molded bricks. Otherwise, these bricks may crack. Firing has three stages:

1. *Smoking or steaming.* This stage will last 12–48 h depending on the quantity. As the quantity of brick increases, the smoking time also increases. In this stage, the temperature is increased from 20°C to 300°C. The atmosphere for this stage should be reducing. Both physisorbed and colloidally held water is removed here.

2. *Decomposition.* The duration for this stage is 10–24 h. The temperature is raised from 300°C to 900°C. Heating is done with an oxidizing flame. The chemisorbed water is removed in this stage. The amorphous residue that remains after the water removal is called *meta-kaolinite.* Above 573°C, α-quartz gets converted to β-quartz. Oxidizing flame gasifies carbon and sulfur, and they escape from the bricks. These elements are harmful in the next stage because they make the brick black-hearted.

3. *Full firing.* This will take place in 12–18 h. The firing temperature is in the range of 1200°C–1400°C. Formation of silicates starts from 1000°C. The firing temperature is fixed such that this vitrification is completed. During the heating to the firing temperature, silica and alumina attain their high temperature forms. They also react to form mullite (above 1100°C). The highest temperature is decided by the amount of fluxes added. As the amount of fluxes increases, the firing temperature can be decreased. Overfiring and underfiring are harmful. Underfiring makes the brick friable and overfiring causes slumping.

The whole operation takes place over three to five days because cooling also has to take place. The microstructure is constituted by Mullite needles in a matrix of glass. If any free silica is present, it may be in the form of quartz, tridymite, or cristobalite.

23.2.2 High Alumina Bricks

High alumina bricks contain more than 46% alumina. The raw materials used for making such bricks are sillimanite, kyanite, and andalusite. These minerals exist as rocks and are used in *as mined* condition to get more alumina content. For a lower percentage of alumina, the as-mined minerals are mixed with clays. On firing, all these bricks contain mullite, glass, and cristobalite or tridymite as the phases. The same phases are also found in firebricks.

The difference here is in the amount of mullite and fluxes. Mullite will be higher and fluxes will be lower in high alumina bricks.

When bricks contain 70%–73% alumina, they are called mullite bricks. These bricks contain very little silica modifications of tridymite or cristobalite. As this much alumina content cannot be obtained from naturally occurring minerals, these bricks are made by mixing in bauxite. When we want 100% alumina bricks, bauxite is used with a small amount of clay. Clay serves as the bonding agent. Bauxite powder is prepared after fusing. The fused bauxite is crushed and ground. Mullite bricks contain mullite and corundum. As seen in Figure 23.1, these bricks remain solid at up to 1810°C.

23.3 Properties

23.3.1 Firebricks

The properties of firebricks depend on the amount of alumina present in them. Therefore, they are classified based on their alumina content. The first class is aluminous firebrick, which contains 38%–45% alumina. When the percentage is between 22 and 38, the brick is called ordinary. The third class is semi-silica firebrick. It contains 10%–22% alumina.

Apart from the main components of silica and alumina, firebrick also contains other oxides. Titania content ranges from 0.5% to 5%. There can be ferric oxide in the range of 0.7% to 4%. The alkali range is 0.1%–2%, but limiting its content to 0.5% is preferred. Otherwise, the refractoriness may be adversely affected. There can be alkaline earths in the range of 0.4%–2%. The silicate bond is formed with the help of the cumulative effect of calcia and magnesia. They cause the fluxing in the brick. The resulting slag after fluxing is useful in preventing any attack on the brick.

As alumina content increases from 25% to 45%, refractoriness of firebricks increases from 1550°C to 1750°C. RUL value ranges from 1200°C to 1620°C when we go from siliceous firebrick to normal brick and then to aluminous brick. Spalling resistance is poor for siliceous firebrick, fair for ordinary brick, and good for aluminous. Thermal expansion increases as temperature rises, and it reaches a maximum of 1.6% at the highest temperature.

Acid slag resistance is good for all types of firebricks. However, resistance to basic slags is poor for siliceous and ordinary firebricks but is good for aluminous ones. The lesser content of alkalies and other fluxes in aluminous bricks is one reason for their good slag resistance. When the alumina percentage is greater than 70%, the brick can absorb 20% of blast furnace slag at 1500°C without the formation of much liquid [1].

Molten silicates are inherently viscous. Once they are formed, they spread on the surface and are retained because of their high viscosity. This viscous layer on the surface delays the failure of the material under corrosive conditions.

Carbon deposition takes place on firebricks if the furnace atmosphere contains carbon monoxide. This happens because of the decomposition of this gas at about 450°C. The decomposition reaction is:

$$2CO = CO_2 + C \qquad (23.3)$$

This temperature prevails at the blast furnace stack. On deposition of carbon, the brick develops a crack and disintegrates. Hence, bricks meant to be used in the construction of a blast furnace stack should be tested for resistance to carbon deposition. This is done by holding a sample of the bricks in a stream of carbon monoxide at 450°C for 200 h. The sample is inspected at frequent intervals for any signs of carbon deposition, cracking, or disintegration.

In order to increase abrasion resistance, firebricks are hard fired. This kind of firing treatment causes partial vitrification. Vitrification reduces the porosity. The brick becomes harder and more compact. But its spalling resistance is reduced. This resistance can be improved by the incorporation of grog. Grog is prefired clay or crushed broken brick. Cracks cannot pass through the grog particles but have to circumvent them in order to propagate. This delays the failure. Grog particles also make the bricks harder.

23.3.2 High Alumina Bricks

As the percentage of alumina increases, the refractoriness and RUL of high alumina bricks increase. The spalling resistance of these bricks is good. Cold crushing strength is extremely high. This makes them useful where mechanical wear or abrasion is encountered. Slagging resistance is fair to good as the alumina content increases. The 50%–60% alumina bricks can resist acid slags and glasses. At 70% alumina, resistance to FeO and CaO is quite good. Bricks containing more than 70% can even resist alkalies. Hence, they are useful in glass tanks. High alumina bricks vary in porosity.

23.4 Uses

23.4.1 Firebricks

Firebricks are commonly used in furnace construction. They are also used in domestic fireplaces and stoves. One reason for their common use is that they are somewhat inexpensive. Also, fireclay is indigenously available. This has contributed to firebricks' traditional use.

In iron-making industries, firebricks are found in the blast furnace stack, bosh, hearth, and ladles. Cowper stoves are also made of firebricks. The blast duct connecting Cowper stoves and the blast furnace is also constructed of firebricks.

In open hearth steel making, firebricks are used in areas where the bricks are not exposed to steelmaking temperatures, such as in checkers of regenerators. The ladles are lined with these bricks and their stoppers are also made of firebrick. In the casting bay, firebrick is used to make runners and guide tubes. The building material used in soaking pits and reheating furnaces is firebrick. Small furnaces in foundries use firebrick. It is also found in steam-raising plants.

The quality of firebrick is decided by the conditions to which it is exposed. For example, blast furnaces use high alumina bricks. Many steel plant applications also use these high quality bricks. Examples include use in the top of Cowper stoves and in checkerwork in regenerators. The bricks used in these applications are exposed to high temperatures and to slagging. The blast furnace stack uses china-clay bricks to resist damage from carbon deposition. These bricks have very low porosity, which results in sufficient abrasion resistance. Ladles should not leak because they are meant to hold liquid metal. Hence, they are made of bricks having a high thermal expansion coefficient. Such bricks are called *bloating firebricks*. On expansion, they are tightly joined.

23.4.2 High Alumina Bricks

High alumina has replaced the better quality firebricks in many applications. Mullite bricks are the most durable of all refractories. They have replaced firebricks used to line the bosh of blast furnaces. Firebricks were found to rapidly slag back almost to the cooling blocks and had to be replaced often. The resistance of high alumina bricks to slag has made them last for a long time. Also, sillimanite and andalusite bricks have also replaced high quality firebricks in the lower blast furnace stack. Mullite and andalusite bricks also are used in the upper courses of checkerwork in regenerators and Cowper stoves, where hot gases enter these heat exchangers. High density is required to achieve good thermal conductivity and heat capacity. The porosity is restricted to 12%. High alumina bricks also find use in soaking pits, reheating furnace walls, roofs, in hearths, and around doors. In the lower parts of soaking pits, attack by FeO is severe. Mullite bricks have sufficient resistance to this attack. These bricks are also used in electric arc steelmaking furnace crowns. They have started replacing silica bricks for wall construction in coke ovens. Here, they are made highly conducting by making them very dense. Heat conduction has to take place through coke oven walls and highly conducting material is desirable.

The applications of high alumina bricks in nonferrous applications include use in brass melting and lead drossing reverberatory furnaces and in aluminum smelting furnaces. These bricks are also used for nonmetallurgical applications in glass, ceramics, cement, and cement industries. Other applications include enameling and in high duty boilers.

Reference

1. W. F. Ford and J. White, Equilibrium relationships in the system $CuO-Cu_2O-SiO_2$, *Trans. Br. Ceram. Soc.*, 50: 461, 1951.

Bibliography

J. D. Gilchrist, *Fuels, Furnaces and Refractories*, Pergamon Press, Oxford, 1977, pp. 258–272.

24

Chrome-Magnesite

24.1 Introduction

The chrome-magnesite group of refractories can be discussed under five headings. They are chromite, magnesite, chrome-magnesite and magnesite-chrome, dolomite, and fosterite. Chromite is a spinel with the formula $FeO \cdot Cr_2O_3$. Magnesite is the raw material used to produce magnesia refractory. Magnesia is one of the most refractory oxides. Its melting point is 2800°C. Chrome-magnesite contains chrome as the major oxide and magnesia as the minor one. It is the opposite for magnesite-chrome refractories.

Dolomite is double carbonate of magnesium and calcium. Once made in the form of a refractory material, it becomes the double oxide of magnesium and calcium and is called doloma. Apart from the carbonates of magnesium and calcium, dolomite contains SiO_2, Al_2O_3, and Fe_2O_3. The calcia in doloma reacts with water according to Equation 24.1. The oxides of calcium and magnesium form a eutectic at 2300°C. The eutectic composition contains 32% calcia. Because the brick crumbles in moist air due to the reaction of calcia with water, calcia in doloma should be stabilized before the brick is made. The stabilization is done by combining calcia with silica. Calcia can form two kinds of silicate with silica—tricalcium silicate ($3CaO \cdot SiO_2$) and dicalcium silicate ($2CaO \cdot SiO_2$). Tricalcium silicate dissociates at 1900°C to produce dicalcium silicate and calcia.

Dicalcium silicate melts at 2130°C. Dicalcium silicate has three allotropic forms. They are α below 678°C, β between 678°C and 1410°C, and γ above 1410°C. The $\alpha \leftrightarrow \beta$ transformation is accompanied by a 10% volume change. This high volume change causes *dusting*. Dusting is powdering. It can be prevented by adding Cr_2O_3, B_2O_3, or P_2O_3. However, any of these additions decreases the slag resistance of the refractory.

Forsterite is also a double oxide. It is a mixture of magnesia and silica forming a silicate.

24.2 Manufacture

24.2.1 Chromite

Chromite is a member of the spinel group. Spinels have the general formula, $R''O \cdot R_2'''O_3$. R'' represents a divalent metal ion, such as Fe^{++}, Mg^{++}, Mn^{++}, and so forth. R''' is a trivalent ion. Some examples are Fe^{+++}, Al^{+++}, and Cr^{+++}. The chemical formula for chromite is $FeO \cdot Cr_2O_3$. The characteristics of this group of minerals are as follows:

a. Cubic crystal structure
b. Hardness
c. Refractoriness
d. Mutual solubility

Each of these minerals can be used as a refractory in their pure form. They can be fused with B_2O_3. This oxide acts as a mineralizer. Once fused, these minerals can be cast into molds. Chromite bricks are made from chromite ore. Chromite ore is a mixture of chromite and picrochromite. The percentage of chromium in the ore varies. Metal extracted from those ores contains the highest chromium content. Bricks are made from the remaining ores and contain high MgO, and low FeO. A typical ore will contain 31% Cr_2O_3, 29% Al_2O_3, 18% MgO, 16% FeO, 5% SiO_2, and 0.4% CaO.

Standard methods are adopted to make chromite bricks. The bond material may be fine lime or hydrated MgO. Tar is added as a temporary bond material. Chromite bricks are arranged on top of magnesite bricks in the kiln. This is required to reduce the height of the chromite stack because green chromite bricks do not have sufficient strength. The firing temperature is 1450°C.

24.2.2 Magnesite

As already mentioned, magnesite ($MgCO_3$) is used to produce magnesia refractories. Because mined magnesite contains calcium and iron carbonates, which are impurities. Ore-dressing techniques such as grinding, washing, calcination, and magnetic separation are used to reduce these impurities. Later, the purified magnesite is fired in a rotary kiln at a temperature above 900°C. This causes magnesite to decompose to magnesia. This magnesia is then fired at a temperature above 1600°C. This produces the compact form of magnesia called periclase, which is stable toward water. This is known as *dead-burned magnesite*.

Another source of magnesia is seawater. Seawater contains 0.2% magnesium chloride. To separate it, dolomite is first calcined to produce doloma.

Doloma is a mixture of calcia and magnesia. This is added to seawater. The oxides are converted to their hydroxides by the following reactions:

$$CaO + H_2O = Ca(OH)_2 \tag{24.1}$$

$$MgO + H_2O = Mg(OH)_2 \tag{24.2}$$

The $Ca(OH)_2$ formed reacts with $MgCl_2$ present in the seawater following Equation 24.3.

$$Ca(OH)_2 + MgCl_2 = CaCl_2 + Mg(OH)_2\downarrow \tag{24.3}$$

Because the $Mg(OH)_2$ is insoluble, it precipitates out and is then dead-burned to periclase.

The periclase is obtained in the form of clinkers that are crushed and separated into various grades depending on particle size. The different grades are mixed with bond materials such as milk of lime, hydrated ferric oxide, or clay. To aid in bonding, a maximum of 5% water is added. A dry press is used to shape the mixture. The pressure applied is 100 MPa. This much pressure results in a good green strength and, after firing, a good fired strength. This is due to the green compact's low porosity. Because of the accompanying high green density, there is also minimal shrinkage in firing. The high fired strength produces good refractoriness under load (RUL).

Periclase contains silica as an essential impurity. This is helpful in forming the silicate bond in the refractory. This bond contains silicate in the form of fine crystals. If it is composed of forsterite ($2MgO{\cdot}SiO_2$), this bond has a melting point of 1890°C. In order to increase its refractoriness, pure magnesia is added in the form of fine particles.

24.2.3 Chrome-Magnesite and Magnesite-Chrome

We have seen that chromite bricks contain as much as 18% magnesite. Sometimes, hydrated magnesia is added as a bond material. In addition to magnesia, chromite is about 5% silica. The formation of $MgO{\cdot}SiO_2$ was found to decrease the RUL value of chromite. To increase the refractoriness of chromite, the percentage of magnesia is increased. Thus, chrome-magnesite bricks are made by adding fine magnesia powder to coarse chromite powder. On firing, the bricks made of this mixture have a continuous bond phase of forsterite in which coarse grains of chromite are distributed. This is the structure of chrome-magnesite brick, which contains a 70/30 chromite-magnesite mixture. When the ratio is 40/60, we get magnesite-chrome refractory.

To manufacture chrome-magnesite and magnesite chrome bricks, the molding is by dry-pressing, and the firing is done in kilns at a temperature of 1450°C–1500°C. At this temperature, a forsterite bond develops between the added fine MgO powder and the silica diffused out of the chromite. Sintering of chromite particles can be enhanced by increasing the firing temperature to 1650°C–1700°C. This happens because of the catalytic effect of MgO, SiO_2, and FeO. The density is also improved for these refractories. Properties are superior to those of low temperature sintered bricks. Structural spalling is less. Because of the higher heating temperature heating and the more time that is taken for firing, the cost is greater.

24.2.4 Dolomite

In manufacturing dolomite bricks, silica is first added to dolomite in the form of serpentine ($3MgO·2SiO_2·H_2O$). This mixture is then calcined, which removes carbon monoxide and water. Oxides of calcium, magnesium, silicon, aluminum, and iron are left behind. This mixture of oxides is then fired at 1600°C in a rotary kiln. The resulting stabilized clinker contains periclase, calcium silicates, aluminates, and ferrates of calcium and magnesium. The clinker is crushed to different sizes, which are then graded. The different grades are then tempered with 4% water. This tempered mixture is then molded by dry pressing. The pressure applied is 70 MPa. The green bricks are then dried. If the calcia in the green compact is not stabilized, it will react with water during drying. The dried bricks are later fired at 1400°C.

After firing, the bricks are checked for stabilization. This is done by keeping the fired brick in boiling water for 24 h. If the brick is not changed by this treatment, then it is stabilized.

A cheaper variety of dolomite brick is known as semi-stabilized dolomite. This is prepared from the calcined dolomite without the separate addition of silica. Fe_2O_3 and Al_2O_3 are added to the calcined dolomite for bonding purposes. Heavy oils are also added that uniformly spread the Fe_2O_3 and Al_2O_3 particles in the mixture. These oils also keep out moisture. Thus, a reaction between moisture and calcia is prevented until the bricks are fired. The firing is done is a kiln at a temperature of 1400°C. If the fired bricks are not put into service immediately, they are coated with a tarry material.

24.2.5 Forsterite

Forsterite is magnesium silicate. The origin of this refractory is olivenes. On decomposition, olivenes gives rise to serpentine, talc, and steatite. All these minerals are used to make forsterite bricks. The magnesia content in the bond is increased by adding very fine MgO. The mixture is then dry-pressed to make the brick. The green bricks are then fired.

24.3 Properties

24.3.1 Chromite

Depending on the ore used, the refractoriness of chromite bricks is 1700°C–1850°C. The RUL is poor—about 1400°C. This is due to the formation of $MgO \cdot SiO_2$ in the bond, which has a low melting point. Spalling resistance is also poor because the thermal conductivity is high. The brick has a distinctive black color. Its density is high at $3g/cm^3$. The brick's only valuable property is its extremely high resistance to acid or basic slag.

24.3.2 Magnesite

The refractoriness of magnesite bricks is about 1800°C. This is dependent on the melting point of the bond. The RUL ranges from 1450°C to 1720°C. Spalling resistance is poor for some grades and good for others. One reason for the poor spalling resistance is a high coefficient of thermal expansion. Magnesite has the highest thermal expansion coefficient among all refractories. Spalling resistance is also dependent on the porosity. To improve the spalling resistance of magnesite bricks, zirconia is sometimes added. Apart from thermal spalling, structural spalling can also take place due to the absorption of FeO, which causes swelling. Because this absorption takes place at the hot face, flaking can result.

Slags containing CaO and FeO do not attack magnesite bricks. Forsterite is formed on reaction with acid slag containing silica. This has a high melting point, so further attack is prevented. The thermal conductivity of magnesite bricks is very high. They are dense, light brown in color, and slightly magnetic.

24.3.3 Chrome-Magnesite and Magnesite-Chrome

Chrome-magnesite and magnesite-chrome bricks are dark brown. Their density is high at $3 \, g/cm^3$. Their porosity varies from 19% to 26%. The higher porosity bricks are useful for the construction of furnace roofs because they are lighter. The bricks' cold crushing strength is low at 27.6 MPa. They get easily damaged at their edges and corners. Their thermal conductivity is lower than that of magnesite bricks. Spalling resistance is excellent in high porous bricks. Refractoriness is also high and RUL values vary. At worst, these bricks are not very satisfactory. The best ones deform at 1600°C.

Excellent slag resistance is obtained against basic slag and iron oxide. The latter can be absorbed to a great extent, which can lead to growth of the brick. Excessive growth can lead to the brick's failure. If the furnace gas contains a high percentage of carbon monoxide and if the brick is high in Fe_2O_3 content, carbon deposition can take place at 450°C. Carbon deposition also leads to brick growth. Finally, spalling can take place.

In bricks containing FeO, the spalling takes place in the following manner. At the high temperatures that prevail at the hot face, the bond phase will be a molten state. This is because the brick's melting point is only 1400°C if it is high in CaO. The pores then act as capillaries to draw in the liquid. That 1400°C temperature is at a depth of about 4 cm from the hot face. The drawn-in liquid freezes below this thickness causing densification of the brick in that zone. FeO in the brick is absorbed by the coarse chromite particles. This causes expansion of these particles, which in turn causes expansion of the brick. This expansion of 4 cm from the hot face is accommodated in the vacant space created by the bond phase. But the expansion of the densified region by the freezing of the liquid cannot be accommodated. Therefore, cracks form and 4 cm thick pieces separate from the bricks. Thus structural spalling takes place.

The remedy for preventing structural spalling is to keep the CaO content in the bond phase below 1%. This can be done by adding pure magnesite while making these bricks.

Properties of magnesite-chrome bricks are similar to those of chrome-magnesite bricks. But the former ones are less liable to evidence structural spalling. These are the most highly developed basic bricks.

24.3.4 Dolomite

As for all basic bricks, refractoriness of dolomite is also very high—above 1750°C. The RUL is 1450°C to 1550°C for stabilized dolomite and 1350°C to 1450°C for semi-stable dolomite. The spalling resistance of stabilized dolomite is poor, but the resistance of semi-stable dolomite is moderately good. Semi-stable dolomite can withstand 20 cycles. Slag resistance is poor for both the types of dolomite. In order to increase this resistance, dolomite is doped with magnesia, which combines with silica to form forsterite. Forsterite has better resistance to slag. Semi-stable dolomite has better resistance than stabilized dolomite.

24.3.5 Forsterite

In olivenes, FeO is an impurity. This impurity tends to reduce the refractoriness of forsterite bricks. The iron content is adjusted to get a refractoriness of 1750°C. The RUL value minimum is 1550°C. Thermal expansion is low, which results in a moderately good spalling resistance. Fosterite is resistant to both acid and basic slags. Its resistance to acid slag is not as much as that of silica, and its resistance to basic slag is not as much as that of chrome-magnesite.

24.4 Uses

24.4.1 Chromite

Chromite is available as a natural refractory. Therefore, it does not require any conversion process for its use. This advantage is exploited in its application—most importantly as a single neutral course between the basic walls and acid roof of a regular open hearth furnace. It is also used to protect silica at the ports of an acid open hearth furnace. Some soaking pits use chromite bricks at the bottom to protect firebrick from FeO attack.

24.4.2 Magnesite

Magnesite bricks are used in basic electric and open hearth furnaces under the hearth, walls, and roofs. Mixers are lined with these bricks. Their slag resistance makes them useful for top runs of checkerwork. Here, the bricks are made resistant to spalling.

Magnesite monolithic walls may be made with mild steel metal casing. On firing, steel oxidizes and becomes integrated with the refractory. Integrated tubing is also made this way. First periclase is rammed onto the tube wall. Afterward, the composite tubing is heated to 1400°C. The structure becomes very strong without the problem of spalling. Kiln walls have been built this way. Later, bricks could be fired in these kilns. Similarly, electric furnace walls have been built in this way.

24.4.3 Chrome-Magnesite and Magnesite-Chrome

Chrome-magnesite and magnesite-chrome bricks had been widely used in basic open hearth furnaces, when those furnaces were in vogue. These bricks were used for gas uptakes, burner zones, back and front walls, and for the roofs.

Other uses of chrome-magnesite and magnesite-chrome bricks are in electric steel furnaces. They may also be used in the hearths of soaking pits and in reheating furnaces. In these furnaces, FeO attack is severe. The roofs of copper reverberatory furnaces are made out of chrome-magnesite and magnesite-chrome bricks. Magnesite-chrome bricks are used to construct the roofs of cement kilns.

Chrome-magnesite and magnesite-chrome are placed in position in one of three ways. One way is by the use of well-shaped bricks that are cemented together in the usual manner. The second way is to keep the material in-between steel sheets. During heating, the steel sheets oxidize and become attached to the refractory. The third way is to encapsulate the material in metal cases, which are then stacked together. The suspended roofs of open hearth furnaces had been made with metal-cased, chemically bonded, magnesite-chrome bricks. For suspension, the hangers were attached in the form of steel rods that were embedded in the bricks. The embedded portion of the rod gets chemically bonded to the refractory on oxidation to FeO. The hook part is at the cold end and is used for suspension.

24.4.4 Dolomite

Dolomite's main application was in the open hearth furnace. This furnace became outdated in the 1960s and has been replaced by Linz-Donawitz and basic oxygen steel converters. In these furnaces, dolomite lining is made up of unfired blocks that are 60 cm long and made to fit precisely within the converters. The blocks are made from crushed and graded doloma. Tar is used for temporary bonding between particles. The density of these blocks is not as high as that of the dolomite bricks. During the converter usage, the tar is burned out and the firing of the bricks takes place. After firing, the structure achieved is the same as that of dolomite bricks. These blocks are backed by magnesite bricks. After about five days of use, the lining is replaced with a new set of unfired blocks. This is the reason why these unfired blocks are not stabilized and are of large size. They are inexpensive because they are less dense and are unstabilized. Quick lining work is made possible by making the bricks large in size and an exact fit. Stabilized bricks are used for occasions when long-term storage or transportation is needed.

After L-D converters, the next most popular use of dolomite is in electric arc furnaces. Here, it replaces costlier magnesite in less severe regions. Dolomite is also used for ladle linings and stopper sleeves. Similar to magnesite, dolomite is crushed to pea size and mixed with tar. This mixture is then rammed to make basic hearths.

A working hearth can be made from crushed magnesite in situ. After crushing, the magnesite is milled to get particles of about 6 mm in size. These are then mixed with tar and rammed onto a template. Afterward, the tar is burned out.

24.4.5 Forsterite

Forsterite has been applied to open hearth back walls with moderate success. It has also been used to construct the downtakes and the top courses for the same furnaces. Presently, its most important application is in copper smelting.

Bibliography

J. D. Gilchrist, *Fuels, Furnaces and Refractories*, Pergamon Press, Oxford, UK, 1977, pp. 284–296.

25

Carbon

25.1 Manufacture

Carbon is available in two varieties: *carbon* and *graphite*. Coke is the raw material for both varieties and is obtained from coal or petroleum. The petroleum coke is more pure and therefore is preferred for making carbon bricks. First, the coke is crushed, and then different grades are separated based on particle size. The required sizes are then mixed with tar or pitch. The shape is made by either extrusion or molding. Carbon bricks are then produced by firing at about 1000°C. The tar in the mixture chars. This char binds the coke particles together. When the coke is fired at 2500°C, we get graphite. This much higher temperature is achieved in a resistance furnace. Here, coke and char develop the crystal structure of graphite. The brick becomes a polycrystalline material.

25.2 Characteristics

Graphite brick exceeds natural graphite in strength and density. The thermal and electrical conductivities are greater than that of a carbon block. Both graphite and carbon have excellent hot strength. Because of their high thermal conductivity, they have good resistance to thermal spalling. They are not wetted by liquid metals or slags. Hence, they are not attacked by these liquids. Only oxidizing environments—such as oxidizing slags and gases—attack them. The latter are air, carbon dioxide, and steam. Graphite is more resistant than carbon to these environments. In order to protect the brick from these environments, a layer of silicon carbide can be added. This can be accomplished by heating the bricks in the presence of silicon tetrachloride. Metals such as iron can attack carbon and graphite, producing stable carbides.

Carbon and graphite bricks can be machined easily. This enables them to be made into complex shapes.

25.3 Uses

Carbon and graphite refractories are used as electrodes in arc furnaces. In aluminum extraction, these electrodes conduct current to the electrolyte. For high current density, graphite is more preferable. This is because this graphite possesses greater conductivity. Where higher resistivity is required, carbon is a better candidate. Graphite is costlier than carbon. Carbon rods, tubes, and granules are used as resistors in electric resistance furnaces.

Iron blast furnaces also use carbon bricks or blocks. This is the principal application of this material. Entire blast furnaces have been lined with carbon. Its use in hearths is well known. Because of its high thermal conductivity, when carbon is used in a hearth, the heat is conducted down to the foundations. The heat developed is extracted through ducts in the hearth. Air is blown through these ducts.

Carbon bricks are also used as linings for furnaces making phosphorus, calcium carbide, aluminum, and magnesium. In some arc furnaces, the arc is struck between the electrodes and the hearth. In those cases, carbon bricks are used to make the hearth conduct heat.

Plumbago is a material made from fireclay and natural graphite. It is used to make crucibles for melting cast iron and other metals. Plumbago's advantageous properties are its high thermal conductivity and the fact that it is nonwetting to metals. A problem with plumbago is a gradual oxidation of graphite with continued use. But it is inexpensive to use under moderately severe conditions.

There are many other uses for carbon in laboratories and in industries. Molds and plungers for hot pressing metallic powders are made from carbon, as are laboratory crucibles. It is a resistor in carbon tube furnaces and is used as an inductor in high frequency furnaces. Carbon is used in rocket engines and in heat transfer systems. Graphite is used as a moderator in some nuclear reactor cores.

Bibliography

J. D. Gilchrist, *Fuels, Furnaces and Refractories*, Pergamon Press, Oxford, UK, 1977, pp. 296–299.

26

Insulating Refractories

26.1 Introduction

Generally, all refractories are insulating materials. But, if we want to use them specifically as insulating refractories, some changes will be needed. Only certain refractories will be receptive to these changes. Those refractories are classified under the category of insulating refractories.

Properties of insulation are imparted to a material by reducing heat transfer through all three modes—conduction, convention, and radiation. Reduction of conductive heat transfer is achieved by selecting a material of low thermal conductivity. Because ceramic materials are generally nonconducting at room temperature, they are usually selected. Gases have even less conductivity, but they can conduct heat by convection. To reduce this mode of heat transfer, the thickness of gas enclosures must also be reduced. In large spaces, this is achieved by separating the space into thin compartments, which reduces the circulation of the gas. Air is the most commonly used gas for insulation purposes. The radiation mode of heat transfer is reduced by radiation shields that are in the form of thin sheets. They function by reflecting the heat back to the source. One shield can reflect back 50% of the heat it absorbs. Two shields in parallel can reflect back 67%, and three can reflect back 75%. Thus, as the number of shields increases, the less heat is absorbed.

An ideal insulator will reduce all three modes of heat transfer to a minimum. An ideal refractory insulator has a honeycomb structure, where the individual cells are minute in size. The cells should be constructed with walls that are not very thick. The material selected should have low thermal conductivity. Porous firebrick is the least expensive insulating refractory. It is useful up to 1500°C.

26.2 Manufacture

In order to make porous fireclay, fireclay powder is mixed with hardwood sawdust. If larger sized pores are required, chips of hardwood can be used.

The mixture is molded and fired. During firing, the sawdust and chips are burned out, leaving pores in the brick.

26.3 Characteristics

The RUL value of porous fireclay is not good. The conductivity of insulating firebrick is 0.3349 W/mK at 700°C, whereas it is 0.8374 W/mK for the normal one. Refractory materials such as kieselguhr, asbestos, and slag wool have lower thermal conductivity than insulating firebrick. They are useful only at temperatures lower than 900°C.

26.4 Uses

Porous fireclay bricks are used to make laboratory furnaces. They can be cut with a hacksaw to the required shape. In furnace construction, low temperature insulating refractories such as slag wool or glass wool are put in a thick blanket of refractory wool on the inside of the furnace casing. Then the bricks are laid on the inside of this blanket. Thus, the hot face is that of the refractory bricks, and the blanket provides sufficient insulating property. For construction of laboratory furnaces, refractory wool made from kaolinite is used. It is very refractory and can be placed behind radiant heaters. Its high reflectivity for both heat and light enables it to reflect most of the heat falling on it. This reflected heat is then thrown to the furnace burden. The more heat that falls on the burden, the faster it heats.

Refractory powders such as alumina, zirconia, and carbon provide very high insulation in laboratory furnaces. Here, the porosity size and amount of the bulk powder are more important than the conductivity of the material as far as the insulating property is concerned. As the size of the pores decreases, the heat transfer due to convection decreases. As the amount of porosity increases, the conduction mode of heat transfer decreases. When using these powdery refractory insulating materials, the particles should not get sintered or they will lose these two important properties. Sintering will be less if the raw material is very pure. Carbon is useful under reducing conditions or in vacuum.

Bibliography

J. D. Gilchrist, *Fuels, Furnaces and Refractories*, Pergamon Press, Oxford, UK, 1977, pp. 300–302.

Appendix 1: Solutions to Problems

SECTION I

Chapter 5

5.1 $\Delta G_{1600}{}^0 = \Delta G_{f,1600}{}^0 (Al_6Si_2O_{13}) + (3/2) \Delta G_{f,1600}{}^0 (Mg_2SiO_4)$

         $- 3\Delta G_{f,1600}{}^0 MgAl_2O_4 - (7/2) \Delta G_{f,1600}{}^0 SiO_2$

         $= 4769.4 + (3/2) \times 1493.7 - 3 \times 1619.6 - (7/2) \times 590.6$

         $= -84.0 \ kJ.mol^{-1}$

SECTION II

Chapter 8

8.1 K^+ ion radius/O^{2-} ion radius = 1.38/1.40 = 0.986

     Therefore, the structure is cubic

8.2

     8.2.1   [2(R + r)] = 2(1.81 + 1.02) = 5.66 A.U.

                                 = lattice constant

     8.2.2   Mass of a NaCl unit cell = $35.453 \times 4 + 22.989 \times 4$

                                     = 233.768

                                     = $233.768 \times 4/6.022 \times 10^{23}$ gm

                                     = 155.250×10^{-26} kg

        Volume of the unit cell = $(5.66 \times 10^{-10})^3 = 181.321 \times 10^{-30} \ m^3$

        Therefore, density of sodium chloride = $155.250 \times 10^{-26}/181.321$

                                     $\times 10^{-30} = 8.562 \times 10^3 \ kg/m^3$

     8.2.3   [$\sqrt{2} \times a$] = 1.81×4

        Therefore, a = 5.12 A.U.

        Therefore, maximum radius of the cation = $(5.12 - 2 \times 1.81)/2$

                                           = 0.75 A.U.

8.2.4   Mass of K^+ = 39.098 a.m.u

Therefore, increase in the mass of the unit cell

$$= (39.098 - 22.989) \times 4 \times 0.0001/6.022 \times 10^{23} = 1.07 \times 10^{-27} \text{ gm}$$

Radius of K^+ = 1.38 A.U.

Therefore, increase in the volume of the unit cell

$$= [(1.38 - 1.02) \times 10^{-8}]^3 \times 10^{-4} \text{ cm}^3$$

Therefore, increase in the density of the unit cell

$$= 1.07 \times 10^{-27}/4.66 \times 10^{-30} = 2.30 \times 10^{-4} \text{ gm/cm}^3$$

Therefore, density of 0.01 mole % KCl solid solution in NaCl

$$= 8.562 + 0.0002$$

$$= 8.5622 \text{ gm/cm}^3$$

8.3   The anion packing in CaO is f. c. c. It has a rock salt structure.

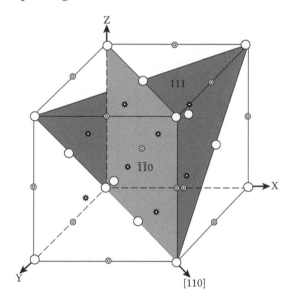

○ Lattice site   ◎ Octahedral site   ✿ Tetrahedral site

8.4   In cubic close packing, there are four and eight octahedral and tetrahedral sites, respectively. Therefore, the ratio of octahedral sites to O^{2-} ions = 1, and the ratio of tetrahedral sites to O^{2-} ions = 2.

8.5   Pauling's first rule says that a coordination polyhedron is formed around each cation in the structure. The second rule states that in a stable structure, the total strength of the bonds reaching an anion from all surrounding cations should be equal to the charge of the anion.

8.5.1 In cubic close packing, there are four octahedral sites, and there are four lattice points. Each cation will be surrounded by six anions and vice versa. Total bond strength reaching an anion = valency of the cation/coordination number = $x/6$

Total strength of the bonds reaching O^{2-} = $6x/6$

$$= 2$$

Therefore, $x = 2$

The example is CaO.

8.5.2 There are eight tetrahedral sites in cubic close packing. This means that each anion will be surrounded by eight cations.

Therefore, $8x/4 = 2$

Therefore, $x = 1$

The example is Li_2O.

8.5.3 Each anion will be surrounded by three cations.

Therefore, $3x/6 = 2$

Therefore, $x = 4$

The example is TiO_2.

8.5.4 Each anion will be surrounded by four cations.

Therefore, $4x/4 = 2$

Therefore, $x = 2$

The example is ZnS.

8.6 The structure is f. c. c. Ba^{2+} at cube corners, Ti^{4+} ions at the cube center, and O^{2-} at the face centers.

The nearest neighbors to the cube corner are the face centers. Hence, CN of Ba^{2+} is 12.

The nearest neighbors to the cube center are the face centers. Hence, CN of Ti^{4+} is 6.

The nearest neighbors to the face center are the cube centers. Hence, CN of O^{2-} is 2.

8.7

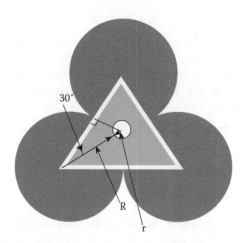

$\cos 30° = R/(R + r)$

$\quad = \sqrt{3}/2$

$\quad = 1.732/2$

Therefore,

$\quad 2R = 1.732\ R + 1.732\ r$

$0.268\ R = 1.732\ r$

Therefore,

$\quad r/R = 0.268/1.732$

$\quad\quad = 0.155$

8.8 Radius of B^{3+}/Radius of O^{2-} = 0.20/1.40 = 0.143

Therefore, the coordination number for B^{3+} = 2

8.9 For tremolite, the total number of O^{2-} ions per molecule = 22

Number of nonbridging O^{2-} ions = 2 + 4 + 10 = 16

O/Si ratio = [16 + (6/2)]/8 = 19/8 ≈ 2.4

For talc, the total number of O^{2-} ions per molecule = 10

Number of nonbridging O^{2-} ions = 2 + 6 = 8

O/Si ratio = [8 + (2/2)]/4 = 9/4 ≈ 2.3

8.10 $E_{latt} = (N_{Av}z_1z_2e^2/4\pi\varepsilon_0r_0)\ (1 - 1/n)\ [(6/1) - (12/\sqrt{2}) + 8/\sqrt{3} - (6/\sqrt{4}) + \cdots]$

$\quad r_0 = (r)\ K^+ + r\ (Cl^-)$

$\quad\quad = 152\ \text{pm (for CN = 6)} + 167\ \text{pm}$

$\quad\quad = 319\ \text{pm}$

Therefore,

$$E_{latt} = (6.02 \times 10^{23})\ (-1)\ (1)\ (1.6 \times 10^{-19})^2\ (1 - 1/\infty)\ \alpha/(4 \times 3.14)$$

$$\times (8.85 \times 10^{-12}) \times (319 \times 10^{-12})$$

$$= -4.35 \times 10^5 [(6/1) - (12/\sqrt{2}) + 8/\sqrt{3} - (6/\sqrt{4}) + \cdots]$$

8.10.1 $E_{latt} = -4.35 \times 10^5 \times 6$

$$= -26.1 \times 10^5$$

$$= -2600\ kJ/mol$$

8.10.2 $E_{latt} = -4.35 \times 10^5 \times [6 - 12/1.414]$

$$= -4.35 \times 10^5 \times 2.49$$

$$= 10.81 \times 10^5$$

$$= 1081\ kJ/mol$$

8.10.3 $E_{latt} = -4.35 \times 10^5 \times [6 - (12/1.414) + 8/\sqrt{3}]$

$$= -4.35 \times 10^5 \times [6 - 8.49 + 4.62]$$

$$= 926\ kJ/mol$$

8.11 $E_{att.} = (z_1 z_2 e^2/4\pi\varepsilon_0 r);\ z_{Ca2+} = 2,\ z_{O2^-} = -2,\ e = 1.6 \times 10^{-19}C,\ \varepsilon_0$

$$= 8.85 \times 10^{-12} C^2/J.\ m$$

$$= -2 \times 2 \times (1.6 \times 10^{-19})^2/(4 \times 3.14) \times (8.85 \times 10^{-12})\ r$$

$$= -(0.09 \times 10^{-26}/r)$$

$$= -(9.21 \times 10^{-19}/r)\ J.\ m,\ r\ in\ nm$$

$E_{rep.} = (B/r^n)$

$$= 0.4 \times 10^{-24}/r^9\ J,\ r\ in\ nm$$

$E_{net} = (z_1 z_2 e^2/4\pi\varepsilon_0 r) + (B/r^n)$

$$-4.61 \times 10^{-18} + 7.8125 \times 10^{-19} = 10^{-19}(-46.1 + 7.8125)$$

$$= -38.2875 \times 10^{-19} = 3.82875 \times 10^{-18}$$

$$= -9.21 \times 10^{-19}/r + 0.4 \times 10^{-24}/r^9$$

r (nm)	$E_{latt.}$ (J), $\times 10^{-18}$	$E_{rep.}$ (J), $\times 10^{-18}$	E_{net} (J), $\times 10^{-18}$
0.20	−4.61	0.78125	−3.8288
0.21	−4.39	0.50360	−3.8864
0.22	−4.19	0.33133	−3.8587
0.23	−4.00	0.22208	−3.7779
0.24	−3.84	0.15141	−3.6886
0.25	−3.68	0.10486	−3.5751

Continued

r (nm)	$E_{latt.}$ (J), $\times\ 10^{-18}$	$E_{rep.}$ (J), $\times\ 10^{-18}$	E_{net} (J), $\times\ 10^{-18}$
0.26	−3.54	0.07367	−3.4663
0.27	−3.41	0.05245	−3.3576
0.28	−3.29	0.03781	−3.2522
0.29	−3.18	0.02757	−3.1524
0.30	−3.07	0.02032	−3.0497

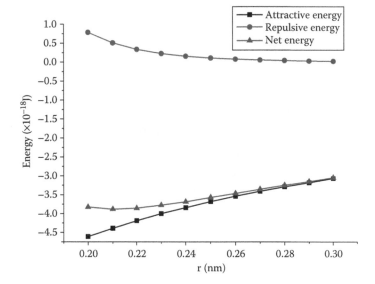

8.12   CsCl has simple cubic structure. Considering the simple cubic made up of Cl⁻ ions, the Cs⁺ ions are located at body centers. For such an arrangement, there will be 8 Cl⁻ ions surrounding a Cs⁺ ion as its first nearest neighbors at a distance of, say, r_0. There will be 6 Cs⁺ ions at a distance of $2r_0/\sqrt{3}$ as its second nearest neighbors. And there will be 24 Cl⁻ ions as its third nearest neighbors at a distance of, say, $\sqrt{11}r_0/\sqrt{3}$. The distances are shown in the following figure.

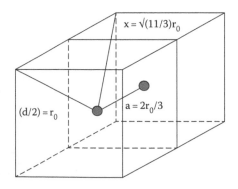

Therefore,

$\alpha = 8 - (6\sqrt{3}/2) - 24\sqrt{3}/\sqrt{11}$

$\quad = 15.31$

8.13

8.13.1 Ba^{2+} corresponds to radon, and O^{2-} to neon.

8.13.2 BaO has a rock salt structure. Therefore, the coordination numbers of both Ba^{2+} and O^{2-} are 6. The ionic radii for this coordination are 149 pm and 126 pm, respectively.

Therefore, the interionic distance = 149 + 126

$$= 275 \text{ pm}$$

8.13.3 $F = -(z_1 z_2 e^2 / 4\pi\varepsilon_0 r^2)$

$\quad = -[(+2) \times (-2) (1.6 \times 10^{-19})^2 / (4 \times 3.14) \times (8.85 \times 10^{-12}) \times (1 \times 10^{-9})^2]$

$\quad = 4 \times 2.56 \times 10^{-38} / 111.156 \times 10^{-30}$

$\quad = 0.09212 \times 10^{-8}$

$\quad = 9.212 \times 10^{-10} \text{ N}$

Assumption = nil

8.14

$$\text{Electronegativity of K} = 0.82$$
$$\text{Electronegativity of Cl} = 3.16$$

Therefore,

Fraction ionic character of K–Cl bond $= 1 - \exp - (0.82 - 3.16)^2/4$

$$= 1 - \exp -1.3689$$
$$= 0.7456$$

$$\text{Electronegativity of Ca} = 1.00$$
$$\text{Electronegativity of O} = 3.44$$

Therefore,

Fraction ionic character of Ca–O bond $= 1 - \exp[-(1.00 - 3.44)^2/4]$

$$= 1 - \exp(-1.4884)$$
$$= 0.7743$$

$$\text{Electronegativity of Ni} = 1.91$$

Therefore,

Fraction ionic character of Ni–O bond $= 1 - \exp - (1.91 - 3.44)^2/4$

$$= 1 - \exp -0.5852$$
$$= 0.4430$$

$$\text{Electronegativity of Pb} = 1.87$$

Therefore,

Fraction ionic character of Pb–O bond $= 1 - \exp - (1.87 - 3.44)^2/4$

$$= 1 - \exp -0.6162$$

$$= 0.5314$$

Electronegativity of Na $= 0.93$

Electronegativity of F $= 3.98$

Therefore,

Fraction ionic character of Na–F bond $= 1 - \exp - (0.93 - 3.98)^{2/4}$

$$= 1 - \exp -2.3256$$

$$= 1 - 0.0977247915824195$$

$$= 0.9022752084175805$$

Chapter 9

9.1

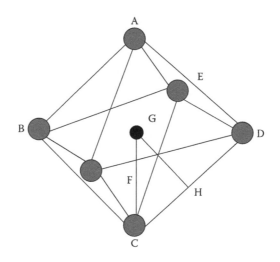

Sin 45° = CH/CG

$\qquad$ = R/(r + R)

$\qquad$ = 1/√2

Therefore,

$\quad$ (r + R) = √2R

$\qquad$ r = 1.414 R – R

$\qquad$ = 0.414 R

$\quad$ r/R = 0.414

9.2

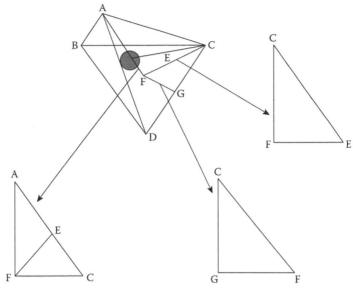

$$\llcorner ECF = 30°$$
$$\tan \llcorner ECF = EF/CE$$
$$= EF/R$$
$$= 1/\sqrt{3}$$
$$EF = R/\sqrt{3}$$
$$\cos 30° = CE/CF$$
$$= R/CF$$
$$CF = R/\cos 30°$$
$$= 2R/\sqrt{3}$$
$$AF^2 + 4R^2/3 = 4R^2$$
$$AF^2 = 4R^2 - 4R^2/3$$
$$= 8\,R^2/3$$
$$EF^2 + CF^2 = CE^2$$
$$[(\sqrt{8}R/\sqrt{3}) - (R + r)]^2 + 4\,R^2/3 = (R + r)^2$$
$$8R^2/3 - (2\sqrt{8}R/\sqrt{3})\,(R + r) + (R + r)^2 + 4R^2/3 = (R + r)^2$$
$$12R^2/3 - (2\sqrt{8}R/\sqrt{3})\,(R + r) = 0$$
$$(2\sqrt{8}R/\sqrt{3})\,r = (12R^2/3) - 2\sqrt{8}R^2/\sqrt{3}$$
$$= (12R^2 - 2\sqrt{24}R^2)/3$$
$$= 2R^2\,(6 - \sqrt{24})/3$$

$$2\sqrt{8}Rr = 2R^2 (6 - \sqrt{24})/\sqrt{3}$$
$$\sqrt{8}r = R (6 - \sqrt{24})/\sqrt{3}$$
$$r/R = 0.225$$

Ratio of the sizes of the tetrahedral to the octahedral sites = 0.225/0.414

$$= 0.543$$

9.3   When oxygen ions are in a hexagonal close packed arrangement, the number of oxygen ions per cell = (1/6) × 12 + (½) × 2 + 3 = 6

   9.3.1   Number of octahedral sites per cell = 6. Therefore, ratio of the octahedral sites to oxygen ions = 1.

   9.3.2   Number of tetrahedral sites per cell = 12. Therefore, the ratio of the tetrahedral sites to oxygen ions = 2.

9.4

   9.4.1   Tetrahedral sites = 8;

             Octahedral sites = 4.

   9.4.2   Octahedral sites to oxygen ions = 4/4

                                            = 1

             Tetrahedral sites to oxygen ions = 8/4

                                            = 2

   9.4.3

        a.   Oxide if one-half of the octahedral sites are filled = MO_2
        b.   Oxide if two-thirds of the octahedral sites are filled = M_2O_3
        c.   Oxide if all of the octahedral sites are filled = MO

   9.4.4

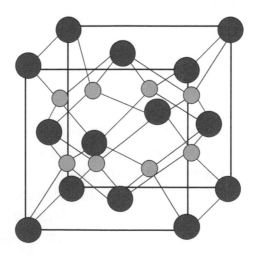

Structure is antifluorite.

Number of O^{2-} ions per cell = 4

Therefore, total negative charges per cell = 8

Number of cations per cell = 8

Therefore, positive charge required per cation = 8/8

= 1

9.4.5

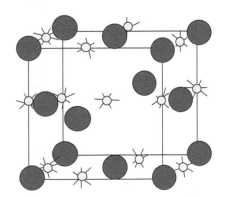

Structure is rock salt.

Number of cations per cell = 4

Therefore, the charge on the cation = 8/4

= 2

9.5

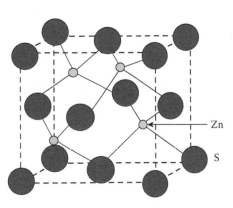

Diamond cubic structure is also same. In it, all the atoms will be carbon.

9.6

9.6.1   Figure same as that given in solution 2.4.4.

Density = Atomic weights × number of atoms per cell/volume of the cell

$$\text{Atomic weight of Li} = 6.941 \times 10^{-3}/6.022 \times 10^{23}$$

$$= 1.153 \times 10^{-26} \text{ kg}$$

$$\text{Number of Li}^+ \text{ ions per cell} = 8$$

$$\text{Therefore, total mass of Li}^+ \text{ ions per cell} = 1.153 \times 10^{-26} \times 8$$

$$= 9.221 \times 10^{-26} \text{ kg}$$

$$\text{Mass of O-atom} = 15.999 \times 10^{-3}/6.022 \times 10^{23}$$

$$= 2.657 \times 10^{-26} \text{ kg}$$

$$\text{Number of O}^{2-} \text{ ions per cell} = 4$$

$$\text{Therefore, total mass of O}^{2-} \text{ ions per cell} = 2.657 \times 10^{-26} \times 4$$

$$= 10.627 \times 10^{-26} \text{ kg}$$

$$\text{Therefore, total mass of all ions per cell} = 9.221 \times 10^{-26} + 10.627 \times 10^{-26}$$

$$= 19.848 \times 10^{-26} \text{ kg}$$

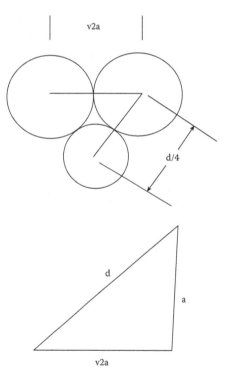

$$\text{CN for Li}^+ = 4$$

$$\text{Ionic radius for Li}^+ \text{ with CN} = 4 \text{ is 73 pm}$$

$$\text{CN for O}^{2-} = 8$$

$$\text{Ionic radius for O}^{2-} \text{ with CN} = 8 \text{ is 128 pm}$$

Therefore, $d/4 = 73 + 128$

$$= 201 \text{ pm}$$

$$a^2 + 2a^2 = d^2$$

That is, $3a^2 = d^2$

Therefore, $a = d/\sqrt{3}$

Volume of the unit cell $= a^3$

$$= d^3/3\sqrt{3}$$

$$= 804^3/5.196$$

$$= 1.000 \times 10^8 \text{ pm}^3$$

$$= 1.000 \times 10^{-28} \text{ m}^3$$

Therefore, density of $Li_2O = 19.848 \times 10^{-26}/1.000 \times 10^{-28}$

$$= 1.98 \text{ g/cm}^3$$

9.6.2 In a tetrahedral interstice, the maximum size of an ion

$$= 0.225 \times \text{radius of lattice ion}$$

$$= 0.225 \times 128$$

$$= 28.8 \text{ pm}$$

9.7

9.7.1

| Cation | Anion | CN | | Ionic Radii, pm | | Ionic Radii Ratio | Cage for Cation |
		Cation	Anion	Cation	Anion		
Ti⁴⁺	O²⁻	4	4	56	124	0.452	Octahedron
-Do-	-Do-	-Do-	6	-Do-	126	0.444	-Do-
-Do-	-Do-	-Do-	8	-Do-	128	0.438	-Do-
-Do-	-Do-	6	4	74.5	124	0.601	-Do-
-Do-	-Do-	-Do-	6	-Do-	126	0.591	-Do-
-Do-	-Do-	-Do-	8	-Do-	128	0.582	-Do-
-Do-	-Do-	8	4	88.0	124	0.710	-Do-
-Do-	-Do-	-Do-	6	-Do-	126	0.698	-Do-
-Do-	-Do-	-Do-	8	-Do-	128	0.688	-Do-
Ba²⁺	-Do-	6	4	149	124	0.832	Cube
-Do-	-Do-	-Do-	6	-Do-	126	0.846	-Do-
-Do-	-Do-	-Do-	8	-Do-	128	0.859	-Do-
-Do-	-Do-	8	4	156	124	0.795	-Do-
-Do-	-Do-	-Do-	6	-Do-	126	0.808	-Do-
-Do-	-Do-	-Do-	8	-Do-	128	0.821	-Do-
-Do-	-Do-	12	4	175	124	0.709	Octahedron
-Do-	-Do-	-Do-	6	-Do-	126	0.720	-Do-
-Do-	-Do-	-Do-	8	-Do-	128	0.731	-Do-

From this table, the most suitable cage for Ti is octahedron.

9.7.2   With respect to Ba^{2+} and O^{2-}, Ba^{2+} forms the polyhedron because it is the larger ion. And O^{-2} has to bond with both Ba^{2+} and Ti^{4+}. So, O^{-2} will be found in a cage formed by its nearest neighbors of Ba^{2+} and Ti^{4+}. There are two Ti^{4+} ions for O^{-2} when Ti^{4+} is put in a cage of O^{-2} forming an octahedron. Hence, there can be four Ba^{2+} ions as the other four nearest neighbors if O^{-2} is to be located inside an octahedron. Such an arrangement is possible in the case of a structure in which Ba^{2+} forms the corner ions, O^{-2} forms the face-centered ions, and Ti^{4+} forms the body-centered ions. The structure is shown in this figure:

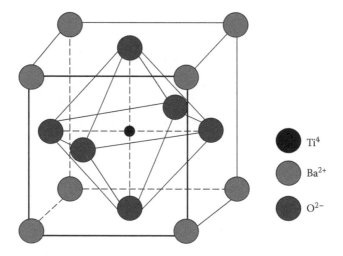

9.7.3   Unit cell contains one Ti^{4+}, three O^{-2}, and one Ba^{2+} ions.

9.8   In FCC structure, there are two kinds of interstices, tetrahedral and octahedral.

9.8.1   Radius of Be^{2+} for tetrahedral coordination = 41.0 pm

When Be^{2+} is in a tetrahedral interstice, O^{2-} will be at the center of a cube. That means, there will be eight nearest Be^{2+} neighbors for O^{2-}.

The ionic radius for O^{2-} for such a coordination = 128.0 pm

Therefore, ionic radii ratio = 41.0/128.0 = 0.320

Because this ratio is greater than 0.225 and less than 0.414, Be^{2+} will occupy the tetrahedral interstice.

Radius of Be^{2+} for octahedral coordination = 59.0 pm

When Be^{2+} is in octahedral interstice, O^{2-} will have 2 as the coordination number.

The ionic radius for O^{2-} for such a coordination = 121.0 pm

Therefore, ionic radii ratio = 59.0/121.0 = 0.488

Because this ratio is greater than 0.414 and less than 0.732, Be^{2+} can occupy the octahedral interstice also. But the number of such interstices in FCC structure is only one per unit cell. In order to satisfy the formula, there should be four such interstices. Because there are eight tetrahedral interstices per cell, Be^{2+} will occupy the tetrahedral interstice.

9.8.2   Half the available interstitial sites will be occupied.

9.9   Consider the simple cubic structure for S^{2-} ions. If all the body-centered interstices are filled, then the CN for both Cd^{2+} and S^{2-} ions will be 8. Then,

Ionic radius of Cd^{2+} for CN 8 = 124 pm

Ionic radius of S^{2-} for CN = 8 ~ 181 pm

Therefore, ratio of ionic radii = 124/181 = 0.685 (not valid for cubic structure).

Consider the base centered cubic (BCC) structure for anion lattice. This structure has three tetrahedral and three octahedral sites per unit cell. Consider the tetrahedral site for Cd^{2+} ions.

For CN = 4, the ionic radius for Cd^{2+} ion = 92.0 pm

The CN for S^{2-} will be 6. Hence, the ionic radius = 170.0 pm

Therefore, ionic radius ratio = 92.0/170.0 = 0.541 pm

Hence, Cd^{2+} cannot be in the tetrahedral interstice. Consider the octahedral interstice.

For CN = 6, the ionic radius for Cd^{2+} ion = 109.0 pm

In this structure, the CN for S^{2-} will be 12.

Ionic radius for S^{2-} ion = ~ 189.0 pm

Therefore, ionic radii ratio = 109/189 = 0.577

Therefore, Cd^{2+} could be located in the octahedral interstice.

Face diagonal length, d = $\sqrt{2}a$

Therefore, a = $d/\sqrt{2}$ = (109 + 189) × 2/1.414 = 421 pm

Volume of the unit cell = 421^3 = 7.471 × 10^{-29} pm^3

$$= 7.471 \times 10^{-23} \text{ cm}^2$$

Mass of Cd^{2+} ion = ~ (112.41/6.022 × 10^{23}) g

Mass of S^{2-} ion = ~ (32.06/6.022 × 10^{23}) g

In BCC structure, there are two lattice ions per cell. Therefore, there should be two Cd^{2+} ions.

Mass of total ions per cell = [(112.41/6.022 × 10^{23}) + (32.06/6.022 × 10^{23})] × 2

$$= 288.94/6.022 \times 10^{23} \text{ g}$$

Therefore, density of CdS = (288.94/6.022 × 10^{23})/7.471 × 10^{-23}

$$= 6.422 \text{ g/cm}^3$$

Because this density is not equal to the actual density, the structure of CdS is not BCC. Now let us consider FCC structure. This structure has eight tetrahedral interstices and four octahedral interstices. First, consider the tetrahedral interstice.

Ionic radius of Cd^{2+} for tetrahedral coordination = 92.0 pm

The coordination number for S^{2-} in this case will be 8.

Ionic radius of S^{2-} for CN = 8 ~ 181 pm

Therefore, ionic radii ratio = 0.508 (but not for tetrahedral coordination).

Now consider the octahedral interstice.

Ionic radius of Cd^{2+} for CN 6 = 109.0 pm

The coordination number for S^{2-} in this case again will be 8.

Ionic radius of S^{2-} for CN 8 = 181.0 pm

Therefore, ionic radii ratio = 0.602 (valid for octahedral coordination).

Mass of total ions per unit cell = [(112.41/6.022 × 10^{23})

$$+ (32.06/6.022 \times 10^{23})] \times 4$$

$$= 577.88/6.022 \times 10^{23} \text{ g}$$

a = d/√3 = (109 + 181) × 2/1.732 = 334.873 pm

Therefore, volume of the unit cell = 3.755 × 10^7 pm^3 3.755 × 10^{-23} cm^3

Therefore, density of CdS = (577.88/6.022 × 10^{23})/3.755 × 10^{-23}

$$= 25.56 \text{ g/cm}^3 \text{ (not the actual density).}$$

These show that CdS will not have a cubic crystal structure.

9.10   Density of unit cell = Mass of unit cell/Volume of unit cell

$$2.1 = \text{Mass of unit cell}/(0.57 \times 10^{-7})^3$$

Therefore, mass of unit cell = 2.1 × 1.85 × 10^{-22} = 3.89 × 10^{-22} g

Total mass of M and X = (28.5 + 30)/6.022 × 10^{23}

$$= 58.5/6.022 \times 10^{23}$$

$$= 9.714 \times 10^{-23} \text{ g}$$

Therefore, the number of molecules of MX per cell

$$= 3.89 \times 10^{-22}/9.714 \times 10^{-23}$$

$$= 4$$

Both NaCl and ZnS can accommodate four molecules per cell. Hence, any one of these two is the possible structure for MX. CsCl can have only one molecule.

9.11 The structure will be based on (O/Network forming atom) ratio.

9.11.1 O/Network forming atom = 22/8 = 2.75 (Double chain structure)

9.11.2 O/Network forming atom = 10/4 = 2.5 (Sheet structure)

9.11.3 O/Network forming atom = 5/2 = 2.5 (Sheet structure)

9.12 Ionic ratio = 49/154 = 0.318 (Tetrahedron)

Ratio of number of cations to number of anions per cell = 2 = Ratio of number of tetrahedral interstices to number of lattice ions per cell.

Therefore, the structure is FCC, in which A ions are in the lattice sites and B ions are in the tetrahedral interstices.

9.13 $(SiO_4)^{3-}$, and $(SiO_4)^{2-}$

9.14 Non-Bridging Oxygens (NBO) per Si atom = Number of charges on the modifying alkali earth ion

$$\text{Number of moles of Si} = 2\eta/y$$

Here, η is the number of moles of alkali earth oxide, and y is the number of moles of SiO_2.

9.15 NBO = $z(\varsigma\eta)/y$, where z = charge on the modifying cation, ς = number modifying cation per molecule of the modifying oxide, η = number of moles of the modifying oxide per formula of the silicate, and y is the number of moles of SiO_2 per formula of the silicate.

$$\text{Therefore, NBO for the given silicate} = [1(2 \times 1)$$

$$+ 2(1 \times 0.5)]/2$$

$$= 1.5$$

Therefore, number of bridging oxygens per Si atom = 4 − 1.5 = 2.5

$$\text{O/Si ratio} = (2.5/2) + 1.5$$

$$= 2.75$$

Therefore, the most likely structure is a double chain.

9.16
$$\text{Molecular formula} = 0.5Na_2O.0.5SiO_2$$

$$\text{Therefore, NBO} = 1 (2 \times 0.5)/0.5 = 2$$

Therefore, number of bridging oxygen = 2

$$\text{Therefore, O/Si ratio} = 2 + 2/2 = 3$$

Therefore, the chains of infinite length would occur for $0.5Na_2O.0.5SiO_2$.

9.17   If X is the number of nonbridging oxygen atoms per Si atom, and Y is
       the number of bridging atoms, then

$$X + 0.5Y = R \qquad (1)$$

and

$$X + Y = 4 \qquad (2)$$

Equation 2 × 0.5 gives

$$0.5X + 0.5Y = 2 \qquad (3)$$

Equations 1 through 3 give

$$0.5X = R - 2 \text{ or}$$

$$X = 2R - 4$$

Substituting the value of X in Equation 2, we get

$$Y = 8 - 2R$$

Chapter 10

10.1   Volume of the unit cell = $0.43^3 = 7.95 \times 10^{-2}$ nm^3

$$= 7.95 \times 10^{-23} \text{ cm}^3$$

Therefore, mass of the unit cell = $5.72 \times 7.95 \times 10^{-23} = 4.55 \times 10^{-22}$ g

Mass of an iron atom = $(55.847/6.022 \times 10^{23})$ g

Mass of an oxygen atom = $(15.999/6.022 \times 10^{23})$ g

Therefore, mass of Fe_yO = $(55.847y/6.022 \times 10^{23})$

$$+ (15.999/6.022 \times 10^{23}) \text{ g}$$

If there is only one molecule per cell, then

$$(55.847y/6.022 \times 10^{23}) + (15.999/6.022 \times 10^{23}) = 4.55 \times 10^{-22}$$

That is, $55.847y + 15.999 = 4.55 \times 10^{-22} \times 6.022 \times 10^{23} = 273.869$

Therefore, y = 4.617

Because y should be a fraction, there should be four molecules
of Fe_yO.

Therefore, $(55.847y + 15.999) \times 4 = 273.869$

Therefore, $55.847 \times 4y = 209.873$

Therefore, y = 0.939

Assumptions:

1. Ferrous oxide is nonstoichiometric.
2. y is a fraction.

10.2

Site fraction = $1.00 - 0.98 = 0.02$

Volume of the unit cell = $0.4352^3 = 8.242 \times 10^{-2}$ nm^3

$$= 8.242 \times 10^{-23} \text{ cm}^3$$

Therefore, mass of the unit cell = $5.7 \times 8.242 \times 10^{-23}$

$$= 46.980 \times 10^{-23} \text{ g}$$

Mass of an iron atom = $(55.847/6.022 \times 10^{23})$ g

Mass of an oxygen atom = $(15.999/6.022 \times 10^{23})$ g

Mass of $Fe_{0.98}O = (55.847 \times 0.98/6.022 \times 10^{23}) + (15.999/6.022 \times 10^{23})$ g

$$= 70.729/6.022 \times 10^{23}) \text{ g}$$

$$= 11.745 \text{ g}$$

Therefore, number of molecules per cell = 46.980×10^{-23}

$$/(70.729/6.022 \times 10^{23})$$

$$= 4$$

10.3 a. V_{Ca}''

b. $V_{Cl}^{\cdot}$

c. $Ca_i^{\cdot\cdot}$

d. Cl_i'

10.4 a. $\frac{1}{2} O_2 (g) + 2M^{2+} \rightarrow 2M^{3+} + O^x{}_O + V''{}_M$

b. $\frac{1}{2} O_2 (g) + 2M^{3+} \rightarrow 2M^{6+} + O^x{}_O + 2V'''{}_M$

Examples: (a) Fe, (b) Cr

10.5 a. When Ca substitutes Zr

```
Ca      O      V
 ↕      ↕
Zr      O      O
CaO     →      Ca_Zr''' + V_O·· + O_O^x
ZrO₂
```

b. When Ca^{2+} goes interstitial

$$Ca^{2+} \rightarrow Ca_i^{\cdot\cdot}$$

10.6 a. $\frac{1}{2} O_2 (g) \rightarrow O_O^x$

b. Null $\rightarrow$ $2V_M''' + 3V_O^{\cdot\cdot}$

c. $Fe_{Fe}^{\cdot\cdot} \rightarrow V_{Fe}'' + 4h\cdot + Fe$

d. Considering only the migration of cations to the interstice, because being smaller in size their migration is easier, we get:

$$Cr_{Cr}^x \quad \rightarrow \quad Cr_i^{\cdot\cdot\cdot} + V_{Cr}'''$$

e. Ca   Ca   V   O   O
$\updownarrow$ $\updownarrow$ $\updownarrow$ $\updownarrow$
Fe   Fe   O   O   O
2CaO $\rightarrow$ $2Ca_{Fe}^{\cdot\cdot} + V_O^{\cdot\cdot} + 2O_O^x$
     Fe_2O_3

f. K   K   O   V
$\updownarrow$ $\updownarrow$ $\updownarrow$
Co   Co   O   O
K_2O $\rightarrow$ $2K_{Co}^{\cdot\cdot} + V_O^{\cdot\cdot} + O_O^x$
     2CoO

g. $\frac{1}{2} O_2 (g) \quad \rightarrow \quad O_O^x$

h. ZnO $\rightarrow$ $V_{Zn}'' + Zn + O_O^x + 2h\cdot$

i. Considering only the migration of cations to the interstice, because their migration is easier being smaller in size, we get:

$$Al_{Al}^x \rightarrow \quad Al_i^{\cdot\cdot\cdot} + V_{Al}'''$$

j. Mg   Mg   V   O   O
$\updownarrow$ $\updownarrow$ $\updownarrow$ $\updownarrow$
Al   Al   O   O   O
2MgO $\rightarrow$ $2Mg_{Al}' + V_O^{\cdot\cdot} + 2O_O^x$
     Al_2O_3

k. Li   Li   V   O
$\updownarrow$ $\updownarrow$ $\updownarrow$
Ni   Ni   O   O
Li_2O $\rightarrow$ $2Li_{Ni}' + V_O^{\cdot\cdot} + O_O^x$
     2NiO

l. Ba   Cl   V   Cl
$\updownarrow$ $\updownarrow$ $\updownarrow$
K   Cl   K   Cl
$BaCl_2$ $\rightarrow$ $Ba_K^{\cdot} + V_K^{\cdot\cdot} + 2Cl_{Cl}^x$
     2KCl

m. Ba   Cl   Cl

    $\updownarrow$ $\updownarrow$ $\updownarrow$

    Ba   Br   Br

    $BaCl_2$ $\rightarrow$ $Ba_{Ba}{}^{x} + 2Cl_{Br}{}^{x}$

                   $BaBr_2$

n. K   Cl   V

    $\updownarrow$ $\updownarrow$

    Ba   Cl   Cl

    KCl $\rightarrow$ $K_{Ba}{}' + Cl_{Cl}{}^{x} + V_{Cl}{}^{\cdot}$

                   $BaCl_2$

o. Crystalline CaO $\rightarrow$ $V_{Ca}{}'' + V_O{}^{\cdot\cdot}$

p. Considering only the migration of cations to the interstice, because their migration is easier being smaller in size, we get:

$$Cu_{Cu}{}^{x} \quad \rightarrow \quad Cu_i{}^{\cdot} + V_{Cu}{}'$$

q. $Pb_{Pb} + Sn_{Sn} \quad \rightarrow \quad Pb_{Sn} + Sn_{Pb}$

r. O_2 can either substitute O^{2-} ions in the lattice or go to the interstice.

    i.   When it substitutes:

         V   V   O   O   O

         Al   Al   O   O   O

         $(3/2)O_2$ $\rightarrow$ $2V_{Al}{}''' + 3O_O{}^{x} + 6h^{\cdot}$

                        Al_2O_3

    ii. When O_2 goes to interstice:

         $\tfrac{1}{2}O_2 \quad \rightarrow \quad O_i{}^{x}$

Chapter 11

11.1

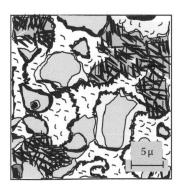

a. Microstructure when fired to achieve phase equilibrium (1450°C for 6 h)

b.   Microstructure when fired to 1300°C for 1 h

11.2 Determination of fractional porosity

   i.   Fractional porosity = 1 – (bulk density/true density)

       Bulk density = Mass of the sample/bulk volume of the sample

       Bulk volume of a regular-shaped sample is obtained from its dimensions.

  ii.   Bulk volume of irregular-shaped sample can be determined by any one of the following ways:

     a.   Measure the weight of the mercury displaced by the sample with a mercury volumeter. Knowing the density of mercury, the volume displaced can be calculated, which will be the same as the bulk volume of the sample.

     b.   Find the force needed to submerge the sample in mercury. This force is proportional to the bulk volume of the sample. The proportionality constant can be determined from the measurements using two samples. Once it is determined, the bulk volume can be calculated.

     c.   Coat the sample with paraffin. Find the weight of the coated sample. Subtract the weight of the sample from this weight to get the weight of the coating. Find the volume of the coating from the knowledge of the density of paraffin. Find the volume of the coated sample by Archemede's method. Subtract the volume of the coating from this volume to get the bulk volume of the sample.

Fractional porosity can also be determine in the following manner:

   i.   Heat the sample in boiling water for 2 h to fill all the open pores completely with water.

ii.   Cool the sample to room temperature.

iii. Find the weight of the sample, and mark it as the saturated weight, W_s.

iv. Suspend the sample in water, and get the suspended weight, W_{sus}.

v.   Find $W_s - W_{sus}$, which gives the bulk volume.

Determination of fractional glass content, and fractional crystal content Polish and etch the sample. Determine the fractional glass content and fractional crystal content by areal analysis, lineal analysis, or point counting method.

11.3   From the micron marker in the figure, we find that 9 mm = 100 μ.

Using a graduated length of 4.5 cm, the following linear intercept lengths of the pores lying on this length were obtained:

Trial Number	Linear Intercept length, mm	Trial Number	Linear Intercept length, mm	Trial Number	Linear Intercept length, mm
1	7	6	8	11	10
2	10	7	3	12	6
3	5	8	5	13	6
4	6	9	6	14	5
5	4	10	9	15	9

Average value of the intercept lengths = 6.6 mm

Therefore, fraction of porosity = 6.6/45 ≈ 0.15

The diameters of 15 pores measured are tabulated here:

Trial Number	Pore Dia, mm	Trial Number	Pore Dia, mm	Trial Number	Pore Dia, mm
1	5	6	4	11	2
2	4	7	2	12	3
3	2	8	1	13	2
4	2	9	2	14	2
5	1	10	3	15	3

Therefore, average pore radius ≈ (2.5/2*9)*100 ≈ 28 μ

Chapter 12

12.2 $\Delta G° = -RT \ln K = -RT \ln p_{CO_2}$

   $= -8.31467 \times T \ln (3 \times 10^{-3} \times 1.013 \times 10^5)$

   $= -8.31467 \times T \ln (3.039 \times 100)$

   $= -47.5325T = 182.50 - 56.16T$

That is, $T = 182.5/8.6275 \approx 21$ K

12.3   Critical speed $= (g/a)^{1/2}/2\pi = (981/5)^{1/2}/(2 \times 3.14)$

   $= 2.23$ revolutions per second

Operating speed $= 2.23 \times 0.75 = 1.25$ revolution per second

12.4   Jander equation is given by:

$$[1-(1-\alpha)^{1/3}]^2 = Kt/r^2 \qquad (1)$$

where α is the fraction of the volume that has already reacted.

For the first 30 min reaction, it is given by:

$$[1 - (1 - 0.25)^{1/3}]^2 = 30K/1^2$$

That is, $8.464 \times 10^{-3} = 30K$

$$\text{or } K = 2.8213 \times 10^{-4} \qquad (2)$$

For all the Al_2O_3 to be reacted, time taken is given by:

$$t = [1 - (1 - 1)^{1/3}]^2/2.8213 \times 10^{-4} = 3.54 \times 10^3 \text{ min}$$

12.5   Volume of HCl gas is 12×22.4 L when $(28.085 \times 3 + 14.007 \times 4)$ g
   of Si_3N_4 is produced at N. T. P.

Therefore, volume of HCl gas when 1 kg of Si_3N_4 is produced

   $= 22.4 \times 1000/140.283 = 159.6772$ L

Chapter 13

13.1   Let the total volume shrinkage be Sv_{total}, shrinkage during sintering be
   Sv_s, and that during binder removal be Sv_b.

Therefore,

$$Sv_{total} = Sv_b + Sv_s$$

Therefore,

$$Sv_b = Sv_{total} - Sv_s = 50\% - 45\% = 5\%$$

13.2 Let Sv_d is the shrinkage during drying.

Then,

$$Sv_{total} = Sv_d + Sv_b + Sv_s$$

Therefore,

$$Sv_b = Sv_{total} - (Sv_d + Sv_s) = 50\% - (3\% + 45\%) = 2\%$$

13.3 Let V_{formed} be the as-formed volume and V_{final} be the final volume.

Then,

$$V_{final} = (1 - Sv_{total})\ V_{formed} = (1 - 0.5) \times 50 = 25\ g/cc$$

13.4 $\Delta L/L_o(\%) = [1-(1 - Sv_{total})^{1/3}]100 = [1-(1 - 0.5)^{1/3}]100 = 20.63\%$

13.5

13.5.1 $Sv_s\ (1) = (1 - D_g/D_f) = (1 - 2.43/3.84) = 0.367$ or 36.7%

$Sv_s\ (2) = (1 - D_g/D_f) = (1 - 2.74/3.90) = 0.297$ or 29.7%

13.5.2 $Sl_s\ (1) = \{1 - [1 - Sv_s\ (1)]^{1/3}\} = \{1 - [1 - 0.367]^{1/3}\} = 0.141$ or 14.1%

$Sl_s\ (2) = \{1 - [1 - Sv_s\ (2)]^{1/3}\} = \{1 - [1 - 0.297]^{1/3}\} = 0.111$ or 11.1%

13.5.3 $D_r\ (1) = D_f\ (1)/D_t = 3.84/3.96 = 0.97$

$D_r\ (2) = D_f\ (2)/D_t = 3.90/3.96 = 0.98$

Chapter 14

14.1 In FCC structure, particles will be in contact along the face diagonal, d. Hence:

$$d = 2\ \mu \tag{1}$$

If $\Delta L/L_o$ is the linear shrinkage, then:

$$1 - (1 - \Delta L/L_o)^3 = 0.44 \tag{2}$$

That is,

$$(1 - \Delta L/L_o)^3 = 0.56 \tag{3}$$

That is,

$$(1 - \Delta L/L_o) = 0.82 \tag{4}$$

That is,

$$\Delta L/L_o = 0.18 \tag{5}$$

Therefore, $\Delta \bar{l} = 2 \times 0.18 = 0.36\ \mu$

14.2 $\Delta V / V_o = 1 - (1 - \Delta L / L_o)^3$ (1)

        $= 1 - (1 - 0.163)^3$ (2)

        $= 1 - 0.586$ (3)

        $= 0.414$

14.3   Grain size $= 5/[(3 + 6 + 5 + 4 + 3 + 5 + 4 + 6 + 3 + 6)/10] \times 100$

               $= 0.0111$ cm $= 111$ μ

14.4   The volume shrinkage, $S_v = (D_f - D_g)100/D_f = (2.47 - 1.48)100/2.47 = 40.1\%$

               $D_f = D_g/(1 - \Delta L / L_o)^3$ or $(1 - \Delta L / L_o)^3 = D_g / D_f$

      or $\Delta L / L_o$ (%) $= [1 - (D_g/D_f)^{1/3}] \times 100 = [1 - (1.48/2.47)^{1/3}] \times 100 = 15.6\%$

      Porosity of the fired material, $\varphi = (1 - D_f/D_t)100 = (1 - 2.47/2.65)100 = 6.8\%$

Chapter 15

15.1 $\sigma_f \sqrt{(\pi c_{crit})} = \sqrt{(2\gamma Y)}$

      $105\sqrt{(3.14 \times c_{crit})} = \sqrt{(2 \times 1 \times 70{,}000)}$

      $105 \times 105 \times 3.14 \times c_{crit} = 2 \times 70{,}000$

      Therefore, $c_{crit} = 2 \times 70{,}000/105 \times 105 \times 3.14 = 4.0$ μ

15.2 $\sigma_f \sqrt{(\pi c_{crit})} = \sqrt{(2\gamma Y)}$

      Therefore, $\sigma_f = \sqrt{(2\gamma Y/\pi c_{crit})} = \sqrt{(2 \times 1 \times 70{,}000/3.14 \times 100)} = 21$ MPa

15.3   MOR $= 3PL/2BW^2$

       $L = 5$ cm

       $B = 1$ cm

      $W = 0.5$ cm

      Therefore, $3L/2BW^2 = 3 \times 5/2 \times 1 \times 0.5^2 = 30$ cm^{-2}

P, kg	MOR, kg/cm²
140	4200
151	4530
154	4620
155	4650
158	4740
165	4950
167	5010
170	5100
173	5190
180	5400

Average MOR $= 4839$ kg/cm²

15.4 Answers to questions 8.4.1 to 8.4.3 are given in the tables that follow:

    a.   HPSN: Mean strength = 433 MPa

Rank, MPa	Deviation from Mean Strength	Square of Deviation	Survival Probability, (S)	ln σ	1/S	ln(1/S)	−ln ln(1/S)
279	−154	23,716	0.970	5.6312	1.0309	0.0304	3.4933
341	−92	8464	0.927	5.8319	1.0787	0.0758	2.5797
360	−73	5329	0.885	5.8861	1.1299	0.1221	2.1029
372	−61	3721	0.842	5.9189	1.1876	0.1719	1.7608
394	−39	1521	0.799	5.9764	1.2516	0.2244	1.4943
398	−35	1225	0.756	5.9865	1.3228	0.2798	1.2737
409	−24	576	0.714	6.0137	1.4006	0.3369	1.0880
422	−11	121	0.671	6.0450	1.4903	0.3990	0.9188
428	−5	25	0.628	6.0591	1.5924	0.4652	0.7653
430	−3	9	0.585	6.0638	1.7094	0.5361	0.6234
439	6	36	0.543	6.0845	1.8416	0.6106	0.4933
441	8	64	0.500	6.0890	2.0000	0.6931	0.3666
444	11	121	0.457	6.0958	2.1882	0.7831	0.2445
448	15	225	0.415	6.1048	2.4096	0.8795	0.1284
452	19	361	0.372	6.1137	2.6882	0.9889	0.0112
453	20	400	0.329	6.1159	3.0395	1.1117	−0.1059
471	38	1444	0.286	6.1549	3.4965	1.2518	−0.2246
474	41	1681	0.244	6.1612	4.0984	1.4106	−0.3440
478	45	2025	0.201	6.1696	4.9751	1.6044	−0.4727
490	57	3249	0.158	6.1944	6.3291	1.8452	−0.6126
500	67	4489	0.115	6.2146	8.6957	2.1628	−0.7714
505	72	5184	0.073	6.2246	13.699	2.6173	−0.9621
521	88	7744	0.030	6.2558	33.333	3.5065	−1.2546

    Sum of squares of deviations = 71,730.

    Mean of the squares ≈ 3119

    Standard deviation ≈ 56

    b.   RBSN: Mean strength = 94 MPa

Rank, MPa	Deviation from Mean Strength	Square of Deviation	Survival Probability	ln σ	1/S	ln(1/S)	−ln ln(1/S)
76	−18	324	0.962	4.3307	1.0395	0.0387	3.2519
78	−16	256	0.908	4.3567	1.1013	0.0965	2.3382
80	−14	196	0.853	4.3820	1.1723	0.1590	1.8389

Continued

Rank, MPa	Deviation from Mean Strength	Square of Deviation	Survival Probability	ln σ	1/S	ln(1/S)	−ln ln(1/S)
80	−14	196	0.799	4.3820	1.2516	0.2244	1.4943
82	−12	144	0.745	4.4067	1.3423	0.2944	1.2228
83	−11	121	0.690	4.4188	1.4493	0.3711	0.9913
84	−10	100	0.636	4.4308	1.5723	0.4525	0.7930
89	−5	25	0.582	4.4886	1.7182	0.5413	0.6138
90	−4	16	0.527	4.4998	1.8975	0.6405	0.4455
93	−1	1	0.473	4.5326	2.1142	0.7487	0.2894
95	1	1	0.418	4.5539	2.3923	0.8723	0.1366
97	3	9	0.364	4.5747	2.7473	1.0106	−0.0105
99	5	25	0.310	4.5951	3.2258	1.1712	−0.1580
103	9	81	0.255	4.6347	3.9216	1.3665	−0.3123
106	12	144	0.201	4.6634	4.9751	1.6044	−0.4727
108	14	196	0.148	4.6821	6.7568	1.9105	−0.6474
120	26	676	0.092	4.7875	10.8696	2.3860	−0.8696
132	38	1444	0.038	4.8828	26.3158	3.2702	−1.1849

Sum of squares of deviations = 3955

Mean of the squares ≈ 220

Standard deviation ≈ 15

c.   CVDSC: Mean strength = 270 MPa

Rank, MPa	Deviation from Mean Strength	Square of Deviation	Survival Probability	ln σ	1/S	ln(1/S)	−ln ln(1/S)
139	−131	17,161	0.957	4.9345	1.0449	0.0439	3.1258
178	−92	8464	0.896	5.1818	1.1161	0.1098	2.2091
199	−71	5041	0.835	5.2933	1.1976	0.1803	1.7131
219	−51	2601	0.774	5.3891	1.2920	0.2562	1.3618
231	−39	1521	0.713	5.4424	1.4025	0.3383	1.0838
248	−22	484	0.652	5.5134	1.5337	0.4277	0.8493
260	−10	100	0.591	5.5607	1.6920	0.5259	0.6426
269	−1	1	0.530	5.5947	1.8868	0.6349	0.4543
279	9	81	0.470	5.6312	2.1277	0.7550	0.2810
290	20	400	0.409	5.6699	2.4450	0.8940	0.1120
296	26	676	0.348	5.6904	2.8736	1.0556	−0.0541
308	38	1444	0.287	5.7301	3.4843	1.2483	−0.2218
327	57	3249	0.226	5.7900	4.4248	1.4872	−0.3969
332	62	3844	0.165	5.8051	6.0606	1.8018	−0.5888
351	81	6561	0.104	5.8608	9.6154	2.2634	−0.8169
386	116	13,456	0.043	5.9558	23.2558	3.1466	−1.1463

Sum of squares of deviations = 65,084

Mean of the squares ≈ 4068

Standard deviation ≈ 64

15.4.4

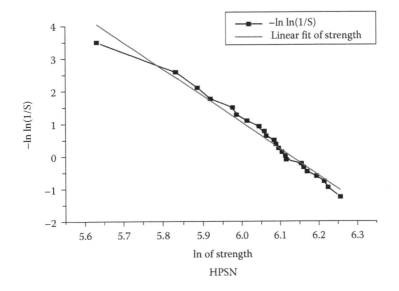

HPSN

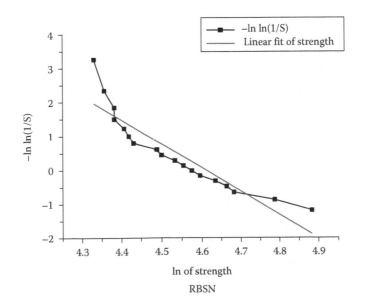

RBSN

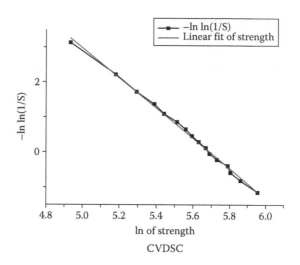

CVDSC

15.4.5   Weibull modulus for HPSN = 8.2, for RBSN = 7.0, and for CVDSC = 4.3.

15.4.6   The design stress for HPSN

$$0.999 = \exp\{-(\sigma/\sigma_o)^{8.2}\}$$

$$\sigma_o \approx 452 \text{ MPa from the table.}$$

Therefore, $-(\sigma/\sigma_o)^{8.2} = \ln 0.999 - 0.0010$

That is, $8.2 \log (\sigma/\sigma_o) = \log 0.0010 = -3$

Therefore, $\log (\sigma/\sigma_o) = -0.375$

That is, $\sigma/\sigma_o = 0.42170$

Therefore, $\sigma = 0.42170 \times 452 = 191 \text{ MPa}$

The design stress for RBSN

$$\sigma_o \approx 97 \text{ MPa from the table.}$$

Therefore, $-(\sigma/\sigma_o)^7 = \ln 0.999 - 0.0010$

That is, $7 \log (\sigma/\sigma_o) = \log 0.0010 = -3$

Therefore, $\log (\sigma/\sigma_o) = -0.4286$

That is, $\sigma/\sigma_o = 0.3727$

Therefore, $\sigma = 0.3727 \times 97 = 36 \text{ MPa}$

The design stress for CVDSC

$$-(\sigma/\sigma_o)^{4.3} = \ln 0.999 - 0.0010$$

That is, $4.3 \times \log (\sigma/\sigma_o) = \log 0.0010 = -3$

Therefore, $\log (\sigma/\sigma_o) = -0.6977$

That is, $\sigma/\sigma_o = 0.2006$

$\sigma_o \approx 295$ MPa from the table

Therefore, $\sigma = 0.2006 \times 295 = 59$ MPa

15.5 The K_I values for the different cracks are tabulated as follows:

c in Meters	K_I, MPa.m$^{1/2}$
1×10^{-2}	3.54
1×10^{-2}	3.54
0.75×10^{-2}	3.07

Hence, $K_{Ic} = 3.54$ MPa.m$^{1/2}$

Assumptions:

1. K_{Ic} is the same as the stress intensity factor (K_I) for which the fracture has taken place.
2. One crack was not perpendicular to the tensile axis. Hence, its effective length perpendicular to the tensile axis was considered.
3. Fracture will take place for the crack having the highest K_I value.

15.6 From the micron marker in the figure, it can be obtained that 1.7 cms = 100 μ.

Also, from the figure, 2c = 3.4 cms, and 2a = 1.3 cms

That is, $2c = (100/1.7)*3.4 = 200$ μ, and $2a = (100/1.7)*1.3 \approx 76$ μ

Given, $K_{Ic} \approx 0.15(H\sqrt{a}) (c/a)^{-1.5}$

Therefore, $K_{Ic} \approx 0.15 (5.5\sqrt{38}) (100/76)^{-1.5} = 122.2248$

15.7 $\quad m \ln \sigma - m \ln \sigma_0 = 0$

That is, $\sigma_0 = \sigma = 100$ MPa

To find σ, when S = 0.95, we have

$$0.95 = \exp [-(\sigma/100)^{10}]$$

That is, $-(\sigma/100)^{10} = \ln 0.95 = -0.0513$

That is, $-10 \log (\sigma/100) = \log (-0.0513) = \log (-2 +1.9487) = -1.0229$

That is, $\log (\sigma/100) = 0.10229$

That is, $\sigma/100 = 1.2656$

That is, $\sigma \approx 127$ MPa

Assumption

Average MOR is taken as the failure stress when $-\ln \ln (1/S) = 0$.

15.8 $\sigma_0 = 435$ MPa

$$0.99 = \exp [-(\sigma/435)^{6.82}]$$

That is, $-(\sigma/435)^{6.82} = \ln 0.99 = -0.0101$

That is, $\log (\sigma/435) = 0.0015$

That is, $\sigma/435 = 1.0034$ or $\sigma \approx 436$ MPa

15.9 D_{gb} at 900K $= 100 \exp (-40 \times 10^3/8.31467 \times 900) = 0.4770$ unit

D_{latt} at 900K $= 300 \exp (-50 \times 10^3/8.31467 \times 900) = 0.3761$ unit

D_{gb} at 1300K $= 100 \exp (-40 \times 10^3/8.31467 \times 1300) = 2.4709$ units

D_{latt} at 1300K $= 300 \exp (-50 \times 10^3/8.31467 \times 1300) = 2.9390$ units

Thus, grain boundary diffusion will dominate at 900 K, and lattice one at 1300 K.

When $D_{gb} = D_{latt}$,

$100 \exp (-40 \times 10^3/8.31467 \times T) = 300 \exp (-50 \times 10^3/8.31467 \times T)$

That is, $(-40 \times 10^3/8.31467 \times T) + 50 \times 10^3/8.31467 \times T = \ln 3 = 1.0986$

That is, $-40 \times 10^3 + 50 \times 10^3 = 1.0986 \times 8.31467 \times T$

Or $T = 10 \times 10^3/1.0986 \times 8.31467 \approx 1095$ K

15.10 $v = A (K_I/K_{Ic})^n$ (1)

$\log v = \log A + n(\log K_I - \log K_{Ic}) = \log A - n \log K_{Ic} + n \log K_I$ (2)

The plot of stress intensity versus crack velocity is given by:

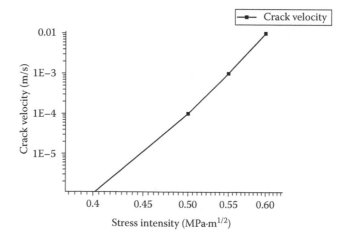

From the plot, intercept = 2.88803, and slope = 22.49311

That is,

$$\log A - n \log K_{Ic} = 2.88803 \text{ or } \log A + 0.1549\, n = 2.88803 \tag{3}$$

$$\text{and } n = 22.49311 \tag{4}$$

Putting the value of n in Equation (3), we get:

$$\log A = 2.88803 - 0.1549 * 22.49311 \text{ or}$$

$$A = 10^{\wedge} (2.88803 - 0.1549 * 22.49311) = 0.2534 \text{ m/s}$$

$$v = A'K_I^{\,n} \text{ or } \log v = \log A' + n \log K_I$$

From the plot, log A' = 2.88803 or

$$A' = 10^{\wedge}2.88803 \approx 773$$

Chapter 16

16.1 $k_{eff} = 4d_p n^2 \sigma e_{eff} T_m^{\,3}$

Because n and e_{eff} are constant for a given material, at a constant temperature, $k_{eff} = Cd_p$, where C is a constant.

For pore size of 5 mm,

$$k_{eff} = C \times 5$$

For pore size of 0.5 mm,

$$k_{eff} = C \times 0.5$$

Therefore, ratio of conductivities at the two pore sizes = 5/0.5 = 10.

16.2 The material having the highest $R_H k_{th}$ value will be the best suited one.

Material	$R_H k_{th}$, W.m$^{-1/2}$
1	1289
2	333
3	889

From the values obtained, it is clear that material 1 is the most well-suited.

16.3 $\partial Q / \partial t = k_{th} A\, \partial T / \partial x$

$$= 1.7 \times 1000 \times 10^{-4} \times 25 / 0.5 \times 10^{-2}$$

$$= 850 \text{ W}$$

16.4 $\partial Q / \partial t = k_{th} A\, \partial T / \partial x$

$$= 1.3 \times 1000 \times 10^{-4} \times 1200 / 0.5 \times 10^{-2}$$

$$\approx 462 \text{ kW}$$

SECTION III

Chapter 18

18.1 In the microscopic domain, entropy is given by Boltzmann equation:

$$S = k \ln \Omega \tag{1}$$

Here, k is the Boltzmann constant and Ω is a measure of the randomness. Let us consider a lattice consisting of N atoms, and n point defects. The number of configurations by which these two species can be arranged is given by:

$$\Omega = (n + N)!/n!N! \tag{2}$$

Substituting this equation in Equation 1, we get Equation 3. Sterling's approximation

$$S = k \ln (n + N)!/n!N! \tag{3}$$

is given by:

$$\ln x! \approx x \ln x - x + 1/2 \ln (2\pi x) \tag{4}$$

For large values of x, the second term on the RHS can be neglected. Hence, Equation 4 becomes Equation 5.

$$\ln x! \approx x \ln x - x \tag{5}$$

Equation 3 can be rearranged as:

$$S = k [\ln (n + N)! - \ln n! - \ln N!] \tag{6}$$

Expanding Equation 6 using Equation 5, we get Equation 7 for configurational entropy.

$$
\begin{aligned}
S_c &= k \{[(n + N) \ln (n + N) - (n + N)] - (n \ln n - n) - (N \ln N - N)\} \\
&= k [(n + N) \ln (n + N) - n \ln n - N \ln N] \\
&= k [n \ln (n + N) + N \ln (n + N) - n \ln n - N \ln N] \\
&= -k [N \ln N - N \ln (n + N) + n \ln n - n \ln (n + N)] \\
&= -k \{N \ln [N/(n + N)] + n \ln [n/(n + N)]\}
\end{aligned}
\tag{7}
$$

18.2 Let us consider a hypothetical equation for the formation of a compound MX.

$$M(s) + \tfrac{1}{2} X_2(g) = MX(s) \tag{1}$$

We have chemical potentials for this reaction as:

$$\mu_M = \mu_M{}^\circ + RT \ln a_M \tag{2}$$

$$\mu_{X_2} = \tfrac{1}{2}\mu_{X_2}{}^\circ + RT \ln P_{X_2}{}^{1/2} \tag{3}$$

$$\mu_{MX} = \mu_{MX}{}^\circ + RT \ln a_{MX} \tag{4}$$

Now, the free energy for the reaction can be written as:

$$\Delta G = \mu_{MX} - \mu_X - \tfrac{1}{2}\mu_{X_2} \tag{5}$$

Writing the values for the chemical potentials of the reactants and product from Equations 2 through 4 in Equation 5, and rearranging the terms, we get:

$$\Delta G = \left(\mu_{MX}{}^\circ - \mu_X{}^\circ - \tfrac{1}{2}\mu_{X_2}{}^\circ\right) + RT \ln\left(a_{MX} / a_M P_{X_2}{}^{1/2}\right) \tag{6}$$

In the lines of Equation 5, standard free energy change for our reaction under consideration can be written as:

$$\Delta G^\circ = \left(\mu_{MX}{}^\circ - \mu_X{}^\circ - \tfrac{1}{2}\mu_{X_2}{}^\circ\right) \tag{7}$$

Also,

$$a_{MX} / a_M P_{X_2}{}^{1/2} = K \tag{8}$$

where K is the equilibrium constant for the reaction. At equilibrium,

$$\Delta G = 0 \tag{9}$$

Therefore, Equation (18.6) becomes:

$$\Delta G^\circ = -RT \ln K \tag{10}$$

18.3 $Si + O_2 = SiO_2$

18.3.1 $\Delta G^\circ = -RT \ln\left(a_{SiO_2} / a_{Si} \times P_{O_2}\right)$

From the standard free energy versus temperature diagram [1], $\Delta G\circ = -652$ kJ.

Therefore,

$$\ln P_{O_2} = -652/8.31467 = -78.4365$$

Therefore,

$$P_{O_2} = \exp(-78.4365) \approx 0$$

18.3.2 $Si + 2H_2O = SiO_2 + 2H_2$ (1)

$$\Delta G^\circ{}_{SiO_2} = -RT \ln\left(a_{SiO_2} \times P_{H_2}{}^2 / a_{Si} \times P_{H_2O}{}^2\right) \qquad (2)$$

$$2H_2 + O_2 = 2H_2O \qquad (3)$$

$$\Delta G^\circ{}_{H_2O} = -RT \ln\left(P_{H_2O}{}^2 / P_{H_2}{}^2 \times P_{O_2}\right) \qquad (4)$$

From the standard free energy versus temperature diagram [1], $\Delta G^\circ{}_{H_2O} = -348kJ$.

Equation 1 + Equation 3 gives:

$$Si + O_2 = SiO_2, \ \Delta G^\circ = -652 \text{ kJ}$$

That is,

$$\Delta G^\circ{}_{SiO_2} + \Delta G^\circ{}_{H_2O} = -652 \text{ kJ}$$

Therefore,

$$\Delta G^\circ{}_{SiO_2} = -652 + 348 = -304 \text{ kJ}$$

$$\ln\ (a_{SiO_2} \times P_{H_2}{}^2 / a_{Si} \times P_{H_2O}{}^2) = -304/8.31467$$

That is, $P_{H_2} / P_{H_2O} = \exp(-18.2914) = 1.14 \times 10^{-8}$

18.3.3   From Equation 4, we have:

$$-348 = -8.31467 \left\{ \ln(P_{H_2O}/P_{H_2})^2 - \ln P_{O_2} \right\}$$

That is, $\ln P_{O_2} - 2\ln\ (P_{H_2O}/P_{H_2}) = -41.8537$

That is, $\ln P_{O_2} = -[41.8537 + 18.2914] = -60.1451$

Therefore, $P_{O_2} = \exp(-60.1451)$ or $P_{O_2} \approx 0$

18.4   Al can reduce Fe_2O_3 at 1200°C if the free energy of formation of Al_2O_3 is less than that of Fe_2O_3.

From the standard free energy of formation of oxides versus temperature diagram [1],

$$(\Delta G_{Al_2O_3}{}^\circ)_{1200} = -799 \text{ kJ/mole of } O_2, \text{ and}$$

$$(\Delta G_{Fe_2O_3}{}^\circ)_{1200} = -83.68 \text{ kJ/mole of } O_2$$

Hence, Al can reduce Fe_2O_3 at 1200°C.

18.5   Oxidation is possible if the free energy change for the oxidation reaction is negative.

Oxidation reaction is given by:

$$Ni + CO_2 = NiO + CO$$

$$\Delta G° = -8.31467 \times 1473 \ln (P_{CO}/P_{CO_2})$$

$$= -12247.50891 \times \ln 0.1$$

$$\approx 28 \text{ MPa}$$

Hence, oxidation is impossible.

18.6 From the standard free energy of formation of oxides versus temperature diagram [1],

$$(\Delta G_{ZnO}°)_{700} = -478 \text{ kJ/mole of } O_2$$

$$(\Delta G_{SiO_2}°)_{700} = -696 \text{ kJ/mole of } O_2$$

Because $(\Delta G_{SiO_2}°)_{700} > (\Delta G_{ZnO}°)_{700}$, SiO_2 will be oxidized rather than be reduced. Hence, SiO_2 will not oxidize Zn.

Reference

1. C. Bodsworth, *Physical Chemistry of Iron and Steel Manufacture*, Longmans, New York, 1963, p. 31.

Bibliography

M. W. Barsoum, *Fundamentals of Ceramics*, Taylor & Francis Group, LLC, New York, 2003, p. 134.

Appendix 2

Conversion from SI to FPS/CGS Units

1	Length	1 m = 3.28 ft
2	Area	1 m^2 = 10.76 ft^2
3	Volume	1 m^3 = 35.32 ft^3
4	Mass	1 kg = 2.205 lb
5	Density	1 kg/m^3 = 6.24 × 10^4 lb/ft^3 1 kg/m^3 = 10^{-3} g/cm^3
6	Force	1 N = 0.2248 lbf
7	Energy	1 J = 9.4844 × 10^{-4} Btu 1 J = 10^7 ergs
8	Power	1 W = 3.414 Btu/h
9	Heat Capacity	1 J/kg.K = 2.39 × 10^{-4} Btu/lb.°F
10	Thermal Conductivity	1 W/m.K = 0.578 Btu/ft.h.°F
11	Pressure (or Stress)	1 Pa = 1.45 × 10^{-4} psi 1 bar = 10^5 Pa 1 atm = 1.013 × 10^5 Pa 1 kg/mm^2 = 9.806 MPa

Appendix 2

Conversion from SI to Imperial/CGS Units

Length	1 m = 3.28 ft	
Area	1 m² =	
Mass	1 kg = 2.2 lb	
Volume	1 m³ =	
Force	1 N = 0.2248 lbf	
Energy	1 J =	
Power	1 W =	
Pressure		
Thermal conductivity		
Temperature		

Index